T0299158

LOCAL GROUP COSMOLOGY

One of the most fascinating unresolved problems of modern astrophysics is how the galaxies we observe today were formed. The Lambda–Cold Dark Matter paradigm predicts that large spiral galaxies such as the Milky Way formed through accretion and tidal disruption of satellite galaxies, a notion previously postulated on empirical grounds from the character of stellar populations found in our Galaxy. The Local Group galaxies are the best laboratory in which to investigate these galaxy formation processes because they can be studied with sufficiently high resolution to exhume fossils of galactic evolution embedded in the spatial distribution, kinematics, and chemical abundances of their oldest stars.

Based on the twentieth Canary Island Winter School of Astrophysics, this volume provides a firm grounding for graduate students and early career researchers working on Local Group cosmology. It presents modules from seven eminent and experienced scientists at the forefront of Local Group research, and includes overviews of observational techniques, diagnostic tools, and various theoretical models.

David Martínez-Delgado is an astronomer at the Max-Planck Institute for Astronomy, Heidelberg, Germany, and at the Instituto de Astrofísica de Canarias, Spain.

Canary Islands Winter School of Astrophysics

Volume XX

Series Editor
Francisco Sánchez, *Instituto de Astrofísica de Canarias*

Previous books in this series

I. Solar Physics
II. Physical and Observational Cosmology
III. Star Formation in Stellar Systems
IV. Infrared Astronomy
V. The Formation of Galaxies
VI. The Structure of the Sun
VII. Instrumentation for Large Telescopes: A Course for Astronomers
VIII. Stellar Astrophysics for the Local Group: A First Step to the Universe
IX. Astrophysics with Large Databases in the Internet Age
X. Globular Clusters
XI. Galaxies at High Redshift
XII. Astrophysical Spectropolarimetry
XIII. Cosmochemistry: The Melting Pot of Elements
XIV. Dark Matter and Dark Energy in the Universe
XV. Payload and Mission Definition in Space Sciences
XVI. Extrasolar Planets
XVII. 3D Spectroscopy in Astronomy
XVIII. The Emission-Line Universe
XIX. The Cosmic Microwave Background: From Quantum Fluctuations to the Present Universe

Participants of the XX Canary Islands Winter School of Astrophysics

LOCAL GROUP COSMOLOGY

Edited by
DAVID MARTÍNEZ-DELGADO
Max-Planck-Institut für Astronomie, Heidelberg

CAMBRIDGE
UNIVERSITY PRESS

University Printing House, Cambridge CB2 8BS, United Kingdom

One Liberty Plaza, 20th Floor, New York, NY 10006, USA

477 Williamstown Road, Port Melbourne, VIC 3207, Australia

314-321, 3rd Floor, Plot 3, Splendor Forum, Jasola District Centre, New Delhi - 110025, India

79 Anson Road, #06-04/06, Singapore 079906

Cambridge University Press is part of the University of Cambridge.

It furthers the University's mission by disseminating knowledge in the pursuit of education, learning and research at the highest international levels of excellence.

www.cambridge.org
Information on this title: www.cambridge.org/9781107023802

First published 2013

A catalogue record for this publication is available from the British Library

Library of Congress Cataloging in Publication data
Canary Islands Winter School of Astrophysics (20th : 2008 : Tenerife, Canary Islands)
Local Group cosmology / [edited by] David Martínez-Delgado, Max-Planck-Institut für Astronomie, Heidelberg.
pages cm
Lectures presented at the XX Canary Islands Winter School of Astrophysics, held in Tenerife, Spain, November 17–18, 2008.
Includes bibliographical references.
ISBN 978-1-107-02380-2 (hardback)
1. Local Group (Astronomy) – Congresses. I. Martínez Delgado, D. (David), editor of compilation. II. Title.
QB858.8.L63C36 2008
523.1′12–dc23 2013012345

ISBN 978-1-107-02380-2 Hardback

Contents

List of contributors

BULLOCK, JAMES, University of California Irvine, USA

FREEMAN, K. C., Australian National University, Australia

IBATA, R., Strasbourg Observatory, France

KROUPA, P., Angelander-Institut für Astronomie, Germany

MAJEWSKI, STEVEN R., University of Virginia, USA

PEÑARRUBIA, J., Institute of Astronomy, University of Cambridge, UK

VALLS-GABAUD, D., Observatoire de Paris, France

List of participants

Alonso García, Javier	University of Michigan – Department of Astronomy (United States of America)
Antoja, Teresa	Universitat de Barcelona – Institut de Ciències del Cosmos Department d'Astronomia i Meteorologia (Spain)
Bakos, Judit	Instituto de Astrofísica de Canarias (Spain)
Beaton, Rachael	University of Virginia (United States of America)
Besla, Gurtina	Harvard-Smithsonian Center for Astrophysics (United States of America)
Bovill, Mia	University of Maryland – Department of Astronomy (United States of America)
Brink, Thomas	University of Michigan – Department of Astronomy (United States of America)
Carballo Bello, Julio Alberto	Instituto de Astrofísica de Canarias (Spain)
Carrera, Ricardo	Osservatorio Astronomico di Bologna (Italy)
Cheng, Judy	UC Santa Cruz (United States of America)
Comerón Limbourg, Sébastien	Instituto de Astrofísica de Canarias (Spain)
Conn, Blair	European Southern Observatory (Chile)
Correnti, Matteo	Universitá degli studi di Bologna (Italy)
Damke, Guillermo	University of Virginia – Astronomy Building (United States of America)
Ebrová, Ivana	Academy of Sciences of the Czech Republic (Czech Republic)
Ferré Mateu, Anna	Instituto de Astrofísica de Canarias (Spain)
Fliri, Jürgen	Instituto de Astrofísica de Canarias (Spain)
Guerras Valera, Eduardo	Instituto de Astrofísica de Canarias (Spain)
Hamden, Erika	Columbia University and Astronomy Department (United States of America)
Hayward, Chris	Harvard-Smithsonian Center for Astrophysics (United States of America)
Hummels, Cameron	Columbia University and Astronomy Department (United States of America)
Ibarra Medel, Héctor Javier	Instituto Nacional de Astrofísica, Óptica y Electrónica (Mexico)
Kang, Aram	University of Edinburgh – Institute of Astronomy (United Kingdom)
Karlsson, Torgny	University of Sidney – Institute of Astronomy (Australia)
Kirby, Emma	Research School of Astronomy and Astrophysics – Mt Stromlo Observatory (Australia)
Klimentowski, Jaroslaw	Nicolaus Copernicus Astronomical Center (Poland)
Kordopatis, George	Observatoire de la Cote d'Azur (France)
Li, Yang-Shyang	Kaptein Astronomical Institute (The Netherlands)
Marks, Michael	Argelander-Institut für Astronomie (Germany)
Mateu, Cecilia	Centro de Investigación de Astronomía (Venezuela)
Narbutis, Donatas	University of Vilnius (Lithuania)
Navarro González, Javier	Instituto de Astrofísica de Canarias (Spain)
Paudel, Sanjaya	Astronomisches Rechen-Institut Zentrum für Astronomie (Germany)

Petrov, Mykola	University of Vienna – Institute of Astronomy (Austria)
Radburn-Smith, David	STScI (United States of America)
Rivero Losada, Illa	Instituto de Astrofísica de Canarias (Spain)
Rubele, Stefano	Universita di Padova – Dipartimento di Astronomia (Italy)
Ruhland, Christine	Max-Planck Institut für Astronomie (Germany)
Sale, Stuart	Astrophysics Group Blackett Laboratory – Imperial College London (United Kingdom)
Salinas, Ricardo	European Southern Observatory (Chile)
Sánchez Gallego, José Ramón	Instituto de Astrofísica de Canarias (Spain)
Simpson, Christine	Columbia University (United States of America)
Sollima, Antonio	Instituto de Astrofísica de Canarias (Spain)
Starkenburg, Else	Kapteyn Astronomical Institute (The Netherlands)
Tapia Peralta, Trinidad	Instituto de Astrofísica de Canarias (Spain)
Tempel, Elmo	Tartu Observatory (Estonia)
Teyssier, Maureen	Columbia University Astronomy Department (United States of America)
Tikhonov, Anton	Sobolev Astronomical Institute, Saint-Petersburg State University – Department of Mathematics and Mechanics (Russia)
Valentini, Marica	Universita di Padova (Italy)

Preface

Background

One of the unresolved problems of modern astrophysics is how the galaxies we observe today were formed. The Lambda–Cold Dark Matter paradigm predicts that large spiral galaxies like the Milky Way formed through the accretion and tidal disruption of satellite galaxies, a notion previously postulated on empirical grounds from the character of stellar populations found in our Galaxy. The Local Group galaxies are the best laboratory in which to investigate these galaxy formation processes as they can be studied with sufficiently high resolution to exhume the fossils of galactic evolution embedded in the spatial distribution, kinematics, and chemical abundances of their oldest stars.

Scientific rationale

This "Galactic archaeology" has recently undergone an unprecedented revolution, brought about by the spectacular increase in the quality and quantity of observations of Local Group galaxies using large-aperture ground-based telescopes and the Hubble Space Telescope, and with the advent of the first large-scale digital sky surveys (such as SLOAN and 2MASS) at the start of the twenty-first century.

The possibility of contrasting these observations with results on a small scale of cosmological simulations has drawn the attention of cosmologists towards the study of Local Group grand design galaxies and their satellites, thus giving rise to new lines of research that have involved numerous resources and a considerable observational and theoretical effort. The disagreement between the results of simulations and observations has also given rise to serious controversies among observers and theoretical cosmologists and is still the subject of active debate in the international community.

There is little doubt that the "golden age" of these investigations will take place in the coming years with the commissioning of ambitious observational projects such as the LSST (Large Synoptic Survey Telescope) in Chile or the launching of the astrometric satellite GAIA. These projects will constitute an enormous qualitative and quantitative leap in data and will require the training of young researchers in a field that has hitherto been the object of study for a relatively small number of scientists. Recognizing the importance of this field, the Instituto de Astrofísica de Canarias organized the XXth in its Winter School series around the topic "Local Group Cosmology."

Outline of the school

The primary aim of the school was to provide a wide-ranging and up-to-date overview of the theoretical, experimental, and analytical tools necessary for carrying out front-line research in the study of the structure, formation, and evolution of Local Group galaxies, based on the results of the latest cosmological simulations. The Winter School was particularly designed to offer young researchers tips and guidelines to help them direct their future research toward these themes, which are among the most important in modern astrophysics.

With this purpose, the school provided to the participants an introduction to the physics of the Local Group galaxies, observational techniques, diagnostic tools, and theoretical codes. In addition, the existing and planned experiments and surveys were reviewed with a description of the analysis methods involved and the constraints imposed on theoretical models. This was complemented by an overview of galactic and extragalactic foregrounds, as well as a summary of other major cosmological probes.

To achieve these goals, the 40 lectures were given by eight eminent and experienced scientists who are actively working on a variety of forefront research projects and who have played a key role in major advances over recent years in the topic of the school. The list of teachers included leading cosmologists and pioneering observers in each area of the subject who were carefully chosen to represent all the leading research teams in each topic covered in the school, especially those topics where there is disagreement among the different teams.

The Editor

Acknowledgments

The organizers of the XX Canary Islands Winter School of Astrophysics would like to express their sincere gratitude, first and foremost, to the lecturers, for making it a great scientific and educational event. The careful preparation of the lectures, the attendance and intense interaction with students, and the subsequent writing up of the manuscripts for this book have been a major commitment in their busy agendas. In particular, we would like to thank Professor Ken Freeman for his entertaining public lecture on "Dark Matter in Galaxies." The students played an important role in the success of the Winter School: their enormous enthusiasm – maintained throughout the entire two weeks – and outstanding human quality resulted in a really pleasant and fruitful event.

The editor wants to express his warmest gratitude to the efficient secretaries, Nieves Villoslada and Lourdes González. Their great knowledge and diligence before, during, and after the Winter School are a key component in its success. The Web page of the Winter School was a vital tool in the preparation and development of the event, and we thank Jorge Andrés Pérez (SIE/IAC) for its care and maintenance. The press room of the Winter School was the responsibility of Nadjejda Vicente and Iván Jiménez, who did a wonderful job interviewing the lecturers. Ramon Castro and Gabriel Pérez (SMM/IAC) designed the posters and additional multimedia support material for the Winter School. Ismael Martínez-Delgado played a major part in the creation of this book, revising in minute detail and with enormous patience and technical editorial skill all the submitted manuscripts.

Each year, the Canary Island Winter School of Astrophysics is a major institutional event at the IAC, whose various departments and support services always actively and enthusiastically contribute to it: Jesús Burgos, Carmen del Puerto, Terry Mahoney, Julio A. Carballo-Bello, Miguel Briganti, Monique Gómez and the technicians of the Servicios Informáticos Comunes (SIC) of the IAC. We thank all concerned for their support and efficiency. We also thank the respective managers and guides who made possible the visits to the Observatories on Tenerife and La Palma.

Finally, we also wish to thank Spain's Ministerio de Ciencia e Innovación, the local governments (Cabildos) of the islands of Tenerife and La Palma, the Puerto de la Cruz Council, and Iberia for their economical support to the organization of this Winter School.

The Editor

Abbreviations

2MASS	Two Micron All Sky Survey
AAT	Anglo Australian Telescope
AGB	Asymptotic Giant Branch
APOD	Astronomy Picture of the Day
APOGEE	Apache Point Observatory Galactic Evolution Experiment
CCD	Charge Coupled Device
CCI	Central Credible Interval
CDM	Cold Dark Matter
CDMS	Cryogenic Dark Matter Search
CMB	Cosmic Microwave Background
CMD	Color Magnitude Diagram
CMSSM	Constrained Minimal Supersymmetric Standard Model
COBE	COsmic Background Explorer
COROT	COnvection ROtation and Planetary Transits
CPU	Central Processing Unit
CR	Credible Region
dE	dwarf Elliptical
DES	Dark Energy Survey
DIRBE	Diffuse Infrared Background Experiment
DM	Dark Matter
DoS	Disk of Satellites
dSph	dwarf Spheroidal
ESA	European Space Agency
ESO	European Southern Observatory
FLAMES	Fibre Large Array Multi-Element Spectrograph
FWHM	Full Width at High Maximum
GR	General Theory of Relativity
GSC	Guide Star Catalog
HDM	Hot Dark Matter
HERMES	High Efficiency and Resolution Multi-Element Spectrograph
HRD	Hertzsprung-Russell Diagram
HSB	High Surface Brightness
HST	Hubble Space Telescope
IMF	Initial Mass Function
IPAC	Infrared Processing and Analysis Center
JASMINE	Japan Astrometry Satellite Mission for INfrared Exploration
LAB	Leiden Argentine Bonn
LG	Local Group
LMC	Large Magellanic Cloud
LSB	Low Surface Brightness
LSR	Local Standard of Rest
LSST	Large Synoptic Survey Telescope
MCI	Minimum Credible Interval
MFL	Mass Follows Light
MIKE	Magellan Inamori Kyocera Echelle
MLT	Mixing Length Convection Theory
MMT	Multi Mirror Telescope
MOG	MOdified Gravity
MOND	MOdified Newtonian Dynamics

MOS	Multi-Object Spectroscopy
MOST	Microvariability and Oscillations of Stars Telescope
MS	Magellanic Stream
MS	Main Sequence
MSP	Missing Satellites Problem
MW	Milky Way
NASA	National Aeronautics and Space Administration
NFW	Navarro-Frenk-White
OLR	Outer Lindblad Resonance
PAndAS	Pan-Andromeda Archeological Survey
Pan-STARRS	Panoramic Survey Telescope & Rapid Response System
PDF	Probability Distribution Function
PM2000	Bordeaux Proper Motion Catalogue
RAVE	RAdial Velocity Experiment
RGB	Red Giant Branch
RHR	Rosenberg-Hertzsprung-Russell Diagram
RMS	Root Mean Square
SAM	Semi Analytical Modelling
SDSS	Sloan Digital Sky Survey
SEGUE	Sloan Extension for Galactic Understanding and Exploration
SEHO	South-East H I Overdensity
SFH	Star-Formation History
SFR	Star-Formation Rate
SMC	Small Magellanic Cloud
SNe	Super Novae
SPH	Smoothed Particle Hydrodynamics
SPM	Southern Proper Motion
STScI	Space Telescope Science Institute
TAMS	Terminal Age Main Sequence
TDG	Tidal Dwarf Galaxies
TO	Turn-Off
UCAC2	Second USNO CCD Astrograph Catalogue
UCD	Ultra Compact Dwarf Galaxy
UKIDSS	UK Infrared Telescope Infrared Deep Sky Survey
USNO	United States Naval Observatory
VHS	Vista Hemisphere Survey
VISTA	Visible and Infrared Survey Telescope for Astronomy
VL II	Via Lactea II Simulation
VLT	Very Large Telescope
WDLF	White Dwarf Luminosity Function
WDM	Warm Dark Matter
WIMP	Weakly Interacting Mass Particle
WMAP	Wilkinson Microwave Anisotropy Probe
ZAMS	Zero Age Mass Sequence
ΛCDM	Dark Energy + Cold Dark Matter

1. The formation of the Milky Way in the CDM paradigm

K. C. FREEMAN

1.1 Introduction

What does our Galaxy look like? We can compare the COBE image of our Galaxy, taken in the near-IR, with the visible image of the edge on spiral NGC 891. Our Galaxy would probably look much like NGC 891 if it were observed in visible light from far away (see Figure 1.1). The Milky Way is very clearly a disk galaxy: its disk is the primary component and is supported almost entirely by its rapid rotation. We also see a small central bulge which contributes about 20% of the total light. Some galaxies have much larger bulges. The small bulge of the Milky Way is a pointer to the events that occurred as it formed and evolved. We would like to understand how our Galaxy came to look like this.

Figure 1.2 shows schematically the five main components of the stellar galaxy. The thin disk and bulge are the main visible components. The thin disk is enveloped in a thicker thick disk which contributes only about 10% of the light of the disk. These thick disks are very common and their formation appears to be part of the formation process of disk galaxies. The stellar halo provides only about 1–2% of the total light but is very important for understanding how the Galaxy was assembled. The stars of the halo are metal-poor, mostly with abundances of $[Fe/H] < -1$. Unlike the disks and the bulge, the stellar halo is not rotating significantly: it is supported against gravity by the random motions of its stars. Currently we believe that the halo represents the debris of small metal-poor galaxies that were accreted by the Galaxy during its formation and evolution. Finally there is the dark halo. It appears to contribute at least 95% of the total mass of the Galaxy. Current opinion is that the dark halo does not contribute much to the gravitational field in the inner few kpc of the Galaxy, but it rapidly becomes the dominant contributor at larger radii. The dark halo appears to extend to a radius of at least 150 kpc.

Each of these components has something to tell us about the formation history of the Galaxy. Our task is to understand how the formation and evolution of the Milky Way took place and to evaluate how the Galaxy compares with the predictions of CDM simulations.

The thin disk is relatively metal-rich and its stars cover a wide range of ages. The other stellar components are all relatively old and more metal-poor. Figure 1.3 summarizes our current belief about the age-metallicity relation for the components of the Galaxy. The similarity of the [Fe/H] range for the thick disk and the globular clusters is worth noting.

The total mass of the Galaxy is about 2×10^{12}. The stellar mass in the bulge is about 1.2×10^{10} $M_{\odot}$, the disk is about 5×10^{10} $M_{\odot}$, and the stellar halo only about 1×10^{9} $M_{\odot}$. The halo and its globular clusters have ages of about 10–12 Gyr and the thick disk stars appear to be older than 10 Gyr. Star formation in the thin disk started about 10 Gyr ago and has continued at a more or less constant rate to the present time.

How did the Galaxy come to be like this? To study the formation and evolution of galaxies observationally, we have a choice. We can observe distant galaxies at high redshift and see them directly as they were long ago at various stages of their formation and evolution. Distant galaxies are faint however, and not much detail can be measured about their chemical properties and motions of their stars. Also, we cannot follow the evolution of any individual galaxy. Alternatively, we can recognize that the main structures of our Galaxy formed long ago, at high redshift. For example, the halo formed at $z > 4$ and the

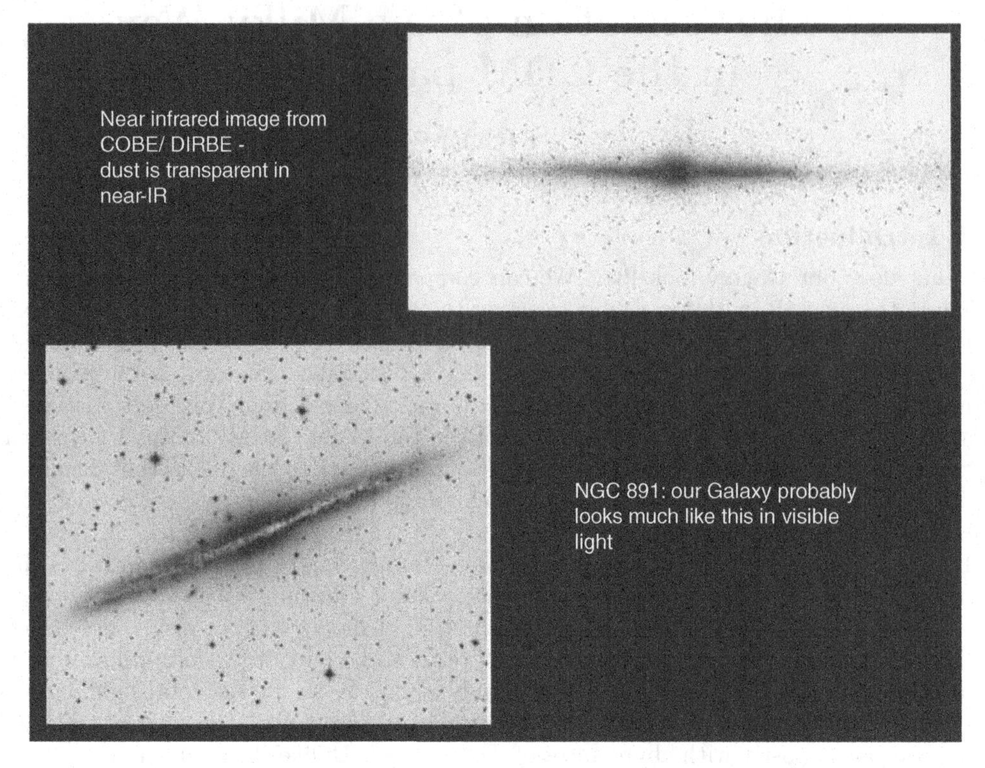

FIG. 1.1. Images of NGC 891 (visible light) and the Milky Way (NIR).

disk at $z \sim 2$. We can study the motions and chemical properties of stars in our Galaxy at a level that is impossible for other galaxies, and we can probe back into the formation epoch of the Galaxy. This approach is now called *near-field cosmology.*

The ages of the oldest stars in the Galaxy are similar to the lookback time for the most distant galaxies we can observe. Both give clues to the sequence of events that led to the formation of galaxies like the Milky Way.

The numerical simulation of galaxy formation (courtesy Sommer-Larsen, 2008) shown in this chapter summarizes our current view of how a disk galaxy like the Milky Way

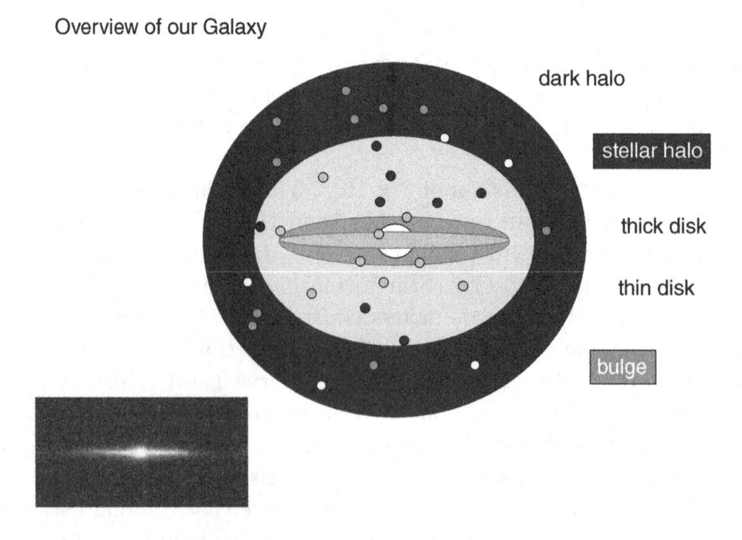

FIG. 1.2. The Galactic components.

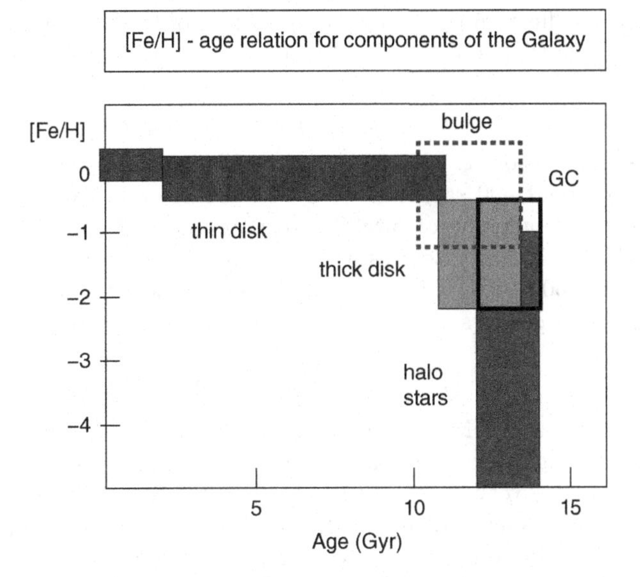

FIG. 1.3. The age-metallicity relation for the Galactic components.

came together from dark matter and baryons. The formation of the oldest halo stars begins at high $z \sim 13$, long before the Galaxy itself became visible. These oldest stars form in small overdensities that ultimately merge to become part of the Galaxy itself.

The formation of the thin rotating disk is a dissipative process that occurs after the dark halo itself has been assembled through the hierarchical merging of smaller structures. Throughout this assembly, much dynamical and chemical evolution is going on. By a redshift of $\sim$3, the Galaxy is partly assembled and is surrounded by hot gas which is cooling to form the disk. At $z \sim 2$, large lumps are falling in, and the Galaxy is already a well-defined rotating system.

The simulation showed the formation and evolution of a large spiral in a ΛCDM simulation. What does each component of the Milky Way contribute to our understanding of the formation and evolution of disk galaxies in the CDM context? Living inside the Milky Way has advantages and disadvantages. The Milky Way will be very good for assessing some CDM issues and not so good for others.

Here are some of the issues with galaxy formation for which CDM has so far not been fully successful. Our Galaxy may be able to contribute to understanding at least some of these difficulties, which include

- the density distribution of the inner dark halo: flat core (as mostly observed) or steep cusp (as predicted)
- the large number of predicted satellites
- the difficulty of forming disks with small bulges
- the related active accretion history within CDM
- the low predicted baryonic angular momentum
- the high predicted fraction of baryons that are converted into stars

1.1.1 *The structure of the inner dark halo*

Simulations consistently predict that the dark halo density distribution has an inner cusp, and observers equally consistently claim that the dark halos have flat inner cores. This argument has been going on for many years. For a density distribution parameterized as $\rho \sim r^{\alpha}$, the halos from simulations have $\alpha \sim -1$ (e.g., Navarro *et al.*, 1996) while rotation curve studies of low surface brightness galaxies (in which the stellar distributions are unlikely to have much effect on the density distribution of the dark halos) typically have α close to zero (e.g., de Blok and Bosma, 2002).

1.1.2 *The number of predicted satellites*

From simulations (e.g., Moore *et al.*, 1999), we would expect a large disk galaxy like the Milky Way to have about 500 satellites with bound masses in excess of 10^8 $M_\odot$. This is much larger than the numbers of satellites detected around the Milky Way optically or in HI. Although new fainter satellites are being discovered, it seems unlikely that the number will approach 500. Are there large numbers of dark satellites? Are some (or all) of the Galactic globular clusters associated in some way with these missing satellites?

1.1.3 *Formation of disks with small or no bulges*

It is currently difficult for ΛCDM to generate galaxies with small or no bulges because of the high continuing merger rate inherent in the CDM context. Most galaxies produced in CDM simulations have very substantial bulges, unlike the Milky Way and the many other giant galaxies with small bulges. Understanding how the bulge of the Milky Way formed could be a contribution toward resolving this problem. Current belief is that the Galactic bulge formed through instabilities of the disk rather than through a merger process. If this turns out to be correct, then we need to understand why the Milky Way escaped the high expected merger rate.

1.1.4 *The high merger rate in CDM*

CDM predicts an active ongoing accretion history, leaving debris of accreted satellites in the stellar disk and halo. The Milky Way stellar halo provides direct evidence of such accretion, of small dense systems that formed before the Milky Way itself. Further evidence comes from the currently disrupting Sgr dwarf galaxy. A very active continuing accretion history of significant sub-halos as expected in CDM is probably inconsistent with the presence of a dominant thin disk. The epoch of last major merger is particularly important for disk survival. We are uniquely located in the Milky Way to evaluate the detailed accretion history of a large spiral and measure the distribution of its first stars.

1.1.5 *The low predicted baryonic angular momentum*

Baryonic angular momentum is lost to the dark halo via hydrodynamical and gravitational effects. This is an old problem, that baryons have been predicted to have less angular momentum than observed. The observational consequences are that disk galaxies are smaller and more rapidly rotating than observed. There is some evidence now that this is less of a problem with higher resolution simulations (e.g., Governato *et al.*, 2007; Kaufmann *et al.*, 2007). This problem is observationally probably better studied in other galaxies.

1.1.6 *The high predicted fraction of baryons converted into stars*

Disk galaxies like the Milky Way appear to have only a small percentage of their baryons in the form of stars. This is not yet seen in most of the simulations, in which most of the baryons are rapidly converted into stars. This problem is probably related to the problem of continuing baryon acquisition needed to fuel ongoing star formation. Without such fueling, the observed star-formation rate would exhaust the current gas supply on a timescale of a few Gyr. Current belief is that there is a substantial reservoir of baryons in the Galactic hot halo, which may come from baryons ejected from the disk or virialized into this hot halo during early baryon infall. The details are poorly understood. Is the baryon acquisition related to the high velocity clouds? Does it come from gas that was previously ejected from the disk, or is gas from the hot halo being entrained by gas lost from the disk and now returning (e.g., Marinacci *et al.*, 2010)? The Milky Way is potentially well suited for investigating these problems of baryon content and acquisition.

1.1.7 *Reconstructing Galaxy formation*

We would like to observationally reconstruct the whole process of galaxy formation as the Galaxy comes together from the CDM hierarchy. What do we mean by the reconstruction of Galaxy formation? We want to understand the sequence of events that led to the Milky Way as it is now. Ideally, we would like to tag or associate the visible components of the Galaxy to parts of the proto-galactic hierarchy: i.e., to the baryon reservoir that fueled the stars in the Galaxy. This seems too difficult. In the process of galaxy formation and evolution from the CDM hierarchy, a lot of information about the proto-galactic hierarchy is lost.

Information about the proto-hierarchy is lost at several phases in the Galaxy formation process:

- as the dark matter virializes
- as baryons dissipate within the dark halo to form the disk
- in the bulge-forming process, whether the bulge forms by mergers or by disk instabilities
- during the subsequent accretion of objects from the environment: information is lost, though some traces remain
- during the evolution of the stellar disk, as orbits are scattered by dynamical processes, including interaction with transient spiral waves and molecular clouds

At each phase, information is lost but some remains. What does the Galaxy remember? What can we hope to discover with Galactic archaeology?

1.1.8 *Signatures remembered from each phase*

We can classify the kinds of information lost as zero order (since dark matter virialized), first-order (since the main epoch of baryon dissipation), and second-order losses (subsequent evolution). Each phase leaves some signatures. In reality, of course, galactic evolution is an ongoing process without such distinct phases. In later chapters, we will look at (1) ways in which we can derive information about the early Galaxy, and (2) some of the processes that cause loss of information or provide bogus information for us to misinterpret.

Zero-order signatures

The virialization phase is dominated by merging and violent relaxation. Early stars form in small elements of the hierarchy, long before the main body of the Galaxy has come together. Some of these stars will become part of the metal-poor halo. The total binding energy E, mass M, and angular momentum parameter $\lambda = J|E|^{1/2}G^{-1}M^{-5/2}$ where J is the angular momentum are more or less established at this phase, although they continue to evolve slowly: E, M, and J determine the gross nature of the galaxy.

The globular cluster system formed around this time: its underlying structure has evolved mainly through the destruction of clusters by evolutionary processes and the changes in the Galactic potential since the clusters formed. Note that the old globular clusters in the Milky Way, LMC, and nearby Fornax dwarf spheroidal galaxy have almost identical ages, within 1 Gyr. Globular clusters in interacting systems like the Antennae (NGC4038/4039) indicate that globular cluster formation is associated with interaction, as in this very early phase.

Some of the properties of the metal-poor stellar halo were probably established in this epoch, as small satellites that had already formed stars were accreted by the virializing halo (more later). The Tully-Fisher law, which relates the rotational velocity and the baryon mass of galaxies, may have been established at this phase, or maybe in the next phase if the loss of baryons associated with star formation was significant.

First-order signatures

What information remains from the epoch when baryons dissipated to form the disk and the bulge? The scale length of the disk may be roughly constant since then. The mass of the disk continues to grow as gas falls in and stars form. Chemical gradients in old components like the thick disk may be conserved but could be affected by radial mixing of stars by transient spiral arms. The vertical scale height of the old disk evolves with disk heating, at least for a few Gyr, but appears to be roughly constant after about 3 Gyr from birth, so we probably see the old disk as it was about 7 Gyr ago.

The old thick disk appears to be a ubiquitous component of disk galaxies, but its formation is not yet well understood. It may represent the early thin disk, dynamically heated by accretion of satellites long ago, or it may have formed much earlier in a gas-rich merger (Brook *et al.*, 2007), or perhaps it is the debris of accreted satellites (Abadi *et al.*, 2003). It is probably now much as it was after it formed and is one of the most important of the Galactic fossils.

The bulge has also probably not changed much since its formation. If it formed by disk instability, then it is probably much as it was about 7 Gyr ago (except for the effects of stellar evolution). The shape of the dark halo may have been affected by the growth of the baryonic component within it, but is probably more or less as it was after the disk began to form stars, except perhaps near the disk plane where the effects of adiabatic compression may flatten and concentrate the dark matter distribution.

Second order signatures

What information remains from the subsequent evolution after the disk began to form? Objects like the Sgr dwarf that are accreted by the Galaxy are tidally disrupted and break up to become part of the stellar halo. Their debris gradually mixes away structurally but their stars conserve some dynamical properties that can in principle be detected in phase space or integral space.

Star-forming events in the disk mostly dissolve and phase-mix around the Galaxy. Some maintain their kinematical identity as moving stellar groups, at least for a few Gyr (e.g., the HR 1614 moving group: De Silva *et al.* (2007). A few survive as open clusters: some of these old open clusters are almost as old as the disk. Old clusters are seen in the Milky Way out to at least 15 kpc in radius. The debris of all of these star-forming events in the disk will maintain their chemical signatures, whether or not the stars stay together in configuration space (old clusters) or phase space (moving groups). Figure 1.4 shows the distribution of [Fe/H], [Mg/Fe], and [Ba/Fe] for the Hyades and Collinder 261 open clusters and for the HR 1614 moving group. These three systems have clearly maintained their chemical identity.

1.1.9 *The metal-poor stellar halo*

Figures 1.5 and 1.6 from Carney *et al.* (1996) show the orbital accentricity and azimuthal velocity against metallicity for a sample of high proper motion stars. These are more recent versions of the famous diagrams from Eggen *et al.* (1962). It shows how the more metal-rich stars that lie in the rapidly rotating thin and thick disks are mostly in orbits of low eccentricity, whereas the metal-poor stars that define the non-rotating stellar halo are in highly eccentric orbits.

Figure 1.7 compares the metallicity distribution of halo stars and globular clusters, as it was known in 1996; now a few halo stars are known with abundances of $[Fe/H] < -5$. The metallicity distribution of the halo globular clusters is narrower than for the halo stars and does not extend below about -2.5.

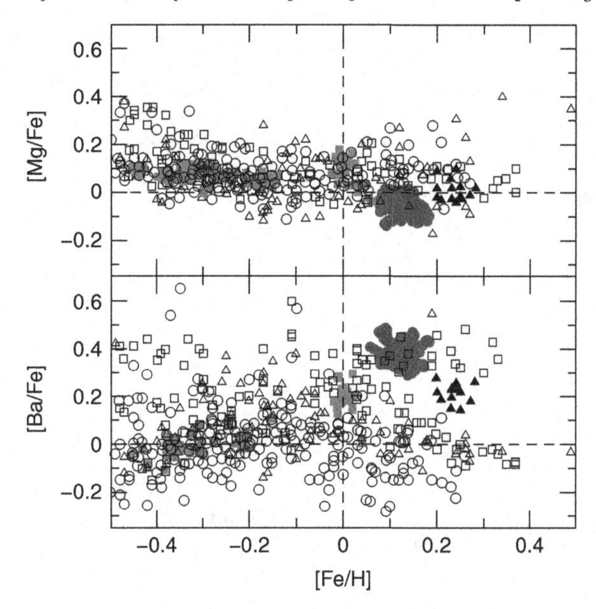

FIG. 1.4. Mg and Ba abundances for stars in the Hyades cluster (full circles), the Collinder 261 cluster (full squares), and the HR 1614 moving group (full triangles), compared with stars in the Galactic disk (open symbols), from De Silva *et al.* (2009).

We discuss the thick disk in more detail later. Most of its stars have abundances between about −0.5 and −1.0, but some of the metal-poor stars ([Fe/H] < −1) have disk-like kinematics and form a metal-poor tail of the thick disk extending down to [Fe/H] = −2. About 25% of the stars with [Fe/H] = −1.5 near the sun belong to the thick disk: this is not so apparent in Figure 1.7 because this sample of stars is kinematically selected and favors stars of the halo. The important point here is that the stellar halo and thick disk are ancient structures and are very significant for galactic archaeology.

Halo streams

The halo extends out beyond a radius of 100 kpc, and its long orbital timescales allow the survival of identifiable debris from accretion events. The Sgr tidal stream, originally discovered behind the Galactic bulge, appears to extend at least twice around the Galaxy. The stars of the stream include some that are younger than those of the bulge and extend out to much redder colors. The Sgr stream is delineated in longitude by M giants.

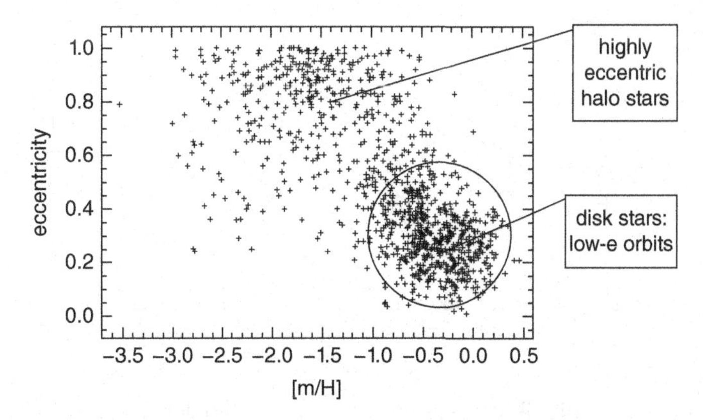

FIG. 1.5. The eccentricity-metallicity relation for stars of the Galactic halo and disk, adapted from Carney *et al.* (1996).

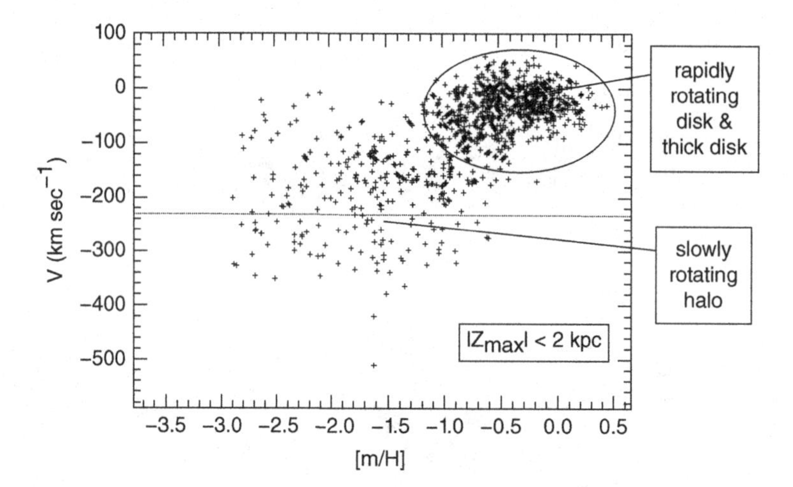

FIG. 1.6. The relation between azimuthal velocity V and metallicity for stars within 2 kpc of the Galactic plane, adapted from Carney *et al.* (1996).

These tidal streams from the currently disrupting Sgr dwarf are interesting for Galactic archaeology, as are the ancient streams from small objects accreted long ago into the halo. The long orbital periods allow these ancient streams to survive in phase space, so the metal-poor halo is the best place to attempt reconstruction of such accretion events. Some are visible in the projection of configuration space on the sky. Even if they are too faint to see in configuration space, they may be visible in projections of phase space, such as position and radial velocity relative to the Galactic center (R_G, V_G), or in the space of integrals of the motion for stellar orbits, like energy and angular momentum (E, L_z).

Accretion is important for building the stellar halo, but it is not clear yet how much of the halo comes from discrete accreted objects (debris of star formation at high z) versus star formation during the baryonic collapse of the Galaxy. At one extreme, simulations of pure dissipative collapse (e.g., Samland and Gerhard, 2003) suggest that the halo may have formed mainly through a lumpy collapse, with only ~10% of its stars coming

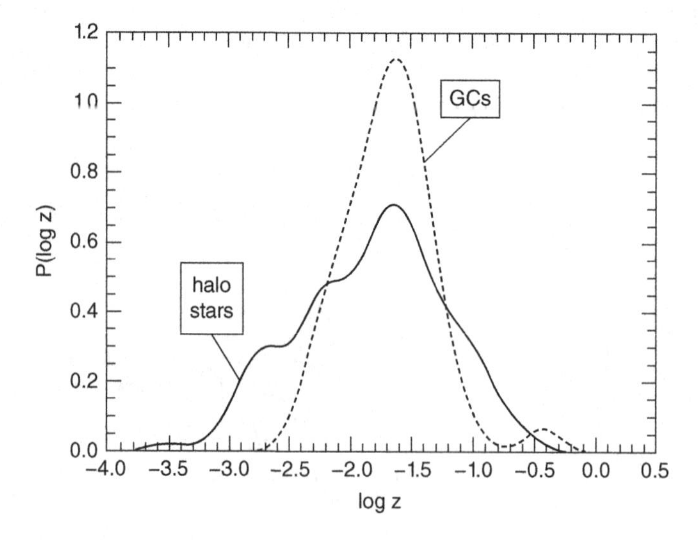

FIG. 1.7. The metallicity distributions of halo stars and globular clusters, from Carney *et al.* (1996). Some halo stars are now known with metallicities below −5.

from accreted satellites. In any case, we can hope to trace the debris of these lumps and accreted satellites from their phase space structure.

The first stars

Where are the first stars now? Simulation indicates that the metal-free stars formed until redshift $z \sim 4$ in chemically isolated sub-halos, far away from the largest progenitor. If these stars survive, they would be spread through the Galactic halo. If they are not found, then their lifetimes are less than a Hubble time, which would imply a truncated IMF for these early stars.

The oldest stars form in the early rare density peaks in the region of the final system. Now they are concentrated to the central region of the Galaxy (Diemand *et al.*, 2005; Brook *et al.*, 2007). The first stars have orbits of fairly high eccentricity (Scannapieco *et al.*, 2006), like the stars of the metal-poor stellar halo.

1.1.10 *General references*

Here are some general references on Galactic dynamics and astronomy.

Galactic Dynamics: Binney and Tremaine (1987, 2008).

Galactic Astronomy: Binney and Merrifield (1998). A more descriptive book, well worth reading for background.

Galaxies in the Universe: Sparke and Gallagher (2007). A good descriptive book about galaxies, including some essential basic theory.

Galactic Populations, Chemistry and Dynamics: Turon *et al.* (2008). This is a very useful and up-to-date compendium of Galactic knowledge, problems, techniques, and surveys.

1.2 Dynamical processes which lose information

In this section, we discuss some of the dynamical processes that lose information or generate potentially misleading information for Galactic archaeology. These processes include the accretion of satellites, dynamical resonances with the bar and spiral structure, heating of the disk by various processes, and radial mixing.

1.2.1 *Accretion and destruction of satellites*

This is an important part of CDM theory: the merging of smaller objects of the hierarchy to form larger objects. Small galaxies are accreted and destroyed by larger galaxies. The debris of the small ones becomes part of the halo, bulge or disk of the larger one.

The orbital energy and angular momentum of the smaller galaxy is absorbed by the dark halo and disk of the larger one. The existence of thin disks constrains the merger history since the disk formed, because disks can be puffed up or destroyed by significant mergers.

The goal here is to describe some of the essential dynamics of merging, accretion and disruption.

Galaxy mergers

The interaction and merging of galaxies and pregalactic fragments is a major element in their formation and evolution. Galaxies are believed to be built up by the merging of a hierarchy of sub-galactic fragments. Mergers of fullyformed galaxies and groups of galaxies are commonly observed, and the end products are believed to be giant elliptical galaxies or large early-type disk galaxies. Accretion of smaller satellites by disk galaxies are believed to contribute to the thickening of the disk, as the disk absorbs energy and angular momentum from the orbit of the satellite. In this section, we discuss the dynamics of the merging and accretion process. Merging usually means the merging of systems of comparable mass, while accretion means the accretion of a small galaxy by a larger one. A nice example of a small galaxy being tidally disrupted by a larger one is seen in the APOD image of the galaxy NGC 5907. The "field of streams" seen in the SDSS star counts for the halo of our Galaxy shows halo streams that are most likely related to one or more accretions of small galaxies into the halo of the Galaxy.

As disk galaxies undergo a close approach, they can interact tidally and merge. The merging stimulates star formation, generates tidal arms and bridges, and disrupts the galaxies. NGC 4038/4039 is a nice example of an ongoing interaction of a pair of spiral galaxies. A fine image of this system can be found on the STScI website.

Much of the discussion follows Binney and Tremaine (1987). When two galaxies interact, direct hits of the stars are unlikely, because the fraction of the area of the galactic disk that is filled by stars is quite small. For example, in the solar neighborhood, the number density of stars is about 20 pc^{-3}, and the radius of a typical star is about $10^{-0.2}$ $R_{\odot}$, so the fractional area covered is about 10^{-14}. But encounters of galaxies do change the dynamical state of stars in the encountering galaxies. Orbital energy is converted into internal energy within the galaxies, and this can lead to merging.

Some encounters lead to mergers and some don't. Consider two interacting galaxies, A and B. A star in orbit about the center (O_A) of galaxy A gains energy at a rate $\mathbf{v} \cdot \mathbf{g}(\mathbf{r})$, where $\mathbf{g}$ is the (gravitational attraction at the position $\mathbf{r}$ of the star) – (the gravitational attraction at O_A) and $\mathbf{v}$ is the stellar velocity relative to O_A. Let the initial relative velocity of galaxy B relative to A be $\mathbf{v}_\infty$. As $\mathbf{v}_\infty$ increases, the time $t_\circ$ to closest approach of the two galaxies decreases, and the total change of energy of our star

$$\Delta E = \int_0^{t_\circ} \mathbf{v} \cdot \mathbf{g}(\mathbf{r}) dt \tag{1.1}$$

decreases, and the star takes less energy from the orbit. There is a critical velocity v_f such that $v_\infty > v_f$ means that the galaxies can escape to infinity after the closest approach. On the other hand, if $v_\infty < v_f$ then the systems will merge. For $v_\infty > v_f$, the orbits and internal structure of the galaxies are relatively weakly affected. However, for galaxies that lie in the tidal field of a cluster of galaxies, even these fast encounters can be quite damaging. They increase the internal energy of the victim which then becomes more loosely bound and prone to disruption by the tidal field of the cluster.

For the Milky Way, with its prominent thin disk and small bulge, accretion of small galaxies is more important for its evolution than major mergers. We look now at the two main processes involved in accretion.

Dynamical friction

Dynamical friction is the frictional effect on a mass M moving through a sea of stars of mass m. Assume that the smaller masses m are uniformly distributed, and adopt the "Jeans Swindle" (i.e., ignore the potential of the uniform distribution of the m objects). Then the motion is determined only by the force of M and the disturbances that M produces to the distribution of m objects.

M raises a response in the sea of smaller objects, and this response acts back on M itself. Summing the effects of the individual encounters of M and m, we see that M suffers a steady *deceleration* parallel to its velocity $\mathbf{v}$. If the velocity distribution of m is Maxwellian

$$f = \frac{n_\circ}{(2\pi\sigma^2)^{3/2}} \exp(-v^2/2\sigma^2) \tag{1.2}$$

then the drag is

$$\frac{d\mathbf{V}_M}{dt} = -\frac{4\pi \ln \Lambda G^2 \rho_m M}{V_M{}^3} \left(\mathrm{erf}\chi - \frac{2\chi}{\sqrt{\pi}} \exp(-\chi^2) \right) \mathbf{V}_M \tag{1.3}$$

for $M \gg m$.

$$\chi = V_M / \sqrt{2}\sigma \tag{1.4}$$

and Λ = (maximum impact parameter) $\times$ (typical speed)$^2/GM$: $\Lambda \gg 1$.

So (i) the drag acceleration is $\propto \rho_m$ and $\propto M$ and (ii) the drag force $\propto M^2$. This comes about because stars deflected by M generate a downstream density enhancement: the enhancement $\propto M$, the force back on $M \propto M^2$.

This estimate neglects the self-gravity of the density enhancement; i.e., it includes the attraction of m and M, but not m and m. The estimate seems to be fairly consistent with the results of N-body simulations, as long as the ratio of M to the total mass of the m objects < 0.2 and the orbit of M is not confined to the core or to the exterior of the larger system. The estimate also neglects resonances between the orbit of M and the orbits of m objects within their system: such resonances enhance dynamical friction, as we shall see.

For example, consider the likely fate of the LMC, now located at about 60 kpc from the Galaxy. New results on the proper motions of the LMC and SMC indicate that they may be unbound to the Galaxy, but for the purposes of this example, we assume that they are in circular orbits around the Galaxy. For circular orbits, the torque from dynamical friction due to the dark halo of our Galaxy gives a decay time

$$t_{fric} = \frac{10^{10}}{3} \left(\frac{r}{60 \text{ kpc}} \right) \left(\frac{V_c}{220 \text{ km s}^{-1}} \right) \left(\frac{2 \times 10^{10}}{\mathrm{M}_\odot} \right) \text{ yr} \tag{1.5}$$

so if the galactic halo extends out beyond a radius of 60 kpc (which appears to be true) and the LMC orbit is approximately circular, then the LMC (and SMC) will sink into the Galaxy in a time less than the Hubble time.

Tidal disruption

Consider a satellite of mass m in a circular orbit around a host of mass M at a distance D. The angular speed around the common center of mass is $\Omega^2 = G(m+M)/D^3$. In this rotating frame, we have the Jacobi integral $E_J = E - \mathbf{\Omega} \cdot \mathbf{L} = \frac{1}{2}v^2 + \Phi_{\mathit{eff}}(\mathbf{r})$, where Φ_{eff} is the effective potential of the gravitational plus centrifugal forces. The contours of Φ_{eff} have a saddle point between the two masses, where the centrifugal and gravitational forces balance and $\partial\Phi_{\mathit{eff}}/\partial x = 0$ (see Binney and Tremaine, 2008:676). For closed contours within the saddle point, particles are bound to one or the other of the two masses. Beyond this saddle point the contours open out and particles can be unbound to either of the two masses. For $m \ll M$, the distance of the saddle point from the smaller mass is

$$r_J = \left(\frac{M}{3M} \right)^{1/3} D \tag{1.6}$$

This is a measure of the tidal radius of the smaller mass m. It is a rough estimate because

(i) the zero velocity surface is not spherical
(ii) orbits do not necessarily escape because the ZV surface is open
(iii) the orbit of m is not usually circular
(iv) m often lies within M, so the point mass approximation is poor

but the main point here is that tidal removal of matter can occur at a radius from m such that the densities $\rho_m(r_J) \sim \rho_M(D)$. For example, we expect an infalling satellite to remain intact to a distance D from the larger galaxy, such that $\rho_M(D) \sim$ the mean density of the satellite.

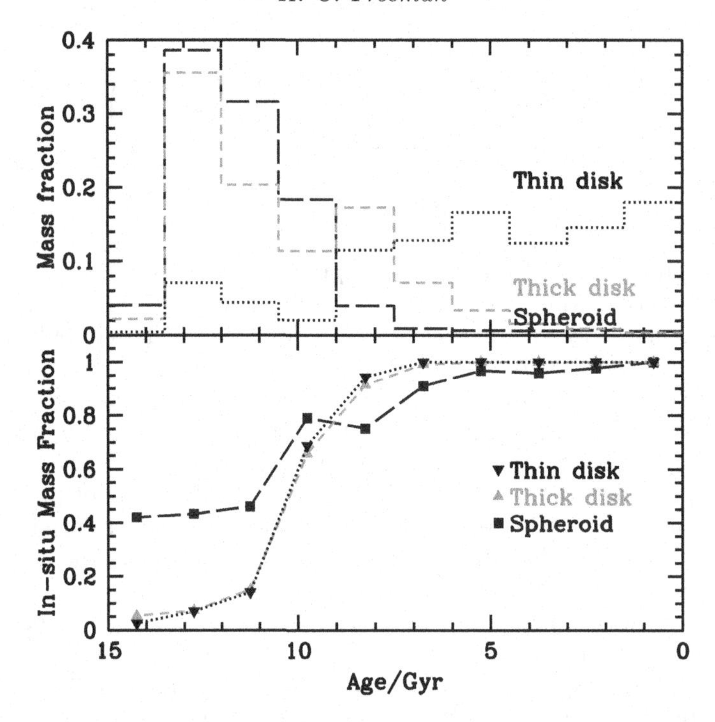

FIG. 1.8. In the simulations of Abadi *et al.* (2003), only a small fraction of the oldest disk stars formed in the disk. Most of these stars come from circularized satellite debris.

To summarize the merger preliminaries:

- Dynamical friction (large particle M in a sea of small particles m): the drag force on $M \propto \rho_m M^2$, neglecting resonances and the self-gravity of the wake.
- Tidal disruption: this occurs for a satellite whose orbit is decaying under dynamical friction, when the mean density of the satellite orbit is same as the mean density of the satellite. Very dense satellites can survive accretion and end up in the inner regions of the larger galaxy, whereas low density satellites are broken up in the outer regions of the parent.

Simulations of accretion of small satellites

During the last 20 years, many simulations have aimed at reproducing the structure of particular interacting galaxies. The antennae system NGC 4038/9 is a favorite. Earlier merger simulations were made with rigid dark halos. Later simulations with live halos showed that the halos can absorb energy and angular momentum, and make the merger process go more rapidly. Similarly, simulations of accretion events by galaxies with live dark halos show that the live halos speed up the accretion process.

A typical model for simulations of the accretion of a small satellite by a disk galaxy has a satellite with mass = 10% of the mass of the large galaxy. The satellite is initially in a circular orbit with a radius of about 6 disk scale lengths and an inclination of 30° to the disk plane, and in a prograde orbit. Typically about 10^6 particles are used in such a simulation; they are distributed in the disk, the live halo of the disk galaxy, and in the satellite.

The satellite orbit decays and its inclination becomes less, due to the effects of dynamical friction. The orbital energy and orbital angular momentum go primarily into the stars of the disk. The heating of the disk is limited by the stripping of the satellite by the tidal field of the large galaxy, and also by the (impulsive) shocks that the satellite receives as its orbit crosses the plane of the disk.

Simulations by Walker *et al.* (1996) show how a satellite in a prograde orbit typically sinks into the plane of a parent disk galaxy in a time of order 1 Gyr. The disk's dynamical friction provides about 75% of the torque on the satellite. Dynamical friction against the dark halo provides the rest. The disk of the parent is thickened by the encounter, and the debris of the satellite becomes part of its disk and halo.

Abadi *et al.* (2003) analyzed an SPH simulation of the formation of a disk galaxy assembled hierarchically in ΛCDM. They identified a large spheroid and two disk components: thin and thick. The spheroid stars are old (>8 Gyr). The disk stars cover a wide range of age, but most of the older (>10 Gyr) disk stars did not form in the disk but came from accreted satellites whose orbits were circularized by dynamical friction before disruption: see Figure 1.8. We need to find observational tests to determine how much of the disk and thick disk of a typical disk galaxy comes from such satellite debris.

How could we evaluate how much of the old disk stars came from outside? Kinematically this would be difficult, because their kinematics would be much like those of disk stars born insitu. Chemical techniques look promising. The overall metal abundance of the satellites depends on their stellar luminosity. Disk stars have abundances [Fe/H]> -1 so the absolute magnitudes of the infalling satellites must have been brighter than about -15. That is consistent with the Abadi *et al.* (2003) satellites that fell into the disk: they were typically more massive than 10^9 $M_\odot$. It makes sense too from dynamical friction theory: only the more massive satellites can suffer significant orbital decay due to dynamical friction.

Chemical properties of dwarf galaxies

The dwarf spheroidal galaxies are now known to show a well-defined relation between their luminosity and their mean metallicity relation, indicating some level of internal chemical evolution. Individual dwarfs show a large range of abundances, with the faintest of the dwarf spheroidal galaxies containing stars with [Fe/H] < -3. The chemical properties of surviving satellites (the dwarf spheroidal galaxies) vary from satellite to satellite and are different in detail from the more homogeneous overall properties of the disk stars (see Venn and Hill, 2008). The faint Hercules dSph galaxy has stars with very extreme ratios of Mg/Ca, suggesting enrichment by just one or two high mass SNII (Koch *et al.*, 2008).

Chemical studies of the old disk stars in the Galaxy might therefore help to identify disk stars that came in from outside in satellites that then disrupted. We can think of a chemical space of abundances of elements O, Na, Mg, Al, Ca, Mn, Fe, Cu, Sr, Ba, Eu, for example. The dimensionality of this space is probably between about 7 and 9. Most disk stars inhabit a sub-region of this space. Stars that came in from satellites may be different enough to stand out from the rest of the disk stars. With this chemical tagging approach (more later), we may be able to detect or put observational limits on the satellite accretion history of the galactic disk.

1.2.2 *The metal-poor stellar halo*

The stellar [Fe/H] abundances for the halo range from about -1 to -5, overlapping with the metal-poor tail of the thick disk. Its density distribution follows approximately a power law distribution $\rho \sim r^{-3.5}$, and the halo extends out to at least 100 kpc. Its mass is about 1×10^9 $M_\odot$, so it provides only a small fraction of the total stellar mass of the Galaxy (about 6×10^{10} $M_\odot$). The halo is believed to be made up at least partly from the debris of low-mass accreted satellites (Searle and Zinn, 1978), and this view is supported by the presence of stellar streams in the halo. Stars in the halo will retain some memory of their dynamics at the time when they were liberated from their parent satellite. When accurate astrometry becomes available from Gaia for large numbers of halo stars, it may be possible to reconstruct these satellites from the distribution of halo stars in integral or action space. Helmi and de Zeeuw (2000) simulated this recovery process, using a time independent gravitational field. More recent simulations (e.g., Gao *et al.*, 2004) indicate that the dark halo has probably doubled its mass since $z = 1$. Even in such a time-dependent situation, the satellite debris retains its identity in integral space (Knebe *et al.*, 2005), although its average position in integral space does change. We could expect the adiabatic invariance of the actions to give a more constant reconstruction of the satellite debris. Dynamically reconstructing at least some of the objects that formed at high redshift and then became part of the Galactic halo seems feasible. Gaia will contribute greatly to this endeavor.

1.2.3 *Some unpleasant stellar dynamical effects*

We now look at some stellar dynamical effects that make Galactic archaeology more difficult. First, a brief discussion of stellar orbits in axisymmetric galaxies is needed.

The equations of motion are

$$\ddot{R} - R\dot{\phi}^2 = -\frac{\partial \Phi}{\partial R} \tag{1.7}$$

$$\frac{d}{dt}(R^2\dot{\phi}) = 0 \tag{1.8}$$

$$\ddot{z} = -\frac{\partial \Phi}{\partial z} \tag{1.9}$$

so $L_z = R^2\dot{\phi} = \text{constant}$.

The motion in the meridional (R, z) plane is then given by

$$\ddot{R} = -\frac{\partial \Phi_{\text{eff}}}{\partial R} \tag{1.10}$$

$$\ddot{z} = -\frac{\partial \Phi_{\text{eff}}}{\partial z} \tag{1.11}$$

where the effective potential is

$$\Phi_{\text{eff}} = \Phi(R, z) + L_z^2/2R^2 \tag{1.12}$$

When these equations of motion are integrated numerically, the typical orbit has the rosette form shown in Figure 1.9. The orbit has an inner and outer radial limit defined by its two integrals, energy and angular momentum. The side-on view in Figure 1.9 shows that the orbit cannot reach all points that are permitted by its energy and angular momentum, so it is affected by a third integral that cannot be written down in analytical form. Galactic disks are built up from such orbits.

Nearly circular orbits

Most stars in the disks of disk galaxies are in orbits that are not far from being circular. We take the system to be axisymmetric. The equations of motion are

$$\ddot{R} = -\frac{\partial \Phi_{\text{eff}}}{\partial R} \tag{1.13}$$

$$\ddot{z} = -\frac{\partial \Phi_{\text{eff}}}{\partial z} \tag{1.14}$$

The circular orbit is defined by

$$\frac{\partial \Phi}{\partial R} = \frac{L_z^2}{R_g^3}, \tag{1.15}$$

where R_g is the radius of the circular orbit.

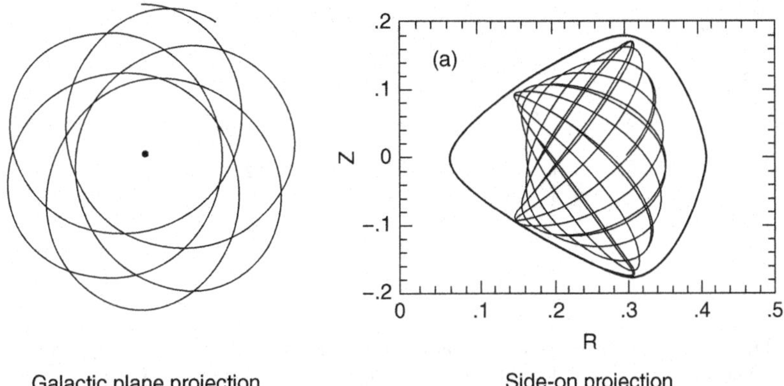

FIG. 1.9. In the plane projection, the orbit has the rosette form, with an inner and outer limit. In the side-on projection, the closed outer curve shows the region within which the orbit can move if limited only by its energy and angular momentum. The orbit cannot reach all points within this region, so a third integral is acting.

Write $x = R - R_g$, so the circular orbit is $(x, z) = (0, 0)$. Expand Φ_{eff} about $(0, 0)$

$$\Phi_{\text{eff}} = \frac{1}{2}x^2 \left[\frac{\partial^2 \Phi_{\text{eff}}}{\partial R^2}\right]_{R_g,0} + \frac{1}{2}z^2 \left[\frac{\partial^2 \Phi_{\text{eff}}}{\partial z^2}\right]_{R_g,0} \tag{1.16}$$

(no linear or xz terms by symmetry).

Write this expansion as

$$\Phi_{\text{eff}} = \frac{1}{2}\kappa^2 x^2 + \frac{1}{2}\nu_z^2 z^2 \tag{1.17}$$

Then $\ddot{x} = -\kappa^2 x$ and $\ddot{z} = -\nu_z^2 z$. These are the equations of motion for a 2-D simple harmonic oscillator. The frequency κ is the *epicyclic frequency* and ν_z is the *vertical frequency.*

The circular frequency Ω is

$$\Omega^2 = \frac{1}{R}\frac{\partial \Phi}{\partial R} \tag{1.18}$$

so

$$\kappa^2 = \left(R\frac{d\Omega^2}{dR} + 4\Omega^2\right)_{R_g} \tag{1.19}$$

In real galaxies, $\sqrt{2}\Omega < \kappa < 2\Omega$. The ratio κ/Ω is usually irrational, so the orbit is an unclosed rosette (see Figure 1.10). Here are values of the periods associated with these three frequencies in the solar neighborhood: $2\pi/\Omega \simeq 2.5 \times 10^8$ yr, $2\pi/\kappa \simeq 1.9 \times 10^8$ yr, and $2\pi/\nu_z \simeq 0.7 \times 10^8$ yr (increasing rapidly with $z-$ amplitude).

Now we have the (R, z)–motion for nearly circular orbits. How about the ϕ-motion? We can write

$$\dot{\phi} = \frac{L_z}{R^2} = \frac{L_z}{R_g^2}\left(1 + \frac{x}{R_g}\right)^{-2} = \Omega_g\left(1 - \frac{2x}{R_g}\right) \tag{1.20}$$

Write $x = X \cos \kappa t$ so

$$\phi - \phi_\circ = \Omega_g t - \frac{2\Omega_g}{\kappa R_g} X \sin \kappa t \tag{1.21}$$

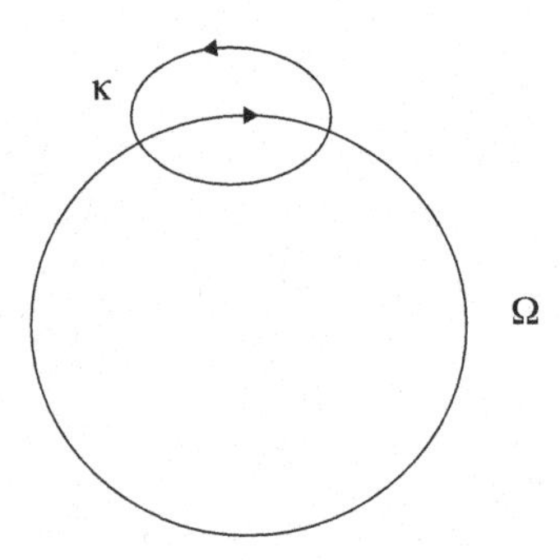

FIG. 1.10. This figure shows the two in-plane components of a near-circular rosette orbit. The rosette is the vector sum of the circular guiding center motion (frequency Ω) and the retrograde elliptical epicyclic motion (frequency κ). The angular momentum is constant on the orbit.

Now write $y = R_g(\phi - \phi_\circ - \Omega_g t)$, so (x, y) are cartesian coordinates with origin at the *guiding center*, which goes around the galaxy at the local circular velocity $\Omega_g R_g$.

In the (x, y) plane, the orbit is an ellipse, with axial ratio $\kappa/2\Omega < 1$, in the retrograde sense. This is called *epicyclic motion.* The velocity dispersion is related to the amplitude of the epicycle: for example, for old disk stars, the velocity dispersion $\sigma_R \simeq 40$ km s^{-1} and the rms x–amplitude is about 1.1 kpc. The radius R_g is about 8.5 kpc. A rosette orbit is the vector superposition of these two components: the epicycle + the circular guiding center motion.

Motion of stars near the sun

Now look at the motions of stars near the sun, in the light of the orbit theory. The Local Standard of Rest (LSR) is a coordinate system going around the Galaxy at the solar radius at the circular velocity of 220 km s^{-1} (recent maser astrometry indicates that a value of 250 km s^{-1} may be closer to the truth). Define the (U,V,W) velocity components of a star relative to the LSR such that U is in the direction of the Galactic center, V is in the direction of Galactic rotation, and W is in the direction of the North Galactic pole. We see that the distribution of stars in the (U,V) plane is lumpy. Are these lumps the remains of star-forming events that would be interesting archaeologically?

Figure 1.11 shows the distribution of stars in the (U,V) plane, from Dehnen (2000) analysis of Hipparcos data for nearby stars. Several lumps in the distribution are identified as stellar moving groups. The Hercules group is believed to be associated with local resonant kinematic disturbances by the inner bar of the Galaxy. The nature of the Sirius and Hyades streams remains contentious but may also be associated with local resonances. The resonant nature of the Hercules group is supported by a chemical study of Hercules stars by Bensby *et al.* (2007). They find that the chemical properties of stars in the Hercules group are indistinguishable from those of a random sample of disk stars, so there is no hint that the Hercules stars are related by birth. On the other hand, we have seen earlier that the stars of the HR 1614 group (which lies near the Hercules group in the (U,V) plane) have a well-defined distribution in chemical space and are very likely to be the debris of a common star-forming event.

Resonant groups

Some stars are in resonance with a rotating gravitational pattern from the bar or spiral structure. The guiding center frequency Ω and the epicyclic frequency κ depend on the radius R. For the rotating pattern, we can take its pattern frequency Ω_p to be constant. The outer Lindblad resonance (OLR) occcurs where $\Omega_p = \Omega(R) + \kappa(R)/2$. Stars with such values of Ω see the same bar potential at the same phase of their κ-oscillation.

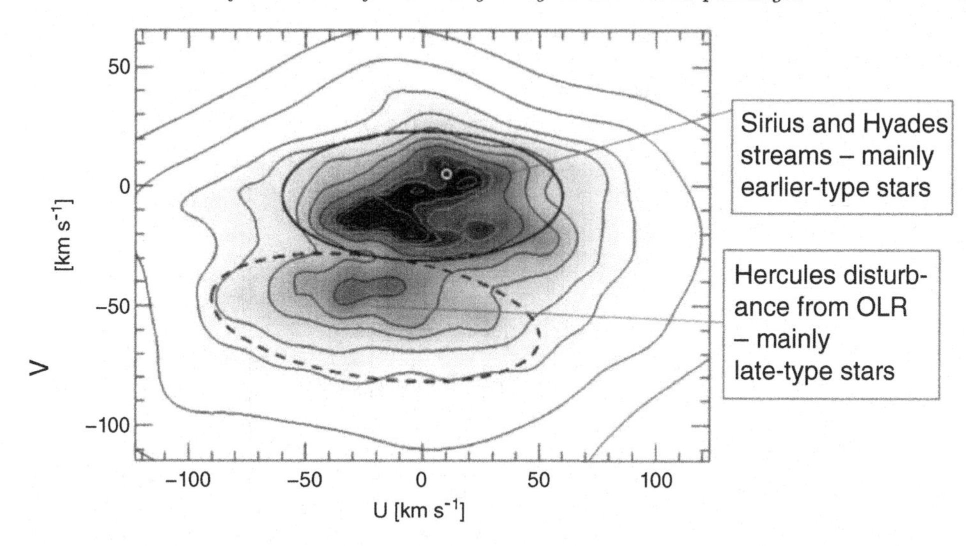

FIG. 1.11. Contours of the (U,V) distribution of nearby stars (Dehnen, 2000). The distribution is not smooth: several concentrations are identified, including the Hercules moving group that is believed to be associated with a local resonant disturbance by the inner bar of the Galaxy.

They are locked into the resonance. For the Galactic bar and disk, the OLR appears to lie near the sun. Stars in OLR have their guiding center radius lying inside the solar radius, so their angular momentum is a little lower than the angular momentum of the LSR and their $V-$velocity is about -50 km s^{-1}, like the Hercules group.

Resonances can occur not only from the bar but also from any azimuthally propagating disturbance like a transient spiral structure that has a well-defined pattern speed. Many (most?) groups of stars observed to be moving together are probably resonant phenomena. There are $\Omega_p = \Omega(R) \pm \kappa(R)/2$ resonances and also $\Omega_p = \Omega(R) \pm \kappa(R)/4$ and $\Omega_p = \Omega(R) \pm \kappa(R)/6$ resonances, etc., all of which can generate dynamical moving stellar groups.

Real coeval chemically homogeneous moving groups do exist. They are potentially very useful for galactic archaeology but are so far relatively rare. The dynamical groups are a nuisance for archaeology. We can expect that Gaia will discover large numbers of moving groups: to identify those that are the debris of some ancient star-forming event, it will be necessary to check their chemical homogeneity.

For small epicyclic amplitudes, resonances occur at a particular guiding center radius, for example, where $\Omega_p = \Omega(R) \pm \kappa(R)/2$. When the epicyclic amplitude A is larger than a few hundred pc, the linear theory breaks down. Orbits can still be represented as a circular guiding center motion plus an epicycle, but the epicycle is no longer a simple ellipse and κ is no longer just a function of R: now $\kappa = \kappa(R, A)$. The resonant condition is still $\Omega_p = \Omega(R) \pm \kappa(R)/2$ and Ω is still $\Omega(R)$, but the non-linearity of κ broadens the resonance in R, so the resonances become yet more widespread.

Some potential metal-poor moving groups

Figure 1.12 shows the histogram of the angular momentum component J_z for Gratton *et al.*'s (2003) sample of nearby metal-poor stars with well-measured chemical abundances; it shows a couple of peaks that may be associated with moving stellar groups. The retrograde ω Centauri feature may come from the accretion event that brought the globular cluster ω Cen into the Milky Way. Recent work by Wylie-de Boer *et al.* (2010) shows that these stars have some of the chemical peculiarities shared with stars of the cluster itself.

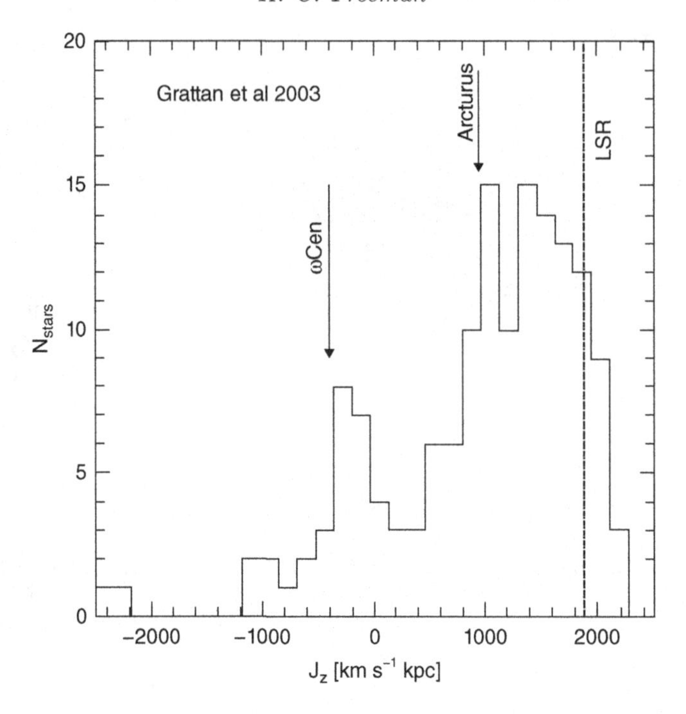

FIG. 1.12. Histogram of angular momentum for a sample of metal-poor stars. The two labelled peaks appear to be associated with the Arcturus moving group and the debris of the ω Cen galaxy.

The Arcturus feature was also thought to be the debris of an accretion event assocated with the thick disk (Navarro *et al.*, 2004). Chemical studies by Williams *et al.* (2009) show no chemical identity among the stars of this feature, so it may turn out to be a resonant feature.

1.2.4 *Radial mixing*

An azimuthally propagating disturbance, like a transient spiral pattern with pattern frequency Ω_p, affects stars near corotation where $\Omega_p = \Omega(R)$. Depending on the relative phase of the disturbance, the star can be flipped from a near-circular orbit at one radius to another near-circular orbit at a different radius (Sellwood and Binney, 2002); see Figure 1.13. The amplitude and sign of the radius change depend on the strength and phase of the disturbance.

This effect adds to the difficulties of Galactic archaeology: we have always believed that stars in near-circular orbits are at radii close to where they were born. This need not be true.

Roškar *et al.* (2008) made SPH simulations of disk formation from cooling gas in an isolated dark halo, including star formation and feedback. The simulation was aimed at understanding the outer break or truncation that is seen in many galactic disks. The break in the radial surface density distribution is seeded by rapid radial decrease in surface density of cool gas. The break is already apparent within 1 Gyr and gradually moves outward as the disk grows. The stars beyond the break radius come almost entirely from the inner regions of the Galaxy via the Sellwood and Binney (2002) spiral arm interactions, so its stars are relatively old. This effect may be very important for understanding the evolution of the abundance gradient in the disks of disk galaxies (e.g., Vlajić *et al.*, 2009).

It seems clear now that stellar kinematics alone cannot reliably show whether a moving group is a resonance effect or the debris of an ancient accretion event or star-formation

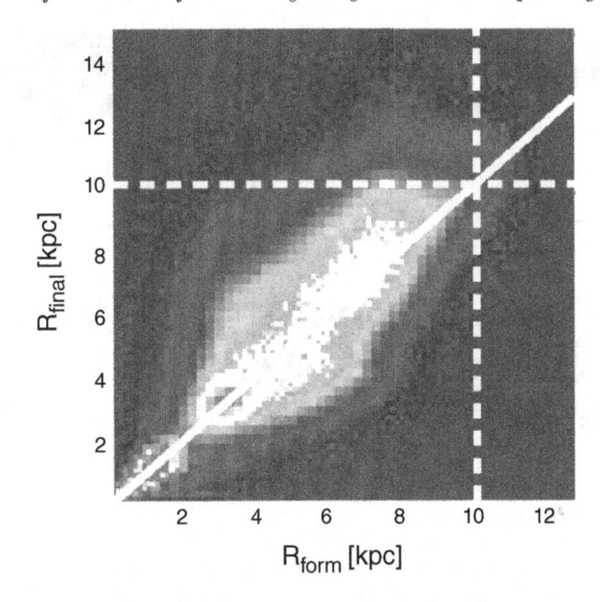

FIG. 1.13. Radius of formation of stars against their final radius, from the simulation by Roškar *et al.* (2008). Most of the stars in the outermost regions of the disk formed at smaller radii and were scattered out to larger radii by the action of the transient spiral structure.

event. Furthermore, we cannot assume that stars are near their birth radius, not even for stars in near-circular orbits. Chemical signatures are essential.

1.3 The Galactic bulge and the globular clusters

1.3.1 *Different kinds of bulges*

Figure 1.14 illustrates the range of bulges seen in galaxies, from the large classical bulge of the Sombrero galaxy to NGC 5907, which appears to have no bulge at all. Recall from Lecture 1 that forming galaxies with small or no bulges is currently difficult in CDM

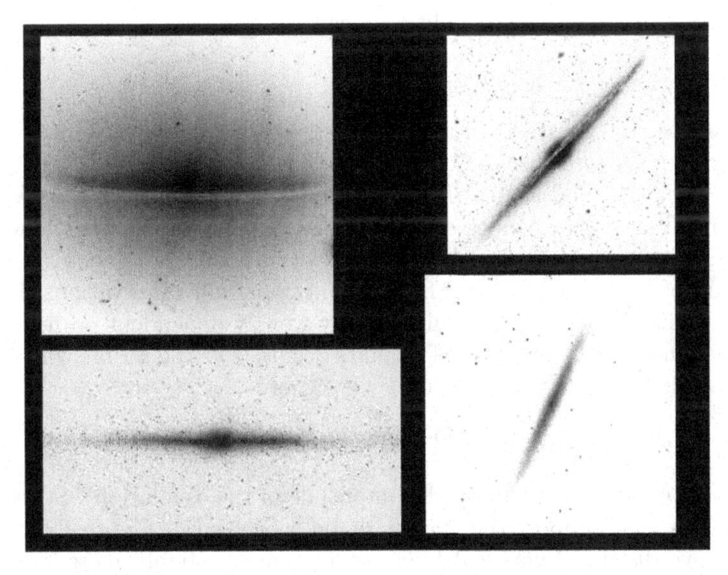

FIG. 1.14. Clockwise from top left: the large classical $r^{1/4}$ bulge of NGC 4594, the boxy bulge of NGC 4565, NGC 5907 with no bulge at all, and our Galaxy with its small boxy bar-bulge.

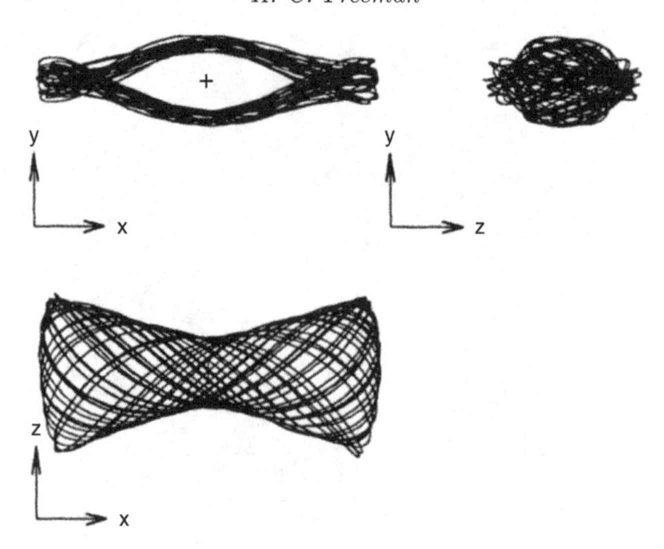

FIG. 1.15. Orbits supporting the peanut bar (from Combes *et al.* 1990).

because of the relatively active ongoing merger history. Establishing the merger history of the Milky Way observationally is a major goal for Galactic archaeology. We need to understand how the Galactic bulge formed: is it even partly a merger product, or did it form entirely through internal processes such as disk and bar instability?

The large classical bulges as in the Sombrero galaxy are believed to be merger products. Merger dynamics and violent relaxation lead to bulges with the characteristic $r^{1/4}$–law light distribution (Sersic index ~4). Classical bulges are common in early-type galaxies but become progressively rarer toward later types. They share some structural, dynamical, and population properties with the lower-luminosity elliptical galaxies.

Later-type galaxies like the Milky Way mostly have small near-exponential boxy bulges rather than $r^{1/4}$ bulges (e.g., Courteau *et al.*, 1996). Boxy bulges, as in our Galaxy, are associated with bars and are believed to form via bar-buckling instability of disk. They are probably not merger products. The observational evidence for this association comes from Kuijken and Merrifield (1995) and Bureau and Freeman (1999). The theoretical basis of forming boxy bulges by bar instabilities goes back to Combes and Sanders (1981).

Bar-buckling to form a boxy/peanut bulge

The process goes in two stages. First, the disk suffers a bar instability, which is very common for relatively cold disks. Then the bar buckles vertically, driven by horizontal and vertical resonances, and forms the boxy/peanut bulge. It takes a few bar revolutions to make this instability go (Combes *et al.*, 1990; Athanassoula, 2008). The whole process takes 2 to 3 Gyr after the formation of the disk.

The rotation of boxy bulges, both from simulations and observations, is cylindrical: the mean rotational velocity of the bulge is only weakly dependent on height above the plane. The maximum vertical extent of peanuts occurs near the radius where the vertical and horizontal Lindblad resonances occur, i.e., where

$$\Omega_p = \Omega - \kappa/2 = \Omega - \nu_z/2 \qquad (1.22)$$

(recall that both κ and ν_z depend on the amplitude of the oscillation). Stars in this zone oscillate on orbits that support the peanut shape, as shown in Figure 1.15.

So far we have discussed classical bulges, which are probably merger products, and boxy/peanut bulges, which are probably disk instability products. There is a third kind of bulge-like structure that looks like an enhancement of the surface brightness profile above the exponential disk, but appears from its shape and kinematics to lie in the disk. These are the pseudobulges.

Pseudobulges

These were discovered by Kormendy (1993) as bulges that lie above the oblate rotator curve in the $V_{max}/\sigma - \epsilon$ plane. Their location in this plane indicates that they are disk-like systems, despite their bulge-like appearance. They are believed to be generated by secular processes associated with the angular momentum transport by bars or weakly oval disks. They often show active star formation within the pseudobulge region (Carollo *et al.*, 1997). (Recall that bars are very common: about two-thirds of disk galaxies show some kind of bar structure in NIR images.)

The terms "pseudobulge" and "secular evolution" are sometimes used to include the boxy/peanut bulges. I think "pseudobulge" is best reserved for these flat disk-like enhancements that look like bulges only in their surface brightness profiles. The boxy/peanut bulges are clearly extended out of the galactic plane: there is nothing pseudo about bulges like the Galactic bulge. The instability process that is believed to generate the boxy/peanut bulges is very different from the secular transport of angular momentum usually associated with building the flat disk-like enhancements.

The different internal kinematics of classical and boxy bulges are nicely illustrated by Falcón-Barroso *et al.* (2004). They compare the cylindrical rotation of the boxy bulge of NGC 7332 with the very clearly non-cylindrical rotation of the classical bulge of NGC 5866.

1.3.2 *The Galactic bar-bulge*

Figure 1.16 shows the boxy structure of the Galactic bulge and the exponential vertical structure, typical of small boxy bulges. The stars of the bulge are believed to be mainly old, from the stellar color-magnitude diagrams: see, for example, Zoccali *et al.* (2003). The stars have a metallicity distribution extending from [Fe/H] $= -1.8$ to $+0.2$ with the mode near the metal-rich end, and the mean metallicity decreases with height above the plane (Minniti *et al.*, 1995; Zoccali *et al.*, 2008). Near the center of the bulge is a younger population on a scale of about 100 pc. It includes a nuclear stellar cluster in the central $\sim$30 pc.

Deprojection of the COBE NIR light distribution shows that the Galactic bar has a length of about 3.5 kpc, has an axis ratio of about 1:0.3:0.3, and points at about 25° from the sun-center line into the first Galactic quadrant (e.g., Bissantz and Gerhard, 2002). There is some evidence from 2MASS counts and red clump star counts for a longer (7.8 kpc $\times$ 1.2 kpc $\times$ 0.2 kpc) flatter bar structure lying in the disk of the Galaxy (López-Corredoira *et al.*, 2007).

The stars of the bulge are old and enhanced in α-elements, which implies a rapid epoch of star formation and chemical evolution. Comparison of the α-enhancement in the bulge and thick disk is currently inconclusive. Some authors (e.g., Fulbright *et al.*, 2007) argue that the α-enhancement of bulge stars persists to higher [Fe/H] values than in the thick disk. Others (e.g., Meléndez *et al.*, 2008) find that the abundance patterns for bulge and thick disk are very similar.

The bar-forming and buckling process takes a few Gyr to act after the disk settles. In this scenario, the bulge *structure* is probably younger than the bulge stars, which would originally have been part of the inner disk. The α-enhancement of the bulge and thick disk is associated with the rapid chemical evolution that took place in the inner disk before the instability had time to act.

The Galactic bulge is rotating, like most other bulges. Beaulieu *et al.* (2000) and Howard *et al.* (2008) show that the mean rotational velocity of stars in the bulge is about 100 km s^{-1}. For comparison, its velocity dispersion near the center is at least 110 km s^{-1}.

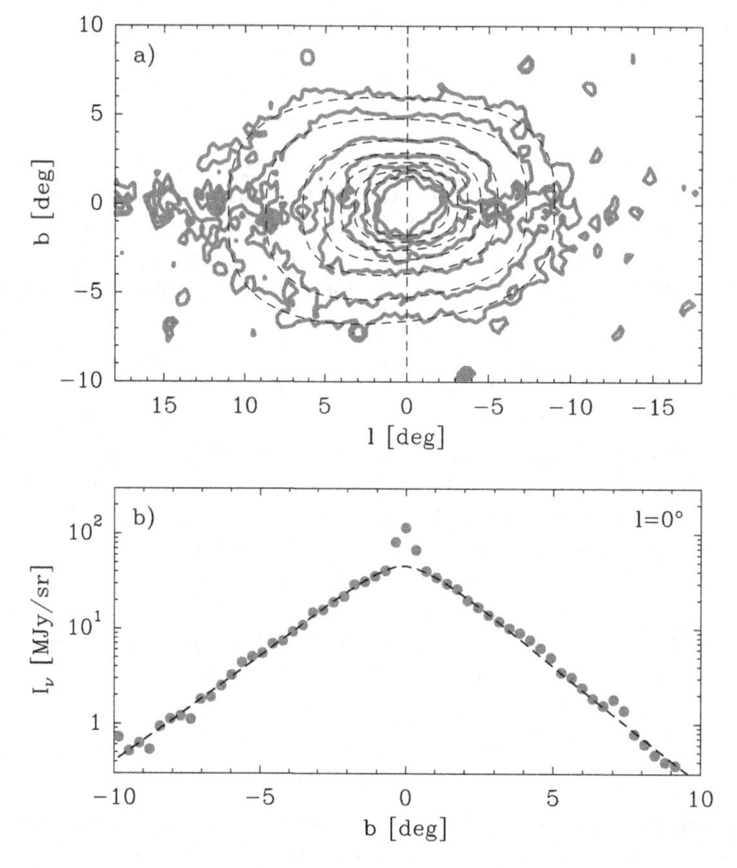

FIG. 1.16. The upper panel shows the boxy structure and higher thickness toward positive longitudes of the NIR isophotes of the Galactic bulge. The lower panel shows the exponential vertical structure of the bulge, which is characteristic of boxy bulges (from Launhardt *et al.*, 2002).

Stars of the inner disk have similar velocity dispersions (Lewis and Freeman, 1989), so it is not easy to separate stars of the inner disk and the bulge kinematically.

Tests of the formation mechanism of the Galactic bulge

One test of the bar-buckling scenario for bulge formation is to compare the observed structure and kinematics of the Galactic bulge with N-body simulations of a disk that has generated a boxy bar/bulge through bar-buckling instability of the disk. Do the simulations match the properties of the Galactic bar/bulge (e.g., exponential structure and the details of its rotation and velocity dispersion as a function of position)? Some large stellar spectroscopic surveys, aimed at these questions, are in progress. I am involved in one such survey, using the AAT and the AAOmega multi-object spectrometer, aimed at acquiring spectra of about 30,000 red clump giants in fields covering the observable southern part of the bulge and extending out into the surrounding disk. We see many metal-poor stars in the bulge region, as found earlier by Harding and Morrison (1993). Are these just the stars of the inner halo, or are they the "first stars" predicted to be concentrated to the Galactic center (e.g., Diemand *et al.*, 2005)?

If the Galactic bulge formed from a bar-buckling instability of the disk, then we might expect some similarities of the stellar populations in the bulge and in the surrounding thin and thick disks.

We can compare the Galactic bulge with the bulge of M31. The M31 bulge has a similar rotational velocity and slightly larger velocity dispersion. It appears to have a

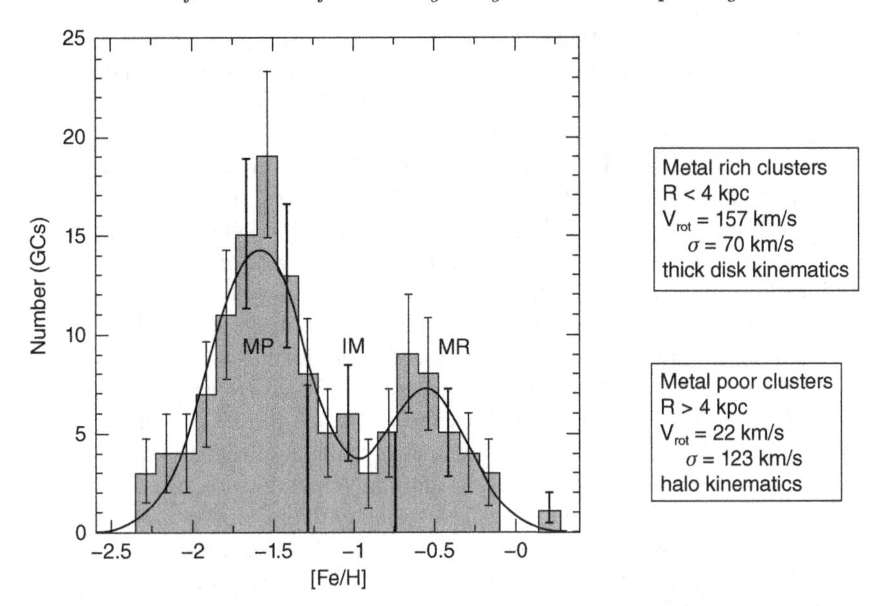

FIG. 1.17. Histogram of globular cluster abundances, from Côté (1999). The clusters in the metal-rich mode have thick disk kinematics, while those in the metal-poor mode have halo kinematics.

classical component plus an inner boxy bar/bulge component, as seen in the studies of Athanassoula and Beaton (2006) and Beaton *et al.* (2007). We can hope that the ongoing surveys of the Galactic bulge will reveal whether it also includes a classical bulge component that might be ascribed to merger activity.

1.3.3 *The Galactic globular clusters*

The globular clusters in our Galaxy are all old. The [Fe/H] distribution shows two modes: the bulge/disk clusters with [Fe/H]> -1 and the halo clusters with [Fe/H]< -1. See Figure 1.17. Zinn (1985) established that

- the more metal-rich clusters form a disk system with a highly flattened spatial distribution and a significant rotational velocity. The scale height and rotational velocity of the system are comparable to that of the thick disk. However, Minniti (1995) argued that the metal-rich globular clusters near the Galactic center were more likely associated with the bulge rather than the thick disk.
- the more metal-poor clusters are part of the halo population, with a nearly spherical distribution about the Galactic center, a small rotational velocity, and a large velocity dispersion.

Globular clusters are potentially very important in galactic archaeology, but we do not understand how they form and therefore what they represent in the process of galaxy formation. Some appear to be associated with the Sgr, Mon, and Can Maj streams so probably formed in the parent objects that were accreted.

Globular cluster formation in the local neighborhood appears to be highly synchronized. The Galaxy contains clusters of almost identical age covering the whole range of abundance, with a slightly younger component that have metallicities [Fe/H]> -1.3. The old globular clusters in the Galaxy, the LMC, and the Fornax dwarf are all coeval within 1 Gyr. The Galactic bulge/disk clusters show no abundance gradient, while the halo clusters show a weak decrease in abundance with galactocentric radius, from about -1.4 for $R < 6$ kpc to -1.8 for $R > 15$ kpc.

In the LMC and M33, some of the globular clusters are very old ($\sim$12 Gyr) and metal-poor, like the clusters in the Galaxy. However, they also contain clusters that are very young (only a few Myr) and others that have ages between 10 Myr and a few Gyr. To understand what globular clusters represent in the process of forming galaxies, we would like to know how these clusters form. The LMC and M33 are able to form them now, but the Milky Way is not: what is the difference?

Some of the Galactic globular clusters show multiple main sequences and abundance anomalies. For example, NGC 2808 has three main sequences, believed to represent three discrete He-levels up to Y = 0.4. Yet its [Fe/H] and other heavy element abundances appear uniform. Pre-enrichment must generate discrete levels of He but must not affect [Fe/H] and [α/Fe]. Pollution by high temperature H-burning in a previous generation of stars is believed to be the cause, but the details are not understood. This is clearly an important issue in understanding how globular clusters form.

Although most globular clusters are homogeneous in their heavy elements, every globular cluster for which there are adequate data shows a spread in its Na and O abundances, sometimes in excess of 1 dex, with a striking anticorrelation of the stellar Na and O abundances. This anticorrelation is seen only in globular clusters, not in the field stars and not in the open clusters. It is somehow related to the globular cluster environment or pre-environment. Again, it requires enrichment from a previous generation of massive stars that must only affect these lighter elements and not the [Fe/H] and [α/Fe] abundances.

The nuclei of low-luminosity bulgeless spirals are similar to the massive globular clusters in velocity dispersion, mass, surface density, and sub-solar metallicity. High-resolution spectra of spiral nuclei indicate continuing episodic star formation (Walcher *et al.*, 2006). The surrounding galaxy environment can provide multiple generations of enrichment, if this material can be funneled into the nucleus. The mechanism is likely to be sporadic and dynamically driven, delivering discrete levels of enrichment at a few particular times.

In summary, globular clusters are potentially useful tracers of the early phases of galaxy evolution but we don't yet understand what physical conditions are traced by their formation, or what they represent in the context of galaxy formation. Some may be the nuclei of accreted galaxies (e.g., Freeman, 1993). We see globular cluster formation in several different situations now: merging galaxies, starburst galaxies, the disks of M33 and LMC. Yet in the large spirals like the MW and M31, the globular clusters are mostly very old, and the oldest clusters have very similar ages throughout the Local Group.

1.4 Data: sources and techniques

1.4.1 *Stellar data*

We discuss here the data needed for galactic archaeology and fossil recovery. What are we trying to do? The goal is to evaluate the state of the Galaxy and understand the formation events, role of mergers, infall history, star-formation history, and dynamical evolution that led to its present state.

We need stellar data to compare with theoretical predictions and to guide the theory. The basic observational data include stellar magnitudes and colors, distances, motions, chemical properties, and ages. We look first at what each kind of data can do for us and then at techniques for acquiring the data.

Stellar photometry

Stellar photometry involves measuring the magnitudes and colors of stars. Photometric catalogs are essential input data for stellar observational programs. They give magnitudes and colors for vast numbers of stars over the whole sky or large fractions of the sky. The catalogs typically have photometry in two or more optical or near-IR bands, at different levels of accuracy. They can be used to estimate stellar parameters like temperature, gravity, chemical abundance, and distance.

From photometric catalogs alone, it is possible to derive useful information about the structure of the Galaxy. For example, the Sloan Digital Sky Survey (SDSS) provides data in five photometric bands from the UV to the near-IR, which can be used to derive distances and photometric estimates of metallicity. Ivezić *et al.* (2008) used SDSS data for 2.5×10^6 F and G stars to demonstrate the planar stratification of the Galactic disk in [Fe/H] out to distances of about 8 kpc. The same study also showed the presence of the Monoceros stream in the outer disk.

A partial list of large ground-based existing and near-future photometric catalogs includes

- 2MASS, whole sky, JHK shallow
- SDSS, 8000 deg^2, mainly northern sky, ugriz
- UKIDSS, 7500 deg^2 of northern sky, YJHK
- SkyMapper, whole southern sky in uvgriz
- PanSTARRS, 30,000 deg^2, mainly northern sky, grizy
- VISTA VHS, whole southern sky, JK
- LSST, whole southern sky deep survey, ugrizy

Stellar distances

We need stellar distances to

- measure transverse velocities from proper motions: distance errors are usually the largest contributor to errors in transverse velocities.
- compute stellar orbits
- map substructure in the halo and disk
- measure structure and kinematics of the galactic components
- calibrate luminosities of different kinds of stars

Stellar motions

How are the different kinds of stars moving in the Galaxy? The data come from radial velocities and proper motions: together they give us the 3D space motions. What insights do we get from the kinematics of stars?

- the sense of their angular momentum: prograde or retrograde
- their orbital properties: how far they are from circular motion
- the consistency of their spatial distributions and kinematic distributions (via Jeans's equations)
- how their orbital properties correlate with their age and metallicity
- detection of kinematic substructure and stellar moving groups
- the total density of matter near the sun (no dark disk?)
- the large-scale distribution of dark matter using distant stars

Stellar motions are usually displayed in cuts through the (U, V, W) space of motions relative to the Local Standard of Rest. The Toomre diagram is a useful diagnostic: $\sqrt{(U^2 + W^2)}$ is plotted against V. The first term is a measure of the orbital energy and the second a measure of the orbital angular momentum. Stars of the thin disk are crowded near the origin, while stars of the thick disk have lower angular momentum and are more energetic: see Figure 1.18. The Lindblad diagram is another very useful diagnostic tool. It shows stars in the space of the two integrals of the motion in an axisymmetric system (E and L_z): see Figure 1.19. The two curves with solid dots superimposed are the circular orbit loci, outside which no star can lie. The sun is located near the prograde circular orbit point marked as 8 kpc. Stars with L_z near zero have low angular momentum. The curve near the bottom, extending from the prograde to retrograde circular orbit loci, is the locus of stars whose orbital apogalactic distance is 6 kpc and therefore do not visit the solar vicinity. The two panels show a striking difference between the orbital characteristics

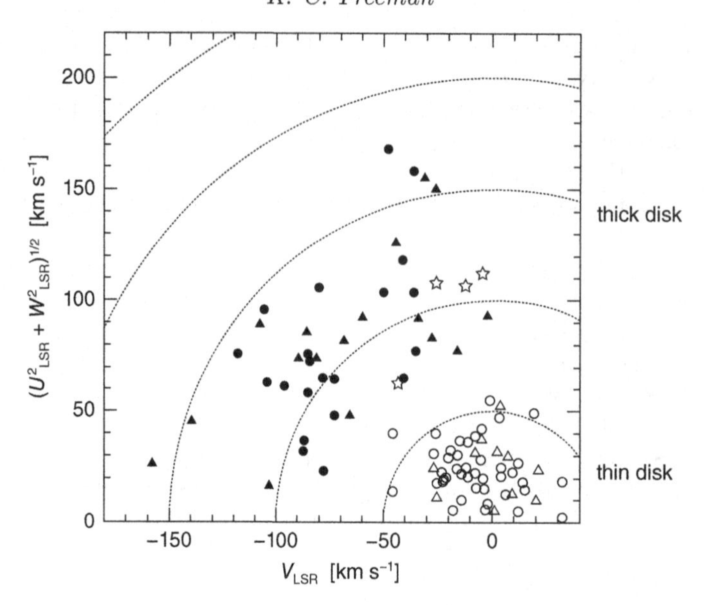

FIG. 1.18. The Toomre diagram for nearby stars of the thin and thick disk (adapted from Bensby *et al.*, 2003).

of the two kinds of stars. The more energetic RR Lyrae stars are concentrated toward orbits of high eccentricity and low angular momentum.

If we know the (U, V, W) motions and have a reliable model for the Galactic potential, then we can compute the stellar orbits (see Figure 1.9). Stars from the inner and outer Galaxy can pass through the solar neighborhood.

Knowing the orbit is not always very useful for Galactic archaeology. Stellar orbits can evolve as the star interacts with spiral structure and giant molecular clouds. Resonances with the bar and spiral structure can flip a star from one near-circular orbit to another: see Figure 1.13. In Figure 1.19, this radial mixing corresponds to moving a star approximately along the prograde circular orbit locus in either direction.

Stellar element abundances

The element abundances of stars come initially from the abundances of the gas from which they formed. This gas has been enriched by previous generations of evolving and dying stars. Different components of the Galaxy (halo, bulge, thick disk, thin disk) each have different characteristic chemical properties. For example, the halo stars are mostly metal-poor, with [Fe/H] values between about -1 and -5 and are enriched in α-elements relative to iron. See Figure 1.3 for a summary.

Different elements come from different progenitors. The Fe-peak elements come mainly from type Ia supernovae, whereas the α- and r-process elements come mainly from the more massive type II SNe. The s-process elements come mainly from thermally pulsing AGB stars. Figure 1.20 indicates the atomic number of the different groups of elements.

The abundances of different element groups in stars can tell us a lot about the star-formation history that led to the formation of these particular stars. For example, α-enrichment relative to Fe indicates that SNII were important for the chemical evolution and the star-formation history was fairly rapid: enrichment from SNIa (which takes $\sim$1 Gyr to evolve) was less important. For most of the heavier elements, stars remember the abundances with which they are born. Groups of stars born together, like open star clusters, appear to have almost identical abundances, reflecting the abundances of the gas

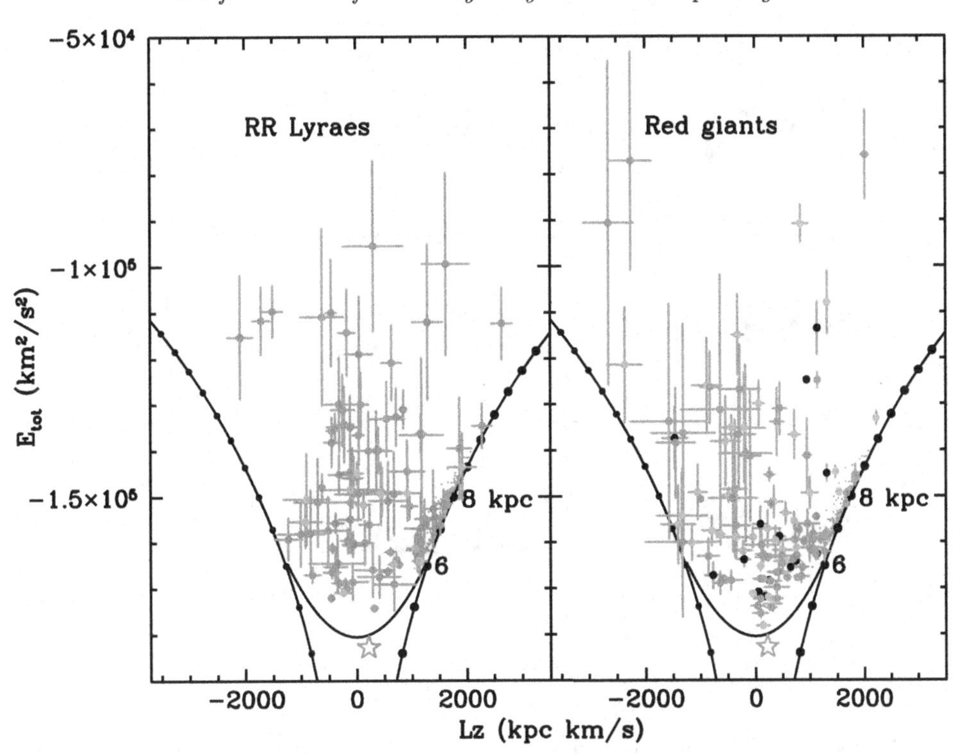

FIG. 1.19. Lindblad diagrams for metal-poor RR Lyrae stars and red giants (from Morrison *et al.*, 2009). See Morrison *et al.* (2009) for a color-coded figure by [Fe/H] values where stars are grouped in four different [Fe/H] ranges: from −1 to −1.5, −1.5 to −2, −2 to −2.5, and < −2.5. See text for more explanation.

from which they formed (see Figure 1.21). Chemical signatures may allow us to recognize groups of stars that were born together but have dispersed and drifted apart (chemical tagging). Not all of the chemical elements vary independently in stars. The number of independently varying elements is probably between 7 and 9. This is the dimensionality of chemical space.

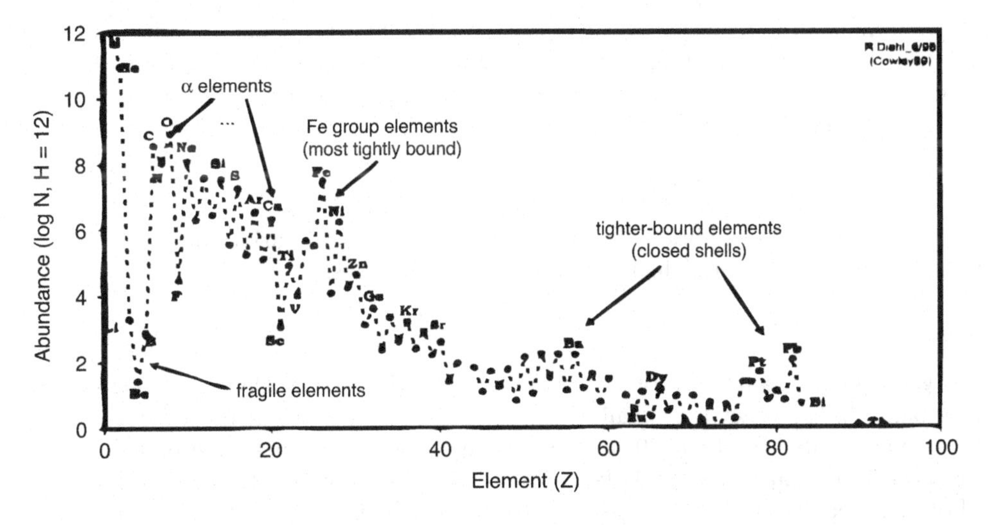

FIG. 1.20. The cosmic abundance distribution. The various groups of elements are indicated.

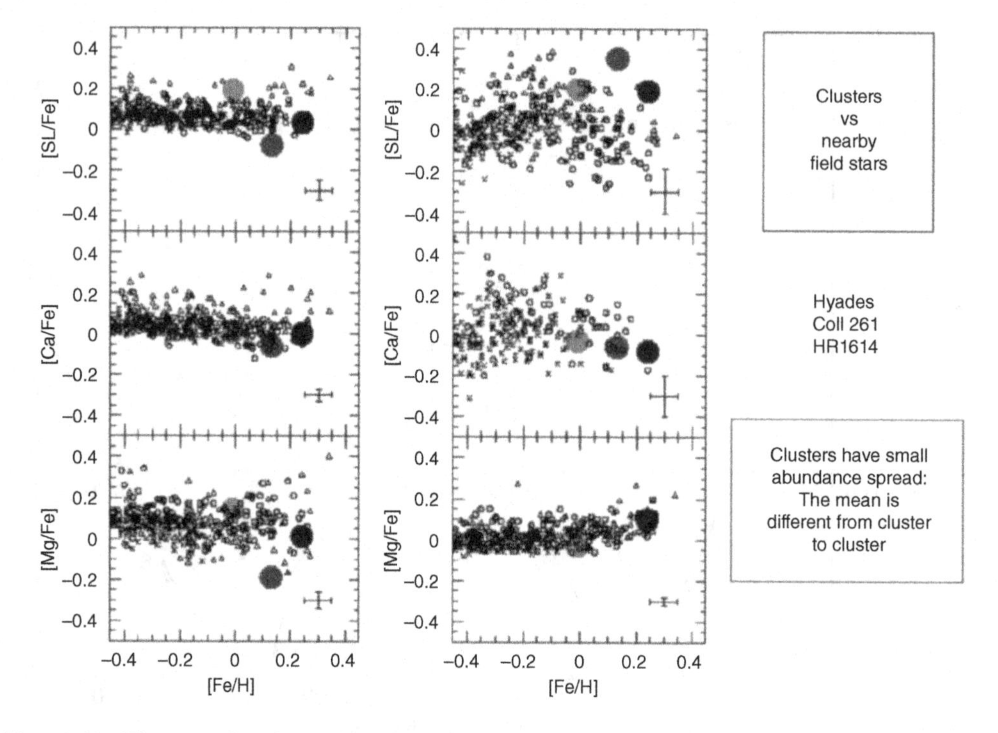

FIG. 1.21. Element abundances in two open clusters and one stellar moving group (HR1614). The abundances in each of these aggregates have an almost unmeasurable dispersion. The black points show stars of the nearby disk (from De Silva *et al.*, 2009).

Stellar ages

Stellar ages allow us to evaluate when events occurred. They are important for measuring the star-formation history and for understanding how the metallicity and dynamics of different groups of stars have evolved: for example, how has the star-formation rate and the metallicity of the thin disk near the sun changed from, say, 10 Gyr ago to the present time? Stellar ages are still difficult to measure and this has led to much uncertainty about the evolution of the Galaxy.

The star-formation history in the Galactic thin disk near the sun appears to have been roughly uniform in the mean over the last 10 Gyr but with some episodic star bursts in which the star formation rises for brief periods to about twice its mean value (e.g., Rocha-Pinto *et al.*, 2000). This conclusion comes mainly from chromospheric ages that may not be very reliable.

1.4.2 *Observational techniques*

In this section, we discuss briefly techniques for measuring the various quantities we need stellar distances, velocities, chemical abundances, and ages.

Stellar distances

Trigonometric parallaxes provide the fundamental distance scale for astronomy. From the ground, the accuracy is typically a few milli-arcsec (mas), so distance errors of about 10% are achievable to about 30 pc. This value of 10% is taken as a standard, because photometric techniques can provide distances with errors of about 10% to 15%.

From space, the Hipparcos mission provided parallaxes with errors ~1 mas, so the distances are useful out to about 100 pc. From the Gaia mission, we expect parallax errors of about 6 μas at $G = 12$, increasing to about 150 μas at $G = 19$, with some dependence

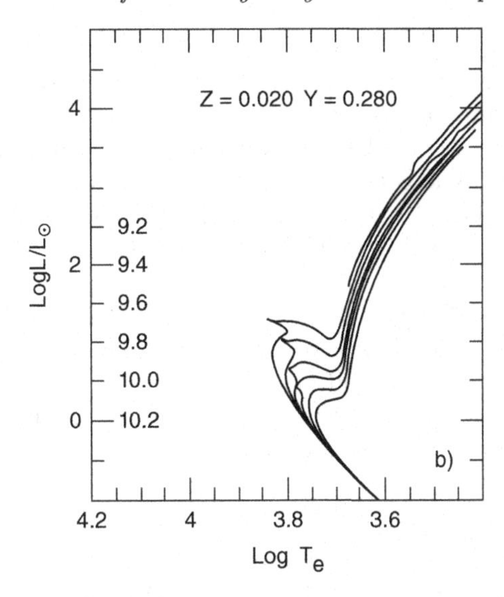

FIG. 1.22. A set of isochrones for metallicity Z and He abundance Y as shown. The log of the age is shown at the left for the six isochrones (from Bertelli *et al.*, 1994).

on stellar color. Gaia will be enormously valuable for determining very accurate distances to the nearer stars. For example, the parallax of a typical disk dwarf with $G = 14$ (distance about 600 pc) will be known to better than 1%. The distance to which the trigonometric parallaxes are better than the photometric parallaxes depends on the kind of star. For example, for disk giants with absolute magnitude near zero, at a distance of 6 kpc, the parallax error will be about 8%. For dwarfs with an absolute magnitude of +4, at 1.6 kpc, the error will be about 5%.

Photometric parallaxes use theoretical or empirical isochrones (see Figure 1.22) to estimate the absolute magnitude and hence the distance of the star. The isochrones depend on chemical abundance. To make this work, one needs the chemical abundance [Fe/H] of the star and an estimate of its temperature: this can come from its color or from spectroscopy. An estimate of the surface gravity is very helpful: otherwise one needs to make an assumption about the age of the star. Gravity estimates can come from photometry and spectroscopy. The more one knows, the better it works.

Distance errors of 10% to 15% can be achieved in the best cases, even for evolved stars, if the abundance, temperature, and gravity are accurately known. If the errors in the stellar parameters are significant, then simple isochrone estimates of the distance can be biased by the shape of the stellar mass function and the metallicity distribution function. Pont and Eyer (2004) have discussed such biases in the context of estimating stellar isochrone ages: their discussion pertains also to isochrone distances in some regions of the luminosity-temperature plane.

For a few kinds of stars, like RR Lyrae stars, cepheids, and blue horizontal stars, accurate absolute magnitudes are known, so the errors in photometric parallaxes can be as low as a few percent.

For star clusters in which the color-magnitude diagram can be measured for evolved and unevolved stars, a fit of the CMD to theoretical isochrones can be used to derive age and distance if the metallicity is known. In globular clusters, the RR Lyrae stars and other horizontal branch stars also provide a distance estimate.

Interstellar reddening and extinction is a problem for photometric parallaxes. This problem is greatly reduced in the near-IR. Multi-color photometry can give reddening-free estimates of the stellar temperature. Galactic reddening models and observed diffuse interstellar bands can give independent reddening estimates.

Stellar velocities

Radial (line of sight) Doppler velocities are measured spectroscopically. Proper (transverse) motions are measured astrometrically.

The typical accuracies of radial velocities range from about 1 m s^{-1} using special techniques at a spectral resolution $R = \lambda/\Delta\lambda > 50,000$ to about 1 km s^{-1} at $R \sim 7000$ (the RAVE survey) to 5 km s^{-1} at $R \sim 2000$ (SEGUE), where λ is the wavelength and $\Delta\lambda$ is the resolution in wavelength units. For most Galactic programs, 5 km s^{-1} is good enough.

Samples of $\sim 5 \times 10^5$ stellar radial velocities are or will soon be available from large fiber spectrograph surveys like RAVE and SEGUE. The spectra also give estimates of the stellar parameters T_e, gravity, and metallicity.

Stellar proper motions μ measured from the ground have accuracies of a few mas yr^{-1}. The transverse velocity of a star at distance Δ is

$$V_t = 4.74\ \mu(\text{mas yr}^{-1})\Delta(\text{kpc}), \tag{1.23}$$

so the velocity error is typically 10 to 20 km s^{-1} at a distance of 1 kpc from the proper motion error alone. Very large samples (10^5 to 10^9) of stellar proper motions are available from several sources (USNO, UCAC2, SPM, SDSS, 2MASS, GSC, PM2000, Starnet2 ...), with more to come from Pan-STARRS, SkyMapper, LSST ....

From space, the Hipparcos/Tycho catalogs provide about 2×10^6 stars with proper motion errors of about 2 mas yr^{-1}. Gaia will measure proper motions for about a billion stars. Again, the accuracy depends on the magnitude: at $G = 14$, the expected accuracy is 7 μas yr^{-1}, which corresponds to 0.5 km s^{-1} at a distance of 15 kpc (if the distance were very accurately known). Gaia will really change Galactic astrophysics, with vast numbers of very precise parallaxes and proper motions. We should be prepared to get the most from this resource. Launch is scheduled for 2013. Other planned astrometric space missions include SIM (a pointed mission with expected proper motion errors of about 4 μas yr^{-1}) and JASMINE (near-IR mission, galactic plane, and bulge, 4 μas yr^{-1})

Stellar element abundances

Intermediate and broad band photometry, like Strömgren photometry and the SDSS/Skymapper 5- and 6-band photometry gives estimates of stellar temperature, gravity, and overall metal abundance [M/H] with an abundance error of about 0.3 dex. These photometric estimates are usually empirically calibrated against stars with known parameters. For example, the Geneva-Copenhagen catalog (Nordström *et al.*, 2004; Holmberg *et al.*, 2007) provides 14,000 FG stars with Strömgren photometry, Hipparcos astrometry, and accurate radial velocities.

Medium resolution spectroscopy ($R \sim$ 2000–7000) measures the strength of spectral features and gives [Fe/H] the α-element to iron ratio [α/Fe] and possibly the abundances of a few other specific elements. The RAVE and SEGUE surveys are examples of large medium-resolution stellar surveys, providing samples of several $\times 10^5$ stars. Typical [Fe/H] errors are about 0.2 dex. Figure 1.23 shows an example of a spectrum of a RAVE star in the Ca-triplet region.

Theoretical stellar atmospheres are widely used for estimating stellar parameters from medium and high resolution spectra by comparing a grid of atmospheres with the observed spectra. They give the flux as a function of wavelength at various spectral resolutions for a wide range of stellar parameters like $T_e, \log g$ and element abundance. For example, the RAVE pipeline works by fitting the observed spectra ($R \sim 7000, SNR \sim 40$) in the Ca triplet region to the grid of Zwitter *et al.* (2004) models. The internal accuracy of this process is about 0.1 in [M/H], 0.2 in $\log g$ and 135 K in T_e.

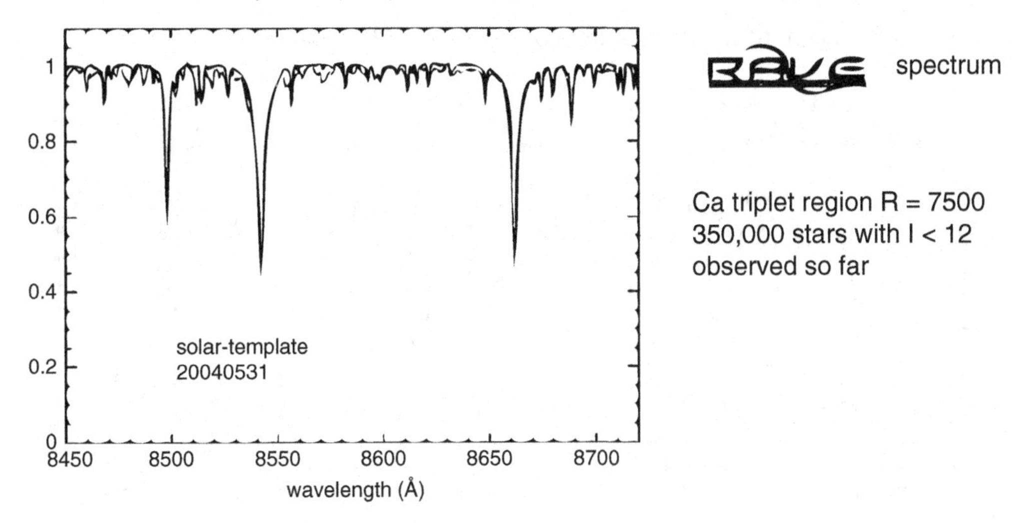

FIG. 1.23. A spectrum from the RAVE survey. The stellar spectrum is in red and a matching synthetic solar template is in black.

High-resolution spectroscopy ($R > 20{,}000$) provides detailed abundances of many different elements (see Figure 1.24). High-resolution spectrographs are usually echelle spectrographs, some with multi-object spectroscopic (MOS) fiber capability. Typically these systems have a few hundred fibers. Hectochelle on MMT, MIKE on Magellan, FLAMES/GIRAFFE on VLT exist already. HERMES on AAT and APOGEE on the Sloan telescope are coming soon and are aimed at large high-resolution surveys of 10^5 to 10^6 stars).

High-resolution spectroscopy is the only way now to measure accurate abundances of many different elements including neutron capture elements (s-process, r-process). The analysis is currently laborious but this is changing, with pipelines for high-resolution surveys of large samples of stars to start soon with HERMES and APOGEE.

The analysis of high-resolution spectra also gives spectroscopic estimates of other stellar parameters, such as effective temperature, surface gravity, and rotational velocity. For large optical high-resolution surveys like HERMES, FGK stars with $T_e = 5000 - 6500$ K will be used: they are cool enough to have plenty of lines and warm enough for analysis to be relatively straightforward. Hotter stars have mostly weaker metallic lines and are often younger and rapidly rotating, which broadens the lines and makes weak lines difficult to measure. Cooler stars have complex atmospheres with molecules: these are easier to study in the near-IR, which is also less affected by interstellar extinction. The near-IR is, however, not so good for the neutron capture elements. I should comment that almost every topic in this paragraph is a major specialty in stellar astrophysics, the subject of much work and individual careers.

Stellar ages

Nuclear cosmochronology compares the abundance ratios of radioactive and stable species to the expected ratios from theory (e.g., $^{238}U/^{232}Th$). For stars other than the sun, it has not been widely used and the age errors are still relatively large.

Asteroseismology uses the dependence of stellar oscillation frequencies on the density profile in the stellar interior, which changes as the star ages. The asteroseismology space missions (MOST, CoRoT, Kepler) are expected to contribute greatly to deriving stellar ages with 5% to 10% errors.

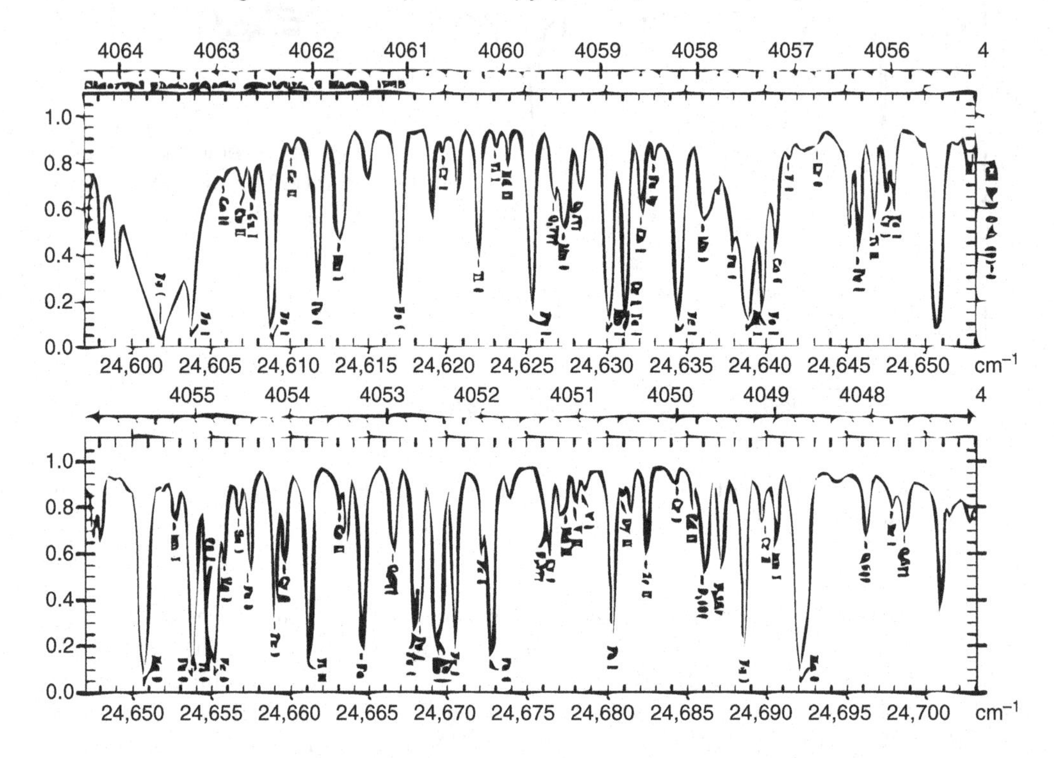

FIG. 1.24. A 17 Å window of the solar spectrum at a resolution of about 80,000. Lines of many elements can be seen, from which abundances can be estimated.

The white dwarf luminosity function (WDLF) is useful for deriving the age of a stellar population (e.g., the Galactic disk) or a globular cluster. The technique uses the theoretical cooling and fading of white dwarfs as an age estimator for the population. For example, Leggett *et al.* (1998) derived an age of 9 Gyr for the stellar disk. Hansen *et al.* (2002) used the WDLF to show that the Galactic disk is clearly younger than the globular clusters M4 and NGC 6397.

Chromospheric emission, as measured, for example, by the emission cores of the Ca II K line, has been used extensively for estimating the ages of dwarf stars, because chromospheric activity appears to decrease with increasing age, at least for a few Gyr. It is still not clear how reliable these estimates are: see, for example, Pace *et al.* (2009) for a recent discussion on chromospheric ages for solar-type stars. Stellar rotation also changes with stellar age and can be used as an age estimator (e.g., Barnes, 2007).

Ages can be derived from theoretical isochrones if the stellar luminosity and temperature (or proxies such as surface gravity and color) and the stellar metallicity are known. For example, trigonometric parallaxes give absolute magnitudes M_V which in turn give the stellar luminosity after bolometric correction. Photometry or spectroscopy give estimates of the effective temperature and surface gravity (which can be used instead of luminosity if no parallax is available). Spectroscopy gives the metallicity.

Isochrone ages can only be derived for evolved stars. On the unevolved lower main sequence, there is no way to distinguish the ages of the stars from the isochrones. On the red giant branch, the ages depend sensitively on color and abundance, and measuring ages for brighter giants is difficult. Looking at Figure 1.22, one can see that isochrone ages for stars close to the main sequence turnoff are difficult to measure, because the isochrones cross. Also, if the stellar parameters have significant errors, then much care is needed to handle properly the biases introduced by the underlying distributions of these parameters, such as the stellar mass (the stellar mass function) and the metallicity distribution function (see Pont and Eyer, 2004). The subgiant region is probably the best region in the L-T plane for measuring isochrone ages.

Gaia will provide a huge increase in accurate isochrone ages. We can expect distance errors of about 1% for turnoff stars and subgiants with V magnitudes of 14.

At this time, there is disagreement about ages derived from different applications of the isochrone technique (see Pont and Eyer, 2004; Reid *et al.*, 2007). This disagreement means that the relationship between age and velocity dispersion, and age and metallicity, are observationally poorly known. It is not even clear whether there is an age-metallicity relation for the nearby stars.

The preceding discussion applies to age estimates for individual stars. For star clusters, stars of a wide range of mass are usually present, and the whole isochrone is observationally delineated. If the metallicity and the reddening are known, then relatively accurate isochrone ages can be derived. As always, these ages are on the system of the particular isochrones adopted. Most isochrone libraries cover a wide range of stellar mass, age, [Fe/H] and [α/Fe], and give magnitudes and colors in several widely used photometric systems. Some include stellar rotation. The list below gives some of the most widely used libraries, with its usual name:

- *Dartmouth:* Dotter and Chaboyer (2002); Dotter *et al.* (2008)
- *Geneva:* Maeder and Meynet (1989, 2008)
- *Padova:* Bertelli *et al.* (1994, 2008)
- *Teramo:* Brocato *et al.* (1999, 2000); Cantiello *et al.* (2003); Raimondo *et al.* (2005)
- *Victoria-Regina:* VandenBerg *et al.* (2006)
- *Yonsei-Yale* (Y^2)*:* Han *et al.* (2009)

Galactic models

Galactic models are constructed to give star counts as a function of color, magnitude, distance, etc., in chosen lines of sight through the Galaxy. They are very useful for simulations of observational programs and for sanity checks on outcomes. The Besançon model[1] is widely used and includes several galactic components, each with prescribed distribution of stellar ages, motions, and chemical properties. A reddening recipe is included. The user specifies the galactic region ($\Delta l, \Delta b, \Delta$ distance), and the age and magnitude range. The model generates a mock catalog of stars in that region and tabulates the stellar parameters: $T_e, \log g$, [Fe/H], components of the kinematics, and photometric magnitudes and colors. Critics argue about some of the basic empirical input parameters for the model, but it seems to work reasonably well.

1.5 The Galactic disk

In this chapter, I discuss the thin and thick disks of the Milky Way, and then describe a large chemical tagging survey we are planning to do with the new HERMES multi-object high-resolution spectrometer on the AAT.

[1] http://model.obs-besancon.fr.

1.5.1 *The thin disk*

The disk is the defining stellar component of disk galaxies. It is the endpoint of the dissipation of most of the baryons, and it contains almost all of the baryonic angular momentum. Understanding its formation is in my opinion the most important goal of galaxy-formation theory. Disks like the large disk of our Galaxy, with its small bulge, are at present difficult to form within the framework of current CDM simulations because of the high level of merger activity, which leads mostly to the formation of large central spheroidal components.

Out of the hierarchical galaxy-formation process come galactic disks with a high level of regularity in the structure and scaling laws. We need to understand the reasons for this regularity.

The exponential structure

Disks have a roughly exponential luminous density distribution in R and z, of the form

$$I(R,z) = I_\circ \exp(-R/h_R)\exp(-z/h_z) \tag{1.24}$$

to a radius R of 3 to 5 radial scalelengths h_R and are then often truncated. The truncation is not abrupt but takes the form of a much-steepened exponential decline in surface density. The radial scale length is usually larger in the larger disk galaxies and ranges from < 1 kpc to at least 5 kpc.

The reason for the exponential form of the radial surface brightness profile is not yet understood. There are many possible mechanisms. Two extreme options are there:

(i) The exponential disk structure comes from the primordial distribution of internal angular momentum of parts of the hierarchy that are destined to become disk galaxies. A torqued element of the hierarchy with the right distribution of internal angular momentum $M(j)$ collapses, conserving its $M(j)$ distribution as in the Fall and Efstathiou (1980) collapse picture, and settles to an exponential disk.

(ii) The exponential disk structure is entirely evolutionary. For example, the gas and stars in the disk are redistributed by viscous or spiral arm torques that drive the disk toward an exponential structure (see Lin and Pringle, 1987).

The vertical exponential structure of disks is directly associated with their star-formation history and dynamical history, including scattering of stellar orbits by molecular clouds, accretion of satellites, disk heating processes, etc. These processes, which are not yet well understood, generate a vertical scale height h_z for the old thin disk that is usually about 200 to 300 pc.

The vertical scaleheight h_z for later-type galaxies is almost independent of radius within the disk. This property gives a constraint on the heating mechanism in late-type disks. For the earlier-type galaxies, the scaleheight increases weakly with radius, as shown in Figure 1.25.

Disk truncation

Whereas most disks show some radial truncation at a radius of a few scalelengths, a fraction do not: their exponential disks continue without break to the limits of the surface brightness measurements. In some cases, like NGC 300 (Bland-Hawthorn *et al.*, 2005) in which it is possible to use deep star-counts to extend the measurements to very low surface brightnesses, the exponential disk continues to a surface brightness level of at least 31 I mag arcsec^{-2}. See also Erwin *et al.* (2005).

The truncation of disks is not understood. It may be associated with

- the star-formation threshold
- angular momentum redistribution by bars and spiral waves
- the hierachical accretion process
- bombardment of the disk by dark matter subhalos (de Jong *et al.*, 2007)

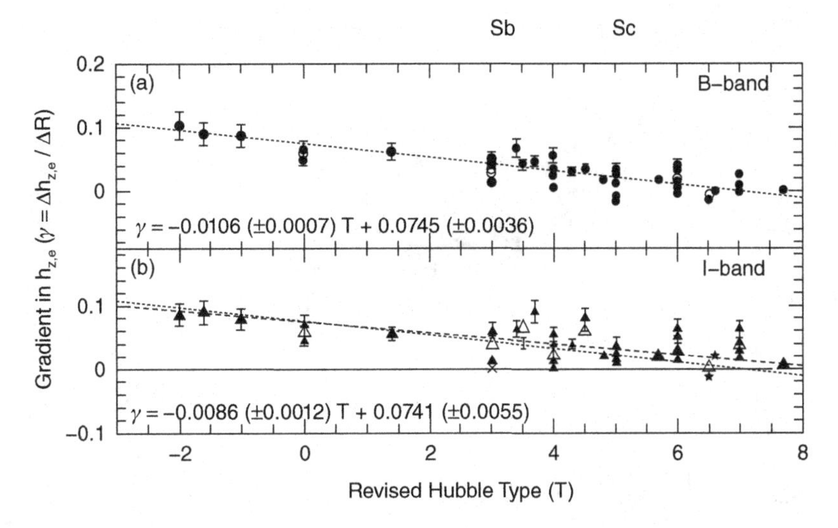

FIG. 1.25. The gradient of the disk scaleheight in the B- and I-bands, as a function of morphological type. In the late-type systems, the disk scaleheight is approximately constant with radius (adapted from de Grijs and Peletier, 1997).

Roškar *et al.* (2008) made an SPH simulation of disk formation from cooling gas in an isolated dark halo. The simulation includes star formation and feedback. The break in surface density is seeded by a rapid radial decrease in surface density of cool gas; the break forms within 1 Gyr and gradually moves outward as the disk grows. Star formation does not extend radially beyond the end of the main exponential. The stars in the outer exponential (at radii larger than the truncation break) are stars from the inner regions, secularly redistributed via the Sellwood and Binney (2002) orbit-swapping mechanism associated with transient spiral structure. Therefore, the stars in the outer exponential are relatively old. In the simulation, few stars form beyond a radius of 10 kpc, but stars are scattered out into the zone with radii between about 10 and 15 kpc to populate the outer exponential.

Chemical properties of the thin disk

The abundance gradient in the younger population of the thin disk is nicely illustrated by the radial distribution of cepheid abundances from Luck *et al.* (2006): see Figure 1.26. Luck *et al.* (2006) found that the abundance gradient is not axisymmetric but shows strong azimuthal structure that may be due to the influence of gas flows associated with the spiral structure of the Galaxy.

Carney *et al.* (2005) and Yong *et al.* (2005) studied the abundances of open clusters and red giants in the outer disk at radii up to about 23 kpc. The clusters have ages between about 1 and 5 Gyr. They found that these objects show an abundance gradient, somewhat steeper than the gradient shown by the cepheids, but that this gradient bottoms out at about 12 kpc. At larger radii, the abundances are roughly constant at $[\mathrm{Fe/H}] \approx -0.5$. These outer open clusters appear to be α-enriched, while the cepheids in the outer Galaxy show a much lower level of α-enrichment. So it appears that the abundance gradient and the $[\alpha/\mathrm{Fe}]$ gradient have flattened with time. This effect may be associated with enriched gas accreted into the outer Galaxy. Alternatively, it may be due to star formation in the outer disk starting only about 6 Gyr ago, so these old outer clusters may be among the oldest objects in this part of the Galactic disk.

M31 shows a very similar bottoming out of the abundance gradient in its disk, with the abundance of disk stars leveling out at a radius of about 20 kpc, again at the same abundance level of $[\mathrm{Fe/H}] \approx -0.5$ (Worthey *et al.*, 2005).

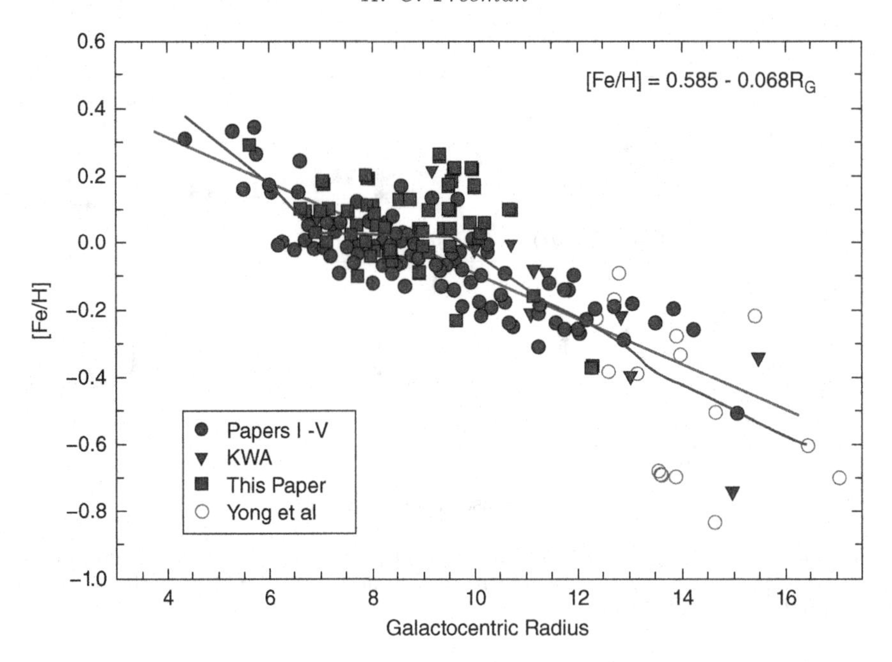

FIG. 1.26. The Galactic abundance gradient as shown by the cepheids (Luck *et al.*, 2006).

The radial abundance distribution in the smaller M33-like galaxy NGC 300 may provide a clue into the reason for this flattening of the abundance gradient in the Milky Way and NGC 300. Vlajić *et al.* (2009) measured the abundance distribution from the colors of giants in this galaxy. They showed that the abundance declines with radius from the inner regions to a radius of about 8 scalelengths and then begins to rise again. We recall that this galaxy shows no radial truncation: its surface brightness follows a single exponential decline over more than 10 scalelengths. It seems possible that the lack of truncation and the reversal of the abundance gradient could both come from the orbit swapping phenomenon, if there was a phase of unusually strong spiral arm activity. Enhanced radial mixing of stars from the inner disk may populate the outer regions with stars from the more metal-rich inner regions, giving an apparently single exponential surface density distribution and a reversal of the abundance gradient.

The evolution of the stellar metallicity distribution of the thin disk in the solar neighborhood is not yet well understood. Observationally, different authors come to different conclusions. Some find a wide range of abundances, with little indication of change with time. Others find less scatter, with a slow increase of abundance with time. The form of the age-metallicity relation in the solar neighborhood urgently needs to be established. With the current uncertainty, it is difficult to (i) assess how much the local metallicity distribution is affected by migration of stars into the solar neighborhood from the inner and outer Galaxy, and (ii) compare the observed chemical evolution of the solar neighborhood with models of the Galactic chemical evolution. It seems likely that errors in stellar ages are contributing to this uncertainty (see Pont and Eyer, 2004, and also the next section).

Disk heating

Disk heating is the secular increase in stellar velocity dispersion with time through various dynamical effects such as the interaction of disk stars with transient spiral waves, giant molecular clouds, and the accretion of small satellites. If the heating is due to interaction with objects of the very young population, which are concentrated to a thin layer in the equatorial plane of the Galaxy, then one might expect this heating to saturate after a few Gyr, when the disk stars have acquired enough vertical energy to spend most of their orbital period away from the Galactic plane. What do the observations show? The facts are not yet clear.

One view is that the stellar velocity dispersion increases steadily with time, like $\sigma \sim t^{0.2-0.5}$, as in Wielen (1977): this study used Ca II emission ages for the McCormick dwarfs. In contrast, accurate individual velocities and ages for a sample of about 200 subgiants studied by Edvardsson *et al.* (1993) showed heating for the first ~2 Gyr, with no significant subsequent heating up to the oldest ages of the thin disk stars (about 10 Gyr), for example, Freeman (1991); Quillen and Garnett (2000). From this sample, it appears that the thin disk heating in the solar neighborhood saturates after about 2 Gyr, at a vertical velocity dispersion of about 20 km s^{-1}. The older stars in the Edvardsson sample show an abrupt increase of vertical velocity dispersion to about 40 km s^{-1}: these older stars are believed to be members of the thick disk, which we discuss later. A more recent study of clump giants in the thin disk by Soubiran *et al.* (2008) has a similar conclusion: their vertical velocity dispersion is approximately constant for stars older than about 3 Gyr at a slightly higher value of about 22 km s^{-1}. The Edvardsson and Soubiran studies used isochrone ages.

The recent Geneva-Copenhagen study measured kinematics, ages, and metallicities for a large sample of Hipparcos stars, using Strömgren photometry, Hipparcos and photometric parallaxes, and accurate radial velocities (Nordström *et al.*, 2004; Holmberg *et al.*, 2007). They derived the change of velocity dispersion with time and found a rather similar steady increase with time to that found by Wielen (1977). Reid *et al.* (2007) have argued that there is some disagreement of stellar ages between the Geneva-Copenhagen study and the isochrone ages from the high resolution spectroscopic study by Valenti and Fischer (2005). It seems possible that errors in ages could transform the stepped young disk – old disk – thick disk age-velocity relation as observed by Quillen and Garnett (2000) into the continuously rising age-velocity relation found in the Geneva-Copenhagen study.

This observational issue needs to be resolved. We need to know how the velocity dispersion of the thin disk changes with time. The age-velocity relation is essential information for understanding the dynamical evolution of the Galactic disk.

1.5.2 *The thick disk*

Most spirals, including our Galaxy, have a second thicker disk component. An extensive study by Yoachim and Dalcanton (2006) shows that the ratio of thick disk stars to thin disk stars depends on the luminosity or circular velocity of the galaxy: it is about 10% for large spirals like the Milky Way, and rises to about 50% for the smallest disk systems. Our Galaxy has a thick disk, first detected in star counts at the South Galactic Pole by Gilmore and Reid (1983). Its scaleheight is about 1000 pc, compared to about 300 pc for the old thin disk. The larger scale height means that its velocity dispersion is higher than for the thin disk (about 40 km s^{-1} in the vertical direction, compared to about 20 km s^{-1} for the thin disk). Its surface brightness is about 10% of the surface brightness of the thin disk. Near the Galactic plane, the rotational lag of the thick disk relative to the LSR is only about 30 km s^{-1} (Chiba and Beers, 2000), but its rotational velocity appears to decrease with height above the plane. Its stars are old (>10 Gyr) and more metal-poor than the thin disk. The metallicity distribution of the thick disk has most of the stars with [Fe/H] between about -0.5 and -1.0, with a tail of metal-poor stars extending to about -2.2. The thick disk stars are enhanced in α-elements relative to the thin disk (Figure 1.27), indicating a rapid history of chemical evolution. The thick disk does not appear to show a significant vertical abundance gradient (Gilmore *et al.*, 1995). It appears to be chemically and kinematically distinct from the thin disk.

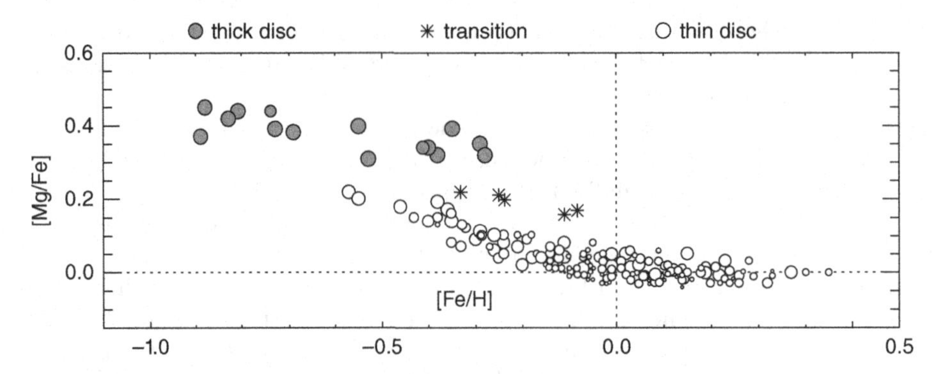

FIG. 1.27. Mg-enrichment of the thick disk relative to the thin disk, indicating that the thick disk is a chemically distinct population (Fuhrmann, 2008).

A decomposition of the kinematics and distribution of stars near the Galactic poles, by Veltz *et al.* (2008), nicely illustrates the three main kinematically discrete stellar components of the Galaxy: a thin disk with a scaleheight of 225 pc and a mean vertical velocity dispersion of about 18 km s^{-1}, a thick disk with a scaleheight of 1048 pc and mean velocity dispersion of 40 km s^{-1}, and a halo component with a velocity dispersion of about 65 km s^{-1}.

The old thick disk presents a kinematically recognizable relic of the early Galaxy and is therefore a very significant component for studying galaxy formation. Because its stars spend most of their time away from the Galactic plane, the thick disk is unlikely to have suffered much secular heating since the time of its formation. Its dynamical evolution was probably dominated by the changing potential field of the Galaxy associated with the continuing growth of the Galaxy since the time at which the thick disk was formed.

We do not yet understand how thick disks form. They appear to be very common, so they are probably a normal part of the early formation of disk galaxies. Possibilities include these:

- thick disks come from energetic early star burst events, maybe associated with gas-rich mergers (Samland and Gerhard, 2003; Brook *et al.*, 2004)
- thick disks are the debris of accreted galaxies that were dragged down by dynamical friction into the plane of the parent galaxy and then disrupted (Abadi *et al.*, 2003; Walker *et al.*, 1996). To provide the observed metallicity of the Galactic thick disk ([Fe/H] ~ -0.7), the accreted galaxies that built up the Galactic thick disk would have been more massive than the SMC and would have had to be chemically evolved at the time of their accretion. The possible discovery of a counter-rotating thick disk (Yoachim and Dalcanton, 2008) would favor this mechanism.
- the thick disk's energy comes from the disruption of massive clusters or star-forming aggregates (Kroupa, 2002), possibly like the massive clumps seen in the high redshift clump cluster galaxies.
- the thick disk represents the remnant of an early thin disk, heated by accretion events. In this picture, the thin disk begins to form at a redshift of 2 or 3 and is partly disrupted and puffed up during the active merger epoch. Subsequently, the rest of the gas gradually settles to form the present thin disk.

1.5.3 *Galactic archaeology: fossil recovery by chemical tagging*

In the process of galaxy formation, much information about the early Galaxy was lost in the dissipation and mixing that occurred as the Galaxy settled. Further information was lost in the dissipation that led to disk formation and the subsequent heating and rearrangement by spiral waves and giant molecular clouds. Some information remains, and the goal of galactic archaeology is to seek signatures or fossils from the epoch of Galaxy formation, to give us insight about the processes that took place as the Galaxy formed.

We would like to be able to reconstruct the star-forming aggregates that built up the disk, bulge, and halo of the Galaxy. Some of these dispersed and phase-mixed aggregates can still be recognized kinematically as stellar moving groups in velocity space or integral (E, L_z) space. For others, the dynamical information was lost through disk heating and mixing processes, but they can still be recognized by their chemical signatures (chemical tagging).

A major goal of Galactic archaeology is to identify how important mergers and accretion events were in building up the Galactic disk and the bulge. CDM predicts a high level of merger activity that conflicts with many observed properties of disk galaxies. The formation of the thick disk is particularly interesting in this context.

Stellar moving groups in the disk

The Galactic disk shows kinematical substructure in the solar neighborhood: these groups of stars, moving together, are usually called moving stellar groups. Olin Eggen made the pioneering studies of these moving groups that include systems with a wide range of age, chemical properties, and likely origins.

Some are associated with dynamical resonances with the Galactic bar or spiral structure. The Hercules group (Dehnen, 2000) is an example. For such dynamical groups, random stars of the disk are collected together in a common motion pattern and one does not expect them to be homogeneous in age or chemical propperties.

Some groups may be debris of infalling objects, as seen in ΛCDM simulations: for example, Abadi *et al.* (2003). We would need to find a way to recognize such systems from their orbital and chemical properties.

Other groups are the debris of star-forming aggregates in the disk. The HR 1614 group is an example (De Silva *et al.*, 2007). Such groups are likely to be homogeneous in age and element abundance, and they could be useful for reconstructing the history of the galactic disk. The stars of the HR 1614 group are scattered all around us but have a common pattern of motion relative to the LSR. The color-magnitude diagram for these stars indicates a common age of about 2 Gyr, and their [Fe/H] value is about +0.2. This group has retained some chemical identity despite its age. De Silva *et al.* (2007) measured accurate differential abundances for many elements in HR 1614 stars and found a very small spread in abundances. This is encouraging for chemical tagging: the stars of this group have a common pattern of chemical abundances over many elements.

Chemical tagging

The goal is to use the detailed chemical abundances of stars to tag or associate them to common ancient star-forming aggregates with similar abundance patterns (Freeman and Bland-Hawthorn, 2002). The detailed abundance pattern of a set of stars reflects the chemical evolution of the gas from which the aggregate formed. Different supernovae provide different yields (depending on mass, metallicity, detonation details, ejected mass, etc.), leading to scatter in detailed abundances, especially at lower metallicities where relatively few SNe contributed to the element enrichment.

Chemical studies of the old disk stars in the Galaxy can help to identify disk stars that came in from outside in disrupting satellites, and also those that are the debris of dispersed star-forming aggregates. The chemical properties of surviving satellites (the dwarf spheroidal galaxies) depend on their star-formation history which varies from satellite to satellite (e.g., Venn and Hill, 2008). Their detailed element abundance distributions also vary from satellite to satellite and differ in detail from the overall properties of the nearby disk stars.

The chemical abundances of elements (O, Na, Mg, Al, Ca, Mn, Fe, Cu, Sr, Ba, Eu, for example) define a chemical space. Although we can measure abundances for about 35 elements in principle, many of the abundances are correlated; the dimensionality of chemical space is believed to be between 7 and 9. Most disk stars inhabit a sub-region of this space. The stars of the homogeneous HR 1614 group fall in a very small region of chemical space, and it seems likely that other examples of debris from star-forming aggregates will be similarly homogeneous (see De Silva *et al.*, 2009). Stars that came in from satellites will not be chemically homogeneous, but their locus in chemical space is likely to be different enough to stand out from the rest of the disk stars. With this chemical tagging approach, we may be able to detect or put observational limits on the satellite accretion history of the galactic disk.

To use chemical tagging to recognize stars that come from common star-forming aggregates, a few conditions are needed.

- stars form in large aggregates: this is believed to be true
- the aggregates are chemically homogeneous
- aggregates have unique chemical signatures defined by several elements that do not vary in lockstep from one aggregate to another. Sufficient variation in abundances from aggregate to aggregate is needed so that chemical signatures can be distinguished with the achievable observational accuracy ($\sim$0.05 dex differentially)

Testing the last two conditions was the goal of De Silva's work on open clusters and moving groups, and they appear to be true. See De Silva *et al.* (2009) for details.

I should stress that chemical tagging is not just assigning stars chemically to a particular *population* (thin disk, thick disk, halo). Chemical tagging is intended to assign stars chemically to *substructure* which is no longer detectable kinematically.

The HERMES chemical tagging survey

HERMES is a new high-resolution multi-object spectrometer under construction for the AAT. The primary science driver for this instrument is galactic archaeology and chemical tagging. HERMES uses the 2dF fiber positioner to feed the light from about 400 stars in a π-deg^2 field to a high-resolution spectrometer with four simultaneous wavelength bands and a resolution of about 28,000. We expect this system to begin operation toward the end of 2012. A comparable near-IR spectrometer, APOGEE, is under construction for the Sloan telescope.

With HERMES, we are planning a stellar survey reaching to a V-magnitude of about 14: at this magnitude, the stellar density matches the fiber density at intermediate latitudes, and we expect to achieve an SN ratio of about 100 per resolution element in a 60-minute integration. Our goal is to observe about one million stars, which should take about 330 clear bright nights.

Most of the stars in this survey will be dwarfs of the thin disk (58%), thick disk (10%), and halo (2%) type. The remaining 30% are giants from these three components. The dwarfs are seen out to distances of about 1 kpc, and the giants are seen to at least 5 kpc.

How many star-formation sites can we expect to detect in this chemical tagging survey? We assume that the debris of each dispersed star-formation event is now mixed around the Galaxy; then the formation debris of about 9% of the thick disk stars and about 14% of the thin disk stars pass through our 1 kpc dwarf horizon. Simulations show that the million-star survey would identify about 20 thick disk dwarfs from each of about 4500 star-formation sites, and about 10 thin disk dwarfs from each of about 35,000 sites. A smaller survey would mean fewer stars from a similar number of sites.

Are there enough cells in the accessible chemical space to detect such a large number of sites? We believe so. We would need about seven independent chemical element groups, each with five measurable abundance levels to get enough independent cells (5^7), and this appears to be within reach. For example, in the studies by De Silva *et al.*, the rms dispersion for most of the element abundance ratios was about 0.03 to 0.04 dex, which provides enough measurable levels within the abundance scatter of the disk stars. From these studies, it appears that there are at least seven element groups that vary at least partly independently:

- light elements like Li, Na, Al, K
- Mg
- other alpha-elements like O, Ca, Si, Ti
- Fe and Fe-peak elements
- light s-process elements like Sr and Zr
- heavy s-process elements like Ba
- r-process elements like Eu

Within several of these groups, there is some variation of element abundances that adds to the dimensionality of the chemical space. Abundance variations in some of these element groups are seen from cluster to cluster.

In addition to the fossil aspects of this survey, we note that chemical tagging may be a way to assess definitively the importance of the Sellwood and Binney (2002) orbit swapping process, in which transient disturbances move stars from one near-circular orbit to another. Recent simulations by Roškar *et al.* (2008) have demonstrated how significant this process could be for redistributing stars in the disk. The spread in the (E, L_z) plane of stars from chemically recovered debris should give a useful indication of how much dynamical evolution has been suffered by the debris via orbit swapping and other dynamical processes.

We should note the synergy of the HERMES survey with data from Gaia. For these bright stars ($V \sim 14$), the parallax and proper motion uncertainties are expected to be about 10 μas and 10 μas yr^{-1}, respectively. This corresponds to a 1% distance error at a distance of 1 kpc (the typical distance of FG dwarfs in the sample) and a 0.7 km s^{-1} velocity error at 15 kpc (the distance of the faintest halo giants in the sample). We could expect to have accurate distances, abundances, and radial and transverse velocities for all the stars in the sample. In addition to the stellar parameters derived spectroscopically, we will also have accurate temperature (absolute magnitude) data using the SkyMapper photometric survey. This will provide an independent check that stars in chemically tagged groups have common ages.

Gaia is likely to detect a vast amount of phase space substructure. The HERMES survey will help to determine which phase space substructures are the debris of remnant star-forming events and which are dynamical artifacts (resonances).

REFERENCES

Abadi, M. G., Navarro, J. F., Steinmetz, M., and Eke, V. R. 2003. Simulations of galaxy formation in a Λ Cold Dark Matter Universe. II. The fine structure of simulated galactic disks. *ApJ*, **597**(Nov.), 21–34.

Athanassoula, E. 2008. Boxy/peanut and discy bulges: formation, evolution and properties. *IAU Symposium*, **245**(July), 93–102.

Athanassoula, E. and Beaton, R. L. 2006. Unraveling the mystery of the M31 bar. *MNRAS*, **370**(Aug.), 1499–1512.

Barnes, S. A. 2007. Ages for illustrative field stars using gyrochronology: viability, limitations, and errors. *ApJ*, **669**(Nov.), 1167–1189.

Beaton, R. L., and 8 colleagues. 2007. Unveiling the boxy bulge and bar of the Andromeda spiral galaxy. *ApJ*, **658**(Apr.), L91–L94.

Beaulieu, S. F., Freeman, K. C., Kalnajs, A. J., Saha, P., and Zhao, H. 2000. Dynamics of the galactic bulge using planetary nebulae. *AJ*, **120**(Aug.), 855–871.

Bensby, T., Feltzing, S., and Lundström, I. 2003. Elemental abundance trends in the Galactic thin and thick disks as traced by nearby F and G dwarf stars. *A&A*, **410**(Nov.), 527–551.

Bensby, T., Oey, M. S., Feltzing, S., and Gustafsson, B. 2007. Disentangling the Hercules stream. *ApJ*, **655**(Feb.), L89–L92.

Bertelli, G., Bressan, A., Chiosi, C., Fagotto, F., and Nasi, E. 1994. Theoretical isochrones from models with new radiative opacities. *A&AS*, **106**(Aug.), 275–302.

Bertelli, G., Girardi, L., Marigo, P., and Nasi, E. 2008. Scaled solar tracks and isochrones in a large region of the Z-Y plane. I. From the ZAMS to the TP-AGB end for 0.15-2.5 M⊙ stars. *A&A*, **484**(June), 815–830.

Binney, J. and Merrifield, M. 1998. *Galactic Astronomy.* Princeton, NJ: Princeton University Press.

Binney, J. and Tremaine, S. 1987. *Galactic Dynamics.* Princeton, NJ: Princeton University Press.

Binney, J. and Tremaine, S. 2008. *Galactic Dynamics.* 2nd ed. Princeton NJ: Princeton University Press.

Bissantz, N. and Gerhard, O. 2002. Spiral arms, bar shape and bulge microlensing in the Milky Way. *MNRAS*, **330**(Mar.), 591–608.

Bland-Hawthorn, J., Vlajić, M., Freeman, K. C., and Draine, B. T. 2005. NGC 300: an extremely faint, outer stellar disk observed to 10 scale lengths. *ApJ*, **629**(Aug.), 239–249.

Brocato, E., Castellani, V., Poli, F. M., and Raimondo, G. 2000. Predicted colours for simple stellar populations. II. The case of old stellar clusters. *A&AS*, **146**(Oct.), 91–101.

Brocato, E., Castellani, V., Raimondo, G., and Romaniello, M. 1999. Predicted HST FOC and broad band colours for young and intermediate simple stellar populations. *A&AS*, **136**(Apr.), 65–80.

Brook, C. B., Kawata, D., Gibson, B. K., and Freeman, K. C. 2004. The emergence of the thick disk in a Cold Dark Matter Universe. *ApJ*, **612**(Sept.), 894–899.

Brook, C., Richard, S., Kawata, D., Martel, H., and Gibson, B. K. 2007. Two disk components from a gas-rich disk-disk merger. *ApJ*, **658**(Mar.), 60–64.

Bureau, M. and Freeman, K. C. 1999. The nature of boxy/peanut-shaped bulges in spiral galaxies. *AJ*, **118**(July), 126–138.

Cantiello, M., Raimondo, G., Brocato, E., and Capaccioli, M. 2003. New optical and near-infrared surface brightness fluctuation models: a primary distance indicator ranging from globular clusters to distant galaxies? *AJ*, **125**(June), 2783–2808.

Carney, B. W., Laird, J. B., Latham, D. W., and Aguilar, L. A. 1996. A survey of proper motion stars. XIII. The halo population. *AJ*, **112**(Aug.), 668–692.

Carney, B. W., Yong, D., Teixera de Almeida, M. L., and Seitzer, P. 2005. Elemental abundance ratios in stars of the outer galactic disk. II. Field red giants. *AJ*, **130**(Sept.), 1111–1126.

Carollo, C. M., Stiavelli, M., de Zeeuw, P. T., and Mack, J. 1997. Spiral galaxies with WFPC2.I. Nuclear morphology, bulges, star clusters, and surface brightness profiles. *AJ*, **114**(Dec.), 2366–2380.

Chiba, M. and Beers, T. C. 2000. Kinematics of metal-poor stars in the Galaxy. III. Formation of the stellar halo and thick disk as revealed from a large sample of sonkinematically selected stars. *AJ*, **119**(June), 2843–2865.

Combes, F., Debbasch, F., Friedli, D., and Pfenniger, D. 1990. Box and peanut shapes generated by stellar bars. *A&A*, **233**(July), 82–95.

Combes, F. and Sanders, R. H. 1981. Formation and properties of persisting stellar bars. *A&A*, **96**(Mar.), 164–173.

Côté, P. 1999. Kinematics of the galactic globular cluster system: new radial velocities for clusters in the direction of the inner Galaxy. *AJ*, **118**(July), 406–420.

Courteau, S., de Jong, R. S., and Broeils, A. H. 1996. Evidence for secular evolution in late-type spirals. *ApJ*, **457**(Feb.), L73–L76.

de Blok, W. J. G. and Bosma, A. 2002. High-resolution rotation curves of low surface brightness galaxies. *A&A*, **385**(Apr.), 816–846.

de Grijs, R. and Peletier, R. F. 1997. The shape of galaxy disks: how the scale height increases with galactocentric distance. *A&A*, **320**(Apr.), L21–L24.

de Jong, R. S., and 14 colleagues 2007. Stellar Populations across the NGC 4244 Truncated Galactic Disk. *ApJ*, **667**(Sept.), L49–L52.

De Silva, G. M., Freeman, K. C., and Bland-Hawthorn, J. 2009. Reconstructing fossil substructures of the Galactic disk: clues from abundance patterns of old open clusters and moving groups. *PASA*, **26**(Apr.), 11–16.

De Silva, G. M., Freeman, K. C., Bland-Hawthorn, J., Asplund, M., and Bessell, M. S. 2007. Chemically tagging the HR 1614 moving group. *AJ*, **133**(Feb.), 694–704.

Dehnen, W. 2000. The effect of the outer Lindblad resonance of the Galactic bar on the local stellar velocity distribution. *AJ*, **119**(Feb.), 800–812.

Diemand, J., Madau, P., and Moore, B. 2005. The distribution and kinematics of early high-σ peaks in present-day haloes: implications for rare objects and old stellar populations. *MNRAS*, **364**(Dec.), 367–383.

Dotter, A. L. and Chaboyer, B. C. 2002. The impact of pollution on stellar evolution models. *Bulletin of the American Astronomical Society*, **34**(Dec.), 1127.

Dotter, A., Chaboyer, B., Jevremović, D., Kostov, V., Baron, E., and Ferguson, J. W. 2008. The Dartmouth stellar evolution database. *ApJS*, **178**(Sept.), 89–101.

Edvardsson, B., Andersen, J., Gustafsson, B., Lambert, D. L., Nissen, P. E., and Tomkin, J. 1993. The chemical evolution of the Galactic disk – Part One – analysis and results. *A&A*, **275**(Aug.), 101–152.

Eggen, O. J., Lynden-Bell, D., and Sandage, A. R. 1962. Evidence from the motions of old stars that the Galaxy collapsed. *ApJ*, **136**(Nov.), 748–766.

Erwin, P., Beckman, J. E., and Pohlen, M. 2005. Antitruncation of disks in early-type barred galaxies. *ApJ*, **626**(June), L81–L84.

Falcón-Barroso, J., and 10 colleagues 2004. A SAURON look at galaxy bulges. *Astronomische Nachrichten*, **325**(Feb.), 92–95.

Fall, S. M. and Efstathiou, G. 1980. Formation and rotation of disc galaxies with haloes. *MNRAS*, **193**(Oct.), 189–206.

Freeman, K. C. 1991. Observational properties of disks. *Dynamics of Disc Galaxies*, ed. B. Sundelius: Göteborgs University and Chalmers University of Technology, p. 15.

Freeman, K. C. 1993. Globular clusters and nucleated dwarf ellipticals. *The Globular Cluster-Galaxy Connection*, ed. G. Smith and J. Brodie (ASP), **48**(Jan.), 608–614.

Freeman, K. and Bland-Hawthorn, J. 2002. The new galaxy: signatures of its formation. *ARA&A*, **40**, 487–537.

Fuhrmann, K. 2008. Nearby stars of the galactic disc and halo IV. *MNRAS*, **384**(Feb.), 173–224.

Fulbright, J. P., McWilliam, A., and Rich, R. M. 2007. Abundances of Baade's Window Giants from Keck HIRES Spectra. II. The alpha and light odd elements. *ApJ*, **661**(June), 1152–1179.

Gao, L., White, S. D. M., Jenkins, A., Stoehr, F., and Springel, V. 2004. The subhalo populations of ΛCDM dark haloes. *MNRAS*, **355**(Dec.), 819–834.

Gilmore, G. and Reid, N. 1983. New light on faint stars III. Galactic structure towards the South Pole and the Galactic thick disc. *MNRAS*, **202**(Mar.), 1025–1047.

Gilmore, G., Wyse, R. F. G., and Jones, J. B. 1995. A determination of the thick disk chemical abundance distribution: implications for galaxy evolution. *AJ*, **109**(Mar.), 1095–1111.

Governato, F., Willman, B., Mayer, L., Brooks, A., Stinson, G., Valenzuela, O., Wadsley, J., and Quinn, T. 2007. Forming disc galaxies in ΛCDM simulations. *MNRAS*, **374**(Feb.), 1479–1494.

Gratton, R. G., Carretta, E., Claudi, R., Lucatello, S., and Barbieri, M. 2003. Abundances for metal-poor stars with accurate parallaxes. I. Basic data. *A&A*, **404**(June), 187–210.

Han, S.-I., Kim, Y.-C., Lee, Y.-W., Yi, S. K., Kim, D.-G., and Demarque, P. 2009. New Yonsei-Yale ($Y^2$2) isochrones and horizontal-branch evolutionary tracks with helium enhancements. *Globular Clusters – Guides to Galaxies.* eds. Richtler, T. and Larsen, S.: Springer, Berlin Heidelberg. p. 33.

Hansen, B. M. S., and 9 colleagues 2002. The white dwarf cooling sequence of the globular cluster Messier 4. *ApJ*, **574**(Aug.), L155–L158.

Harding, P. and Morrison, H. 1993. The Bulge/halo interface: rotational kinematics from [Fe/H]=-:3.0 to Solar. *Galactic Bulges*, **153**, 297–298.

Helmi, A. and de Zeeuw, P. T. 2000. Mapping the substructure in the Galactic halo with the next generation of astrometric satellites. *MNRAS*, **319**(Dec.), 657–665.

Holmberg, J., Nordström, B., and Andersen, J. 2007. The Geneva-Copenhagen survey of the Solar neighbourhood II. New uvby calibrations and rediscussion of stellar ages, the G dwarf problem, age-metallicity diagram, and heating mechanisms of the disk. *A&A*, **475**(Nov.), 519–537.

Howard, C. D., Rich, R. M., Reitzel, D. B., Koch, A., De Propris, R., and Zhao, H. 2008. The Bulge Radial Velocity Assay (BRAVA). I. Sample selection and a rotation curve. *ApJ*, **688**(Dec.), 1060–1077.

Ivezić, Ž., and 52 colleagues 2008. The Milky Way tomography with SDSS. II. Stellar metallicity. *ApJ*, **684**(Sept.), 287–325.

Kaufmann, T., Mayer, L., Wadsley, J., Stadel, J., and Moore, B. 2007. Angular momentum transport and disc morphology in smoothed particle hydrodynamics simulations of galaxy formation. *MNRAS*, **375**(Feb.), 53–67.

Knebe, A., Gill, S. P. D., Kawata, D., and Gibson, B. K. 2005. Mapping substructures in dark matter haloes. *MNRAS*, **357**(Feb.), L35–L39.

Koch, A., McWilliam, A., Grebel, E. K., Zucker, D. B., and Belokurov, V. 2008. The highly unusual chemical composition of the Hercules Dwarf Spheroidal Galaxy. *ApJ*, **688**(Nov.), L13–L16.

Kormendy, J. 1993. Kinematics of extragalactic bulges: evidence that some bulges are really disks. *Galactic Bulges*, **153**, p. 209.

Kroupa, P. 2002. Thickening of galactic discs through clustered star formation. *MNRAS*, **330**(Mar.), 707–718.

Kuijken, K. and Merrifield, M. R. 1995. Establishing the connection between peanut-shaped bulges and galactic bars. *ApJ*, **443**(Apr.), L13–L16.

Launhardt, R., Zylka, R., and Mezger, P. G. 2002. The nuclear bulge of the Galaxy. III. Large-scale physical characteristics of stars and interstellar matter. *A&A*, **384**(Mar.), 112–139.

Leggett, S. K., Ruiz, M. T., and Bergeron, P. 1998. The cool white dwarf luminosity function and the age of the Galactic disk. *ApJ*, **497**(Apr.), p. 294.

Lewis, J. R. and Freeman, K. C. 1989. Kinematics and chemical properties of the old disk of the Galaxy. *AJ*, **97**(Jan.), 139–162.

Lin, D. N. C. and Pringle, J. E. 1987. The formation of the exponential disk in spiral galaxies. *ApJ*, **320**(Sept.), L87–L91.

López-Corredoira, M., Cabrera-Lavers, A., Mahoney, T. J., Hammersley, P. L., Garzón, F., and González-Fernández, C. 2007. The long bar in the Milky Way: corroboration of an old hypothesis. *AJ*, **133**(Jan.), 154–161.

Luck, R. E., Kovtyukh, V. V., and Andrievsky, S. M. 2006. The distribution of the elements in the Galactic Disk. *AJ*, **132**(Aug.), 902–918.

Maeder, A. and Meynet, G. 1989. Grids of evolutionary models from 0.85 to 120 solar masses. Observational tests and the mass limits. *A&A*, **210**(Feb.), 155–173.

Maeder, A. and Meynet, G. 2008. Massive Star Evolution with Mass Loss and Rotation. *Revista Mexicana de Astronomia y Astrofisica Conference Series*, **33**(Aug.), 38–43.

Marinacci, F., Binney, J., Fraternali, F., Nipoti, C., Ciotti, L., and Londrillo, P. 2010. The mode of gas accretion on to star-forming galaxies. *MNRAS*, **404**(May), 1464–1474.

Meléndez, J., and 10 colleagues 2008. Chemical similarities between Galactic bulge and local thick disk red giant stars. *A&A*, **484**(June), L21–L25.

Minniti, D. 1995. Metal-rich globular clusters with R less than or equal 3 kpc: Disk or bulge clusters. *AJ*, **109**(Apr.), 1663–1669.

Minniti, D., Olszewski, E. W., Liebert, J., White, S. D. M., Hill, J. M., and Irwin, M. J. 1995. The metallicity gradient of the Galactic bulge. *MNRAS*, **277**(Dec.), 1293–1311.

Moore, B., Ghigna, S., Governato, F., Lake, G., Quinn, T., Stadel, J., and Tozzi, P. 1999. Dark matter substructure within Galactic halos. *ApJ*, **524**(Oct.), L19–L22.

Morrison, H. L., and 11 colleagues. 2009. Fashionably late? Building up the Milky Way's inner halo. *ApJ*, **694**(Mar.), 130–143.

Navarro, J. F., Helmi, A., and Freeman, K. C. 2004. The extragalactic origin of the Arcturus Group. *ApJ*, **601**(Jan.), L43–L46.

Navarro, J. F., Frenk, C. S., and White, S. D. M. 1996. The structure of cold dark matter halos. *ApJ*, **462**(May), 563–575.

Nordström, B., and 8 colleagues. 2004. The Geneva-Copenhagen survey of the Solar neighbourhood. Ages, metallicities, and kinematic properties of ~14 000 F and G dwarfs. *A&A*, **418**(May), 989–1019.

Pace, G., Melendez, J., Pasquini, L., Carraro, G., Danziger, J., François, P., Matteucci, F., and Santos, N. C. 2009. An investigation of chromospheric activity spanning the Vaughan-Preston gap: impact on stellar ages. *A&A*, **499**(May), L9–L12.

Pont, F. and Eyer, L. 2004. Isochrone ages for field dwarfs: method and application to the age-metallicity relation. *MNRAS*, **351**(June), 487–504.

Quillen, A. C. and Garnett, D. R. 2000. The saturation of disk heating in the solar neighborhood and evidence for a merger 9 Gyrs ago. (Apr.), arXiv:astro-ph/0004210.

Raimondo, G., Brocato, E., Cantiello, M., and Capaccioli, M. 2005. New optical and near-infrared surface brightness fluctuation models. II. Young and intermediate-age stellar populations. *AJ*, **130**(Dec.), 2625–2646.

Reid, I. N., Turner, E. L., Turnbull, M. C., Mountain, M., and Valenti, J. A. 2007. Searching for earth analogs around the nearest stars: the disk age-metallicity relation and the age distribution in the Solar neighborhood. *ApJ*, **665**(Aug.), 767–784.

Rocha-Pinto, H. J., Scalo, J., Maciel, W. J., and Flynn, C. 2000. Chemical enrichment and star formation in the Milky Way disk. II. Star formation history. *A&A*, **358**(June), 869–885.

Roškar, R., Debattista, V. P., Stinson, G. S., Quinn, T. R., Kaufmann, T., and Wadsley, J. 2008. Beyond inside-out growth: formation and evolution of disk outskirts. *ApJ*, **675**(Mar.), L65–L68.

Samland, M. and Gerhard, O. E. 2003. The formation of a disk galaxy within a growing dark halo. *A&A*, **399**(Mar.), 961–982.

Scannapieco, E., Kawata, D., Brook, C. B., Schneider, R., Ferrara, A., and Gibson, B. K. 2006. The spatial distribution of the Galactic first stars. I. High-resolution N-body approach. *ApJ*, **653**(Dec.), 285–299.

Searle, L. and Zinn, R. 1978. Compositions of halo clusters and the formation of the galactic halo. *ApJ*, **225**(Oct.), 357–379.

Sellwood, J. A. and Binney, J. J. 2002. Radial mixing in galactic discs. *MNRAS*, **336**(Nov.), 785–796.

Sommer-Larsen, J. Numerical simulation of galaxy formation shown in this chapter.

Soubiran, C., Bienaymé, O., Mishenina, T. V., and Kovtyukh, V. V. 2008. Vertical distribution of Galactic disk stars. IV. AMR and AVR from clump giants. *A&A*, **480**(Mar.), 91–101.

Sparke, L. S. and Gallagher, J. S., III. 2007. *Galaxies in the Universe: An Introduction.* 2nd ed. Cambridge UK: Cambridge University Press.

Turon, C., Primas, F., Binney, J., Chiappini, C., Drew, J., Helmi, A., Robin, A. C., and Ryan, S. G. 2008. ESA-ESO Working Group on Galactic Populations, Chemistry and Dynamics. *ESA-ESO Working Group reports*(Sept.)

Valenti, J. A. and Fischer, D. A. 2005. Spectroscopic properties of cool ctars (SPOCS). I. 1040 F, G, and K Dwarfs from Keck, Lick, and AAT Planet Search Programs. *ApJS*, **159**(July), 141–166.

VandenBerg, D. A., Bergbusch, P. A., and Dowler, P. D. 2006. The Victoria-Regina stellar models: evolutionary tracks and isochrones for a wide range in mass and metallicity that allow for empirically constrained amounts of convective core overshooting. *ApJS*, **162**(Feb.), 375–387.

Veltz, L., and 18 colleagues 2008. Galactic kinematics with RAVE data. I. The distribution of stars towards the Galactic poles. *A&A*, **480**(Mar.), 753–765.

Venn, K. A. and Hill, V. M. 2008. Chemical signatures in dwarf galaxies. *The Messenger*, **134**(Dec.), 23–27.

Vlajić, M., Bland-Hawthorn, J., and Freeman, K. C. 2009. The abundance gradient in the extremely faint outer disk of NGC 300. *ApJ*, **697**(May), 361–372.

Walcher, C. J., Böker, T., Charlot, S., Ho, L. C., Rix, H.-W., Rossa, J., Shields, J. C., and van der Marel, R. P. 2006. Stellar populations in the nuclei of late-type spiral galaxies. *ApJ*, **649**(Oct.), 692–708.

Walker, I. R., Mihos, J. C., and Hernquist, L. 1996. Quantifying the fragility of galactic disks in minor mergers. *ApJ*, **460**(Mar.), 121–135.

Wielen, R. 1977. The diffusion of stellar orbits derived from the observed age-dependence of the velocity dispersion. *A&A*, **60**(Sept.), 263–275.

Williams, M. E. K., Freeman, K. C., Helmi, A., and RAVE Collaboration 2009. The Arcturus moving group: its place in the Galaxy. *IAU Symposium*, **254**(Mar.), 139–144.

Worthey, G., España, A., MacArthur, L. A., and Courteau, S. 2005. M31's heavy-element distribution and outer disk. *ApJ*, **631**(Oct.), 820–831.

Wylie-de Boer, E., Freeman, K., and Williams, M. 2010. Evidence of tidal debris from ω Cen in the Kapteyn Group. *AJ*, **139**(Feb.), 636–645.

Yoachim, P. and Dalcanton, J. J. 2006. Structural parameters of thin and thick disks in edge-on disk galaxies. *AJ*, **131**(Jan.), 226–249.

Yoachim, P. and Dalcanton, J. J. 2008. The kinematics of thick disks in nine external galaxies. *ApJ*, **682**(Aug.), 1004–1019.

Yong, D., Carney, B. W., and Teixera de Almeida, M. L. 2005. Elemental abundance ratios in stars of the Outer Galactic Disk. I. Open clusters. *AJ*, **130**(Aug.), 597–625.

Zinn, R. 1985. The globular cluster system of the galaxy. IV. The halo and disk subsystems. *ApJ*, **293**(June), 424–444.

Zoccali, M., and 9 colleagues. 2003. Age and metallicity distribution of the Galactic bulge from extensive optical and near-IR stellar photometry. *A&A*, **399**(Mar.), 931–956.

Zoccali, M., Hill, V., Lecureur, A., Barbuy, B., Renzini, A., Minniti, D., Gómez, A., and Ortolani, S. 2008. The metal content of bulge field stars from FLAMES-GIRAFFE spectra. I. Stellar parameters and iron abundances. *A&A*, **486**(July), 177–189.

Zwitter, T., Castelli, F., and Munari, U. 2004. An extensive library of synthetic spectra covering the far red, RAVE and GAIA wavelength ranges. *A&A*, **417**(Apr.), 1055–1062.

2. Dark matter content and tidal effects in Local Group dwarf galaxies

STEVEN R. MAJEWSKI

2.1 Somewhat historical: overview of the Local Group, dwarf galaxies, and their observed structures

Before taking on a discussion of the dynamics of Local Group (LG) galaxies and the contributing and competing effects of dark matter and tides, it is useful to have an understanding of the spatial distribution of these galaxies, the distribution of their types and masses, and their morphologies – all of which play critical roles in defining how dark matter and tides play out their dynamical tug-of-war. The most common types of galaxies – the dwarfs – which are the most dark matter dominated as well as those among LG galaxies to show the greatest evidence for tidal effects, are the primary focus of this chapter.

2.1.1 *The Local Group in context*

Large-scale galaxy redshift surveys over the past decades (e.g., Davis *et al.*, 1982; Geller and Huchra, 1989; Shectman *et al.*, 1996; York *et al.*, 2000; Colless *et al.*, 2001; Strauss *et al.*, 2002; Abazajian *et al.*, 2009; Jones *et al.*, 2009) have revealed clearly the filamentary structure of the distribution of galaxies in the Universe. The nearest 100 Mpc shows vast voids but several large mass concentrations, such as the Perseus-Pisces, Pegasus, Pavo, Coma, Hydra-Centaurus, and Virgo Superclusters. The Milky Way (MW) and the LG of galaxies live on the outskirts of the Virgo Supercluster, whose center lies about 15 Mpc away.

Closer to home, within several Mpc, the LG appears to be the most densely populated of several loose aggregates of galaxies, including the NGC 3109 and Maffei Groups (lying about an Mpc closer to the Virgo Supercluster than us), and the Sculptor Group (lying several Mpc farther away from Virgo). The LG itself contains two primary concentrations of galaxies, centered on each of the two large spirals, M31 and the MW. But the looseness of the distribution of galaxies in this neighborhood of scattered galaxy groups and "subgroups" makes it somewhat tricky to define the actual limits of the LG and to make an accurate census of its membership. Thus, the definition of the "Local Group" is fuzzy and can be made in various ways.

The three-dimensional distribution of galaxies in the nearest several Mpc (e.g., the nice plots given in the Encyclopedia Galactica)[1] gives one the impression of an apparent minimum in density at a radius of about 1 Mpc from a point roughly centered between M31 and the MW (the approximate barycenter of the LG). Compilations of nearby galaxies circa the year 2000 (e.g., Mateo, 1998; Sparke and Gallagher, 2000) enumerated about three dozen within 1 Mpc of the Sun – a number that is now known to be at least a factor of two (and probably several more factors than this) too small, given the recent discovery of the ultrafaint class of dwarf galaxies (see Section 2.1.4). Increasing this "by-eye" definition of the LG to a 1.6 Mpc radius from the LG barycenter only adds a few (non-ultrafaint) galaxies (mainly from the NGC 3109 "subgroup"). Distance-based accountings of LG galaxies hold the extra complication that accurate distances to some potential members are still not in hand, so that these lists may be incomplete or too inclusive. This is particularly problematical for low luminosity dSphs where use of tip of the red giant branch (RGB) magnitudes – the most accessible and reliable means for using stellar population information to gauge distances in far-off but resolved systems – breaks

[1] http://www.astro.utu.fi/EGal/elg/ELG3Dl.html.

down simply from the paucity of stars on the RGB, a problem that makes it difficult to determine the location of the RGB tip in a color-magnitude diagram (CMD). By the way, resolvability itself has been a traditional way to at least get a sense of distance for stellar systems; for example, Baade (1944) used his ability to resolve individual stars in several dE galaxies (M32 and NGC 205) located near M31 on the sky to determine that they were also at a similar distance to, and therefore associated with, M31. Through Hubble Space Telescope (HST) imaging we are now able to resolve stars in galaxies well past what are the commonly accepted limits of the LG, which means that color-magnitude information and isochrone fitting can be brought to bear not only on gauging the distances but also the ages and metallicities of LG galaxies. However, current state-of-the-art CMDs of more distant LG systems are typically limited to the brightest parts of RGBs even with HST (though there are some notable exceptions – for example, Brown *et al.*, 2003; Dalcanton *et al.*, 2009), so that there is still some uncertainty in defining LG members this way.

Dynamics provide an alternative means to establish LG membership. For example, galaxies in equilibrium with the LG should have lower relative velocities with respect to the LG barycenter than systems participating in the Hubble flow. To estimate the *zero-velocity surface*, which separates the LG from the Hubble flow, requires an estimate of the total mass of the LG (which is dominated by the masses of M31 and the MW) and the LG velocity dispersion (which is about 60 km s^{-1}). Defining the zero-velocity surface to be the edge of the LG yields a typical radius of about 1.8 Mpc (Sandage, 1986; Mateo, 1998). Note that this inferred LG size and observed velocity dispersion implies a crossing time longer than a Hubble time, so that not all "LG galaxies" would be "infalling" yet – i.e., virialized. There are also some nuances to such dynamical definitions of LG membership. For example, two dSphs – And XII (Martin *et al.*, 2006; Chapman *et al.*, 2007) and And XIV (Majewski *et al.*, 2007) – have been discovered that look to be within the LG spatially but that may not presently be bound to any large galaxy (they lie near, but apparently outside the escape velocity for, M31; Figure 2.1). These two dSph systems may be falling into the LG for the first time and presently have large radial velocities that might exclude them from dynamical definitions of LG membership. Nevertheless, these two galaxies may well represent precursors to the very types of dwarfs we see tidally disrupting in the LG today.

Why do we care to have a reasonable census of the LG? Apart from the obvious reasons of wanting to assess its net mass and stellar population/galaxy content, an understanding of what's in the LG is useful because the LG is still growing, as predicted by Cold Dark Matter (CDM) models of the growth of structure in the universe, and it is through this growth that smaller systems are brought closer to large systems via hierarchical formation. Identifying systems like And XII and And XIV (which may be part of an infalling *group* of galaxies – Martin *et al.*, 2006; Majewski *et al.*, 2007) and taking a census of the current satellite galaxy retinue allows us to gauge the past, present, and future rates of growth of LG galaxies, estimate the number of tidally disrupting systems we expect to see, and develop an understanding of how galaxies grow through accretion of subsystems.

2.1.2 *Galaxy content of the Local Group and the origin of morphological differences*

The mass and luminosity of the LG is dominated by its spiral systems, M31 (morphological type SAb or SBb), the MW (likely SBbc), M33 (Sc or Sd), and the Large Magellanic Cloud ("LMC," classified either as a dIrr type galaxy or as the prototype for "Magellanic spirals," SBm). Together, MW+M31+M33 emit 90% of the light in the LG, and, interestingly, all three show evidence of warps in their disks – typically thought to be formed from gravitational encounters. Obviously, these disks are rotationally supported, with $V_{\rm rot}/\sigma_v \sim$ (220–240 km s^{-1}) / (30 km s^{-1}) $\sim$ 7–8 for the MW. M31 is thought to

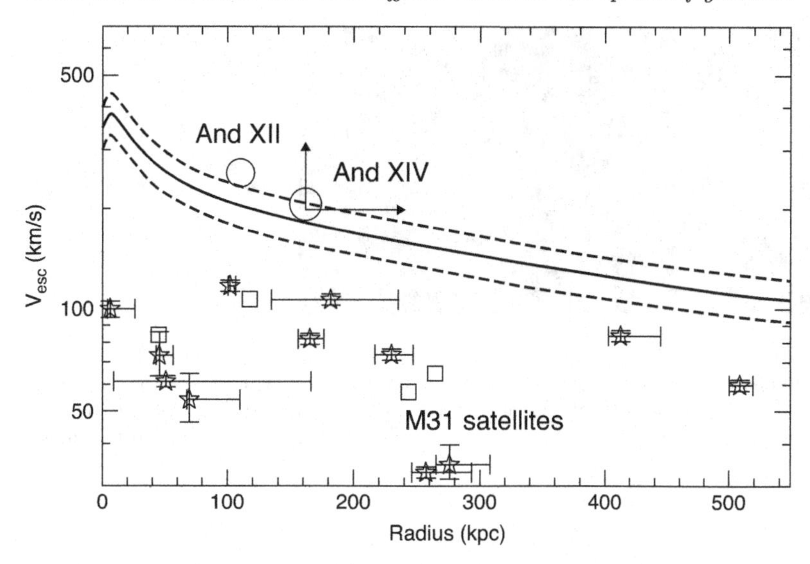

FIG. 2.1. Escape velocity versus three-dimensional distance from the center of M31 for an M31 mass model by Seigar *et al.* (2008). The star symbols and squares represent the positions and radial velocities relative to M31 (reduced to a common projection parallel to the M31-MW vector) for bound Andromeda satellites. The circles represent the conservative positions and velocities of And XII and XIV based on using the *projected* radial distance from M31. Given the conservative limits used to place the objects in the plot, both the And XII and XIV dwarf galaxies appear to be above the median model escape velocity of M31. Figure modified from Majewski *et al.* (2007).

be more luminous and more massive than the MW,[2] with $V_{\rm rot} = 260$ km s^{-1}, double the scalelength of the MW and an earlier morphological type. It has only recently become certain that M31 and the MW are barred spirals (Gerhard, 2002; Merrifield, 2004; Beaton *et al.*, 2007), as is the LMC. All of these spirals contain gas and dust and are actively forming stars. M31 and the MW are the primary centers of satellite galaxy and globular cluster systems in the LG.

The LG contains only one system, M32, that could nominally be categorized as an elliptical galaxy. But M32 itself is in many ways an odd elliptical system that often shows up as extreme or exceptional in trends of galaxy properties. Unlike other LG systems, M32 does not have a flat inner light profile, but one that instead rises to a cusp. The central surface brightness of M32 is the brightest known for any galaxy. Both of the latter properties suggest that M32 may possess a central, million-solar mass black hole (e.g., van der Marel *et al.*, 1998). M32 is partly rotationally supported, with $V_{\rm rot}/\sigma_v \sim 1$. It contains no young stars or gas, and yet has enriched to about solar metallicity, just like larger ellipticals. Although M32 shows isophotal distortion that has been attributed to tidal interaction with M31, Howley *et al.* (2011) find no obvious sharp velocity gradients over the "tidal distortion" region, a feature often expected for tidal disruption (but see Section 2.4.2). Because of the compactness of M32 compared with other dEs, it is considered the prototype of the compact elliptical or cE morphological type (e.g., Graham, 2002; Chilingarian *et al.*, 2009).

Apart from the small elliptical M32, the LG contains three other systems that can truly be considered *dwarf ellipticals* (dE); like M32, these three systems, NGC 147, NGC 185, and NGC 205, are all satellites associated with M31. These dE are diffuse systems that are apparently susceptible to tidal stripping – tidal tails emanating from NGC 205

[2] This point, however, is still debated (Evans and Wilkinson, 2000; Gottesman *et al.*, 2002; Klypin *et al.*, 2002; Karachentsev *et al.*, 2009).

FIG. 2.2. (a): Local Group (LG) dIrr galaxy Sextans A. Image from SD. Van Dyk (IPAC/Caltech) *et al.* and National Optical Astronomy Observatories, at http://apod.nasa.gov/apod/ap981103.html. (b): "Transitional" type LG galaxy, Pegasus. Image by D. A. Hunter (Lowell Observatory). (c): LG dSph galaxy Sculptor. Image by D. Malin (Anglo-Australian Observatory).

are now well documented (Howley *et al.*, 2008; Saviane *et al.*, 2010). Internally, these dEs show no ordered rotation, but rather they are "pressure supported" ($V_{rot}/\sigma_v \sim 0.1$). Clearly, these dE systems are "later" in morphological type than M32 in that they contain trace amounts of gas and dust and, though dominated by intermediate to old stellar populations, two contain small numbers of younger stars seen near their nuclei (NGC 205 and NGC 185). Unlike M32, which has no known globular clusters, all three LG dE systems have their own globular clusters, and NGC 185 even contains relatively young star clusters.

More abundant in the LG are the Irr and dIrr systems, with the two classes often divided at about 10^8 solar luminosities. In this case, the LG Irr systems would be the Small Magellanic Cloud ("SMC"), IC 10, and NGC 6822, and in some morphological systems the LMC and NGC 3109 would also be included. The dIrr include such systems as Sextans A (Figure 2.2), IC 1613, WLM, Leo A, Aquarius and the Sagittarius dIrr galaxy (not to be confused with the Sagittarius dSph galaxy). The irregular systems are characterized by large gas fractions and current star formation featuring bright, young, blue stars, and H II regions. It is the presence of bright, patchy star formation regions that is primarily responsible for the irregular shapes of these galaxies, although interactions between LG members may have also played a role (e.g., in the LMC/SMC/MW system). The LG dIrr systems are generally rotationally supported, with $V_{rot}/\sigma_v \sim 1$–6, and not as chemically evolved, with metallicities typically 1/10 solar or less.

Dwarf spheroidal (dSph) galaxies (e.g., the Sculptor system shown in Figure 2.2) are by far the most prominent type of galaxy in the LG (and presumably in the universe), especially if you consider the "classical" dSph and the ultrafaint dwarf galaxies (Section 2.1.4) as a group. These are the main type of LG galaxy still being discovered today. The dSph are extremely low surface brightness systems, with total luminosities not too dissimilar in range from those of globular clusters, but central surface brightnesses nearly 10 magnitudes arcsec^{-2} fainter due to the extremely diffuse, extended structure of dSphs, with hundreds of times larger core radii than those of globular clusters. This observed structure alone hints at a quite different distribution of matter from that of other star systems, for the distended shapes of dSphs require "hidden mass" for these shapes to be maintained, if these systems are in dynamical equilibrium. The dSphs are clearly pressure-supported systems ($V_{rot}/\sigma_v < \sim 0.1$). They also have no detectable hydrogen and no signs of current, nor (generally) recent star formation (at least within the past 10^8 years or so). On the other hand, while all dSph seem to contain an old ($\sim$13 Gyr) population, also most show at least one other generation of younger stars, from 1 to 10 Gyr old. Like dIrr systems, the dSph have intermediate to low metallicities.

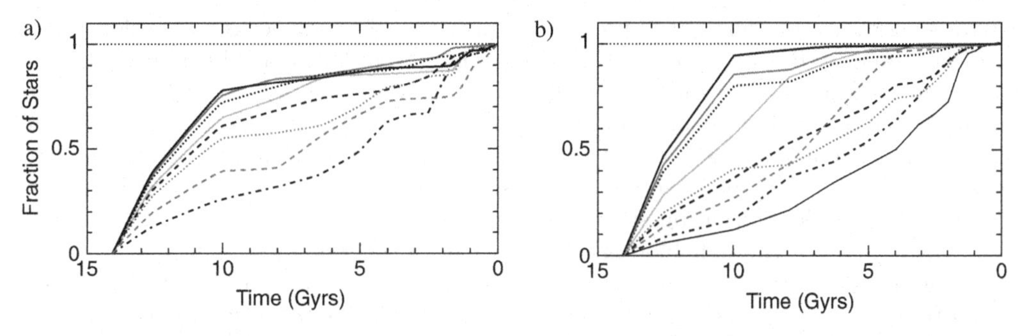

FIG. 2.3. (a): Holtzman diagram for different dIrrs. (b): Same for different dSph. The greater similarity and more typically constant SFHs among dIrr galaxies compared to those of dSph galaxies is obvious. Figure courtesy of Jon Holtzman.

Indeed, a number of similarities between dSph and dIrr suggest that they may have an evolutionary connection. The two classes of galaxies have similar masses, with dIrr (e.g., the LG systems GR8, SagDIG, Leo A) having estimated masses as low as $< \sim 10^7$ $M_\odot$, and the dSphs (at least the "classical" dSphs), having about the same mass (see Section 2.2.1). dSph and dIrr also have similar total luminosities, especially if you strip away the pyrotechnics of the recent star formation in the latter. Indeed, because the dIrr are also fairly diffuse, the surface brightnesses in between their star-forming regions are low, and similar to those of dSph galaxies. Indeed, the primary difference between the dSph and dIrr seems to lie in their relative star-formation histories (SFHs): Obviously a major distinction is that dIrr show active star formation now while dSph do not, and one can't help but wonder whether this alone biases the classification – for example, where one of the current dSphs observed during an active star formation phase, it seems highly plausible that it would be classified as a dIrr type system.

There is even another morphological subclass of LG galaxies, the so-called transitional or dIrr/dSph, that belies a potential semantical bias in our categorization. Local Group dIrr/dSph systems like Pegasus (Figure 2.2) and Pisces (also known as LGS3) have properties intermediate between those of dIrr and dSph galaxies, including just a little gas, modest recent star formation with a few young stars but no H II regions, and a $V_{rot}/\sigma_v \sim 1.7$ in Pegasus (the one such system measured so far). The existence of these transitional type systems strengthens the notion of an evolutionary link between dIrr and dSph, and one related to relative SFHs.

Systematically evaluated SFHs using homogeneous data from HST by, for example, Dolphin *et al.* (2003) reveal an apparent overall trend in the SFHs of LG dwarf galaxies, with dIrr tending to show more or less continuous and steady SFH; transition types showing a declining SFH; and dSph, while showing a wide variety of SFHs, typically showing an overall more rapidly declining SFH than the transition types, with perhaps some bursts at later times. These differences are most easily seen in "Holtzman Diagrams" (Figure 2.3), which are cumulative distributions of the fraction of stars created as a function of time.[3] These global trends suggest that an important distinction in the morphological types is primarily correlated with how much gas is left and how quickly it was used up.

What could drive such differences in SFHs and gas consumption among the LG dwarf galaxies? Suppressing star formation is one mechanism for prolonging it, and some dIrr, like Sextans A, have apparently had very delayed star formation requiring an effective

[3] Comparing star formation rates (SFR) in *differential* form can be tricky because different mass systems require different normalizations. Moreover, the derivation of star formation histories using color-magnitude distributions of resolved stars relies on *population synthesis models*, and in such models the errors in derived SFR from time bin to time bin can be highly correlated; these problems are mitigated by comparing cumulative SFHs.

suppressing agent. Heating gas to prevent it from collapsing can occur from prior generations of stars, particularly populous ones. Extreme UV radiation fields from strong star bursts, or even an intense ionizing background radiation, such as was experienced due to the first round of quasar and supernova formation (from "Population III" stars) at the epoch of reionization (about 600 million years after the Big Bang), could effectively destroy cold molecular clouds, squelching for some time thereafter the formation of stars. Especially energetic internal bursts of star formation can even remove gas permanently from small galaxies through blowout. Perhaps this is an appropriate model for those dSph galaxies that are dominated by a single early star-formation burst and not much star formation later on. As we shall discuss in Section 2.5, the blowout model may be an active process in the LMC and aiding in the formation of the Magellanic Stream.

Alternatively, variations in dwarf SFHs could come about through mechanisms that *hasten* star formation in some systems relative to others. Environmental triggers are strongly suggested by the relative spatial distributions of dwarf types in the local universe: dE and dSph galaxies are strongly clustered around M31 and the MW, nearby transitional type dwarfs have typical distances from nearest spirals (MW, M31, or NGC 3109) that are a bit higher, whereas dIrr tend to be found distributed in all environments but are the principal morphological type in the most isolated realms of the local universe (Skillman *et al.*, 2003; Grebel, 2004, see Figure 2.4). This clear *morphology-density relationship* strongly implies that environment drives dwarf galaxy SFHs. Because tidal shocking fosters star formation bursts, more frequent gravitational encounters with larger systems would be expected to accelerate the net SFR in a dwarf galaxy. In addition, or alternatively, the gaseous fuel for dwarf galaxy star formation might be stripped out by ram pressure interactions with the hot coronal gas around larger galaxies like M31 and the MW (another traditional model for the formation of the Magellanic Stream; see Section 2.5); thus more frequent or more intense encounters would result in a faster depletion of gas and a gradual (or even dramatic) squelching of star formation in an affected dwarf galaxy.

The morphology-density relationship and the similarities cited earlier between dIrr and dSph systems provide strong, if circumstantial, evidence of their evolutionary connection and differentiation by environmental influences. Even in the case of the LMC, where self-created blowout may be a key role in the loss of fuel in the system, environment has its role to play as well (Section 2.5).

Obviously, environmental influences strongly shape the stellar populations in a dwarf galaxy, but what about their dynamics? If one believes that dIrr and dSph are just two phases in the life of a dwarf galaxy dictated by gravitational encounters, then those encounters might also account for a transition from rotation to pressure support in the dwarf galaxy. This can occur through the process of *tidal stirring*, which may be a key mechanism at play in the LG. Simulations by Mayer *et al.* (2001), see Figure 2.5, of tidal stirring acting on the evolution of a high surface brightness, rotating dwarf disk galaxy inside a MW-sized dark matter halo shows how the tidal field from the latter induces the formation and subsequent buckling of a bar in the dwarf; this process dynamically heats the system and, over the course of order 6 Gyr or so, transforms it into a pressure-supported galaxy. The strong ellipticities now seen in some of the dSph satellites of the MW may be a residual feature of the bars formed in this tidal stirring process (e.g., Łokas *et al.*, 2010).

To summarize this discussion, and to connect it to the main theme of this chapter, we can see how, on the simple basis of dwarf galaxy morphologies and the distribution of these different morphologies in relative numbers and locations within the LG, one is led naturally to the conclusion that many LG dwarf galaxies – dSphs in particular – have experienced strong tidal effects. The degree to which these tides have shaped the morphologies of these systems likely depends on their masses (and therefore dark matter contents) and orbits, as well as the operation of a variety of processes that may be

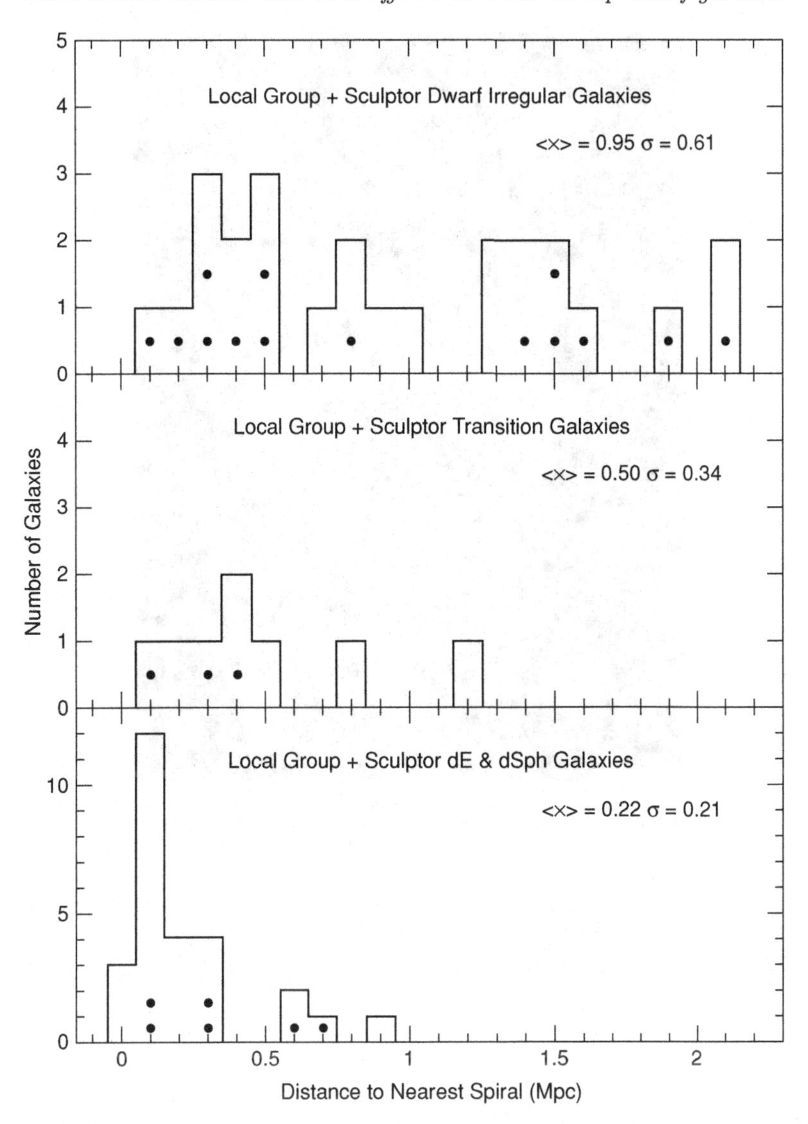

FIG. 2.4. Morphology-density relationship for the Local and Sculptor Groups, from Skillman *et al.* (2003). The mean distances and standard deviations are given in each panel. Filled circles represent the distribution of Sculptor Group dwarf galaxies.

involved in the transformation – for example, starbursts, ram pressure stripping, tidal stirring etc.

Three points, two of them caveats, are worth adding to the oft-discussed paradigm for the origin of the morphology-density relation, the evolution of dwarf galaxies, and the dIrr-dSph connection:

(1) Technically, the LG morphology-density relationship could be considered even stronger than observed at present if one accounts for the possibility that the dIrr systems closest to spirals in the LG are the Magellanic Clouds, recent proper motion measurements made using Hubble Space Telescope images suggest that they may be on their first trip into the MW system (Kallivayalil *et al.*, 2006a,b; Besla *et al.*, 2007); that is, a few Gyr ago the Clouds may have been well outside the MW, in the outskirts of the LG, where most dIrr lie today.

(2) Mass loss through tidal stripping (Section 2.4) will act preferentially to remove stars from the outer parts of dSphs, while new star formation will usually occur at

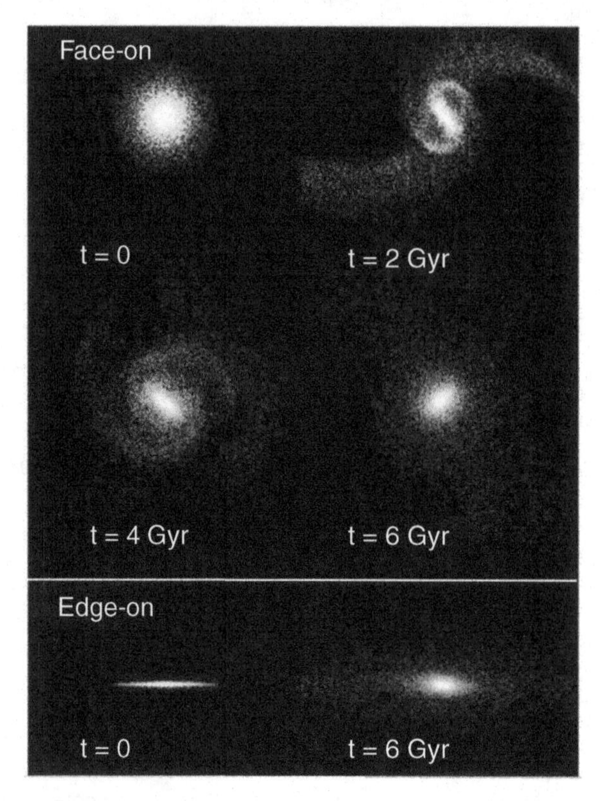

FIG. 2.5. Face-on and edge-on views of the evolution of a high surface brightness, dwarf disk galaxy into a dwarf spheroidal due to tidal stirring induced while in an orbit around a MW-sized halo, from simulations by Mayer *et al.* (2001). Heating induced by bar buckling is found to be more important than direct heating by the tidal field. The boxes are about 30 kpc on each side. Modified from Mayer *et al.* (2001).

the center of the system's gravitational potential, to where gas will sink and be most concentrated. Thus, we can expect the proportions of stellar populations recorded in the distribution of stars in present dSphs to not necessarily give an accurate assessment of the true SFH of a dwarf galaxy: older populations of stars will be underrepresented if they have been preferentially stripped away (Majewski *et al.*, 2005; Chou *et al.*, 2007; Muñoz *et al.*, 2008; Law and Majewski, 2010) compared to more tightly bound, younger populations. At apparent present rates of tidal mass loss, dSphs that today are dominated by intermediate-aged populations are calculated to have had a much more significant older stellar population if one includes lost stars (Majewski *et al.*, 2002; Chou *et al.*, 2007; Muñoz *et al.*, 2008; Law and Majewski, 2010); this effect differentiates the SFHs of dSphs even more from those of the more remotely situated, less tidally affected dIrr systems.

(3) There are detractors to the dIrr-dSph evolutionary connection described earlier. Grebel *et al.* (2003) argue that (a) once dIrr systems stop forming stars and fade in both luminosity and surface brightness, this fading would result in galaxies with too low a surface brightness as compared with presentday dSphs, and (b) "dIrrs have too low a metallicity for their luminosity as compared with dSphs." Grebel *et al.* (2003) suggest that the class of "transitional" dwarfs more closely resemble dSphs and may be their more direct and principal progenitors, distinct from dIrr, which follow a different evolutionary course. Nevertheless, despite this difference in evolutionary connections between dSphs and other dwarf systems, Grebel *et al.* (2003) make the case that environment (tidal forces) still drive the evolution in dwarf galaxy morphology and SFH.

2.1.3 *The classical dwarf spheroidal galaxies*

In a series of lectures on the topic of dark matter and tidal effects in LG galaxies, the dSph systems – the galaxies exhibiting the highest dark matter content, and those also most suspected of having endured tidal effects – must receive prominent attention. dSphs are also the most numerous galaxies in the LG, despite the extreme challenge of finding these very faint ($M_V > -12$), low surface brightness ($\Sigma_V > 22$ mag arcsec^{-2}) systems compared to the more flashy dwarfs with current star formation.

The dSph galaxies are so faint and spread out that the first one known, Sculptor (Figure 2.2), wasn't discovered until 1938 on a deeply exposed, photographic plate of "abnormal sensitiveness" by Harlow Shapley, who noted in referring to this "large rich cluster" that "nothing quite like it [was] known" at the time. Shapley (1938, 1939) mapped Sculptor's broad stellar density distribution and estimated its distance with variable stars. Sculptor's nature was further revealed by a study using the 60-inch telescope at Boyden Observatory and the 100-inch telescope at Mt. Wilson by Baade and Hubble (1939), who discovered both (Population II) Cepheid variables (i.e., W Virginis stars) and dozens of RR Lyrae stars that they could use to estimate a distance modulus for the system (19.62 mag, remarkably close to the most recent estimates using near infrared observations of RR Lyrae stars, 19.68±0.08 mag, by Pietrzyński *et al.*, 2008) and therefore a luminosity, confirming that it is a faint, dwarf galaxy, but one with "remarkably uniform" distributions of stars over a large angle, and "no indications of a nucleus, of clusters, or of diffuse nebulosity, either luminous or dark," as Shapley earlier described. After an exhaustive survey of Harvard Observatory plates, Shapley (1939) was able to find a second example of one of these unusual "Sculptor-type" galaxies, the Fornax system, which he recognized to be much more distant than Sculptor. Four additional "extreme dwarf ellipticals" (Hodge, 1971) near the MW were discovered in the 1950s: Leo I, Leo II (Harrington and Wilson, 1950), and Ursa Minor and Draco (Wilson, 1955). By 2005, despite many deliberate searches, only three more dSph satellites of the MW were found (Carina, Sextans and Sagittarius (Sgr);[4] Cannon *et al.*, 1977; Irwin *et al.*, 1990; Ibata *et al.*, 1994), but seven were found around M31, where the greater distance makes the systems more compact and therefore easier to see in contrast to their surroundings. We refer to these first nine found MW dSphs as the "classical dSphs," to distinguish them from the subsequently discovered slew of "ultrafaint dSphs" (Section 2.1.4).

The first systematic, quantitative morphological studies of the dSphs didn't happen until decades after their discovery with the photographic starcount studies by Hodge (1961a,b). These studies made clear that these systems did not follow density laws at all like those conventionally fitted to other elliptical galaxy types, for example, (1) the Hubble Law, where the density of stars (or, equivalently, the luminosity profile of the system, under the assumption that the stellar luminosity function is constant with radius) I goes with radius R as $I(R) = I_0/(R/a+1)^2$, where a and I_0 are size and luminosity scale parameters, or (2) the de Vaucouleurs Law, where $\ln I(R) = \ln I_e + 7.67[1 - (R/R_e)^{1/4}]$, where R_e is the effective radius in which lies half the light of the system and I_e is the surface brightness at that radius.

On the other hand, the density profiles of dSphs did seem to match *King profiles.* This is surprising, given that King (1962) originally derived this analytical shape to describe the density distributions of the much more compact *globular clusters.* In 1962, Ivan King noticed that the density distributions of globular clusters seemed to follow a simple family of curves (Figure 2.6) described by only two free parameters, a core (or half-light) radius, r_c, and a limiting radius, r_t, where the density drops dramatically (often referred to as the "King radius" or "tidal radius"). For a typical globular cluster, $r_c \sim 1$ pc (or 20 arcsec for a cluster at 10 kpc) and $r_t \sim 30$ pc (or 10 arcmin at 10 kpc). From visual analysis,

[4] The discovery of the Sgr dSph was actually serendipitous, not the result of a deliberate search (Ibata *et al.*, 1994).

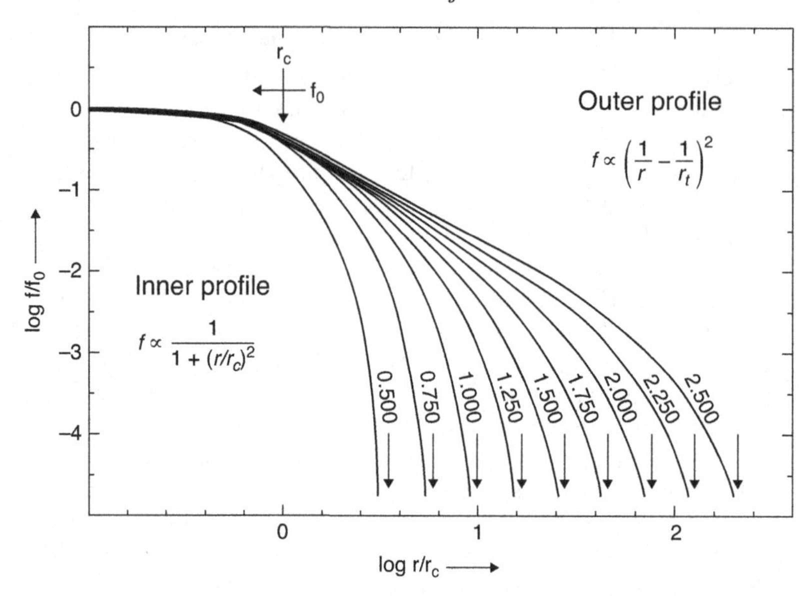

FIG. 2.6. Family of analytical King profiles normalized to the core radius, with the concentration parameter and King limiting radius shown for each curve. King's (1962) separate empirical relations for the inner and outer parts of the profile are shown. Figure modified from that given in the lecture by Elson (1999), who presented this topic at the Tenth Canary Islands Winter School in 1998, shortly before her untimely passing.

he formulated a single analytical form that encapsulates the empirical distributions with these parameters:

$$f = k\left(\frac{1}{[1+(r/r_c)^2]^{1/2}} - \frac{1}{[1+(r_t/r_c)^2]^{1/2}}\right)^2, \tag{2.1}$$

where the first term in the brackets dictates the shape of the inner, flatter part of the profile and the second term dictates the steeper, rapidly declining shape of the outer profile (Figure 2.6). With this formulation, the relative *concentration* of the cluster can be defined by $c = \log(r_t/r_c)$; for MW globulars the concentration spans approximately $0.5 < c < 3$ (a range of shapes slightly more diverse than that shown in Figure 2.6). In a remarkable sequel paper to his 1962 empirical studies, King (1966) showed theoretically how his empirically chosen density law arises naturally in the physics of *tidally truncated*, spherically symmetric, isotropic, self-gravitating systems made of a "gas" of stars of a single mass. By his analysis, the stars in the gas are found to relax via repeated two-body interactions, eventually resulting in a family of models that look just like the *empirical* King models, but now characterized by one free parameter, the central potential (W_0). In the theoretical models, r_t – the limiting radius – is the *tidal radius*, i.e., the radius of tidal truncation. The "two-body relaxation" timescale, defined as the time it takes cumulative star-star encounters to result in a net deflection of 90 degrees in the motion of a typical star, follows $t_{2body} \sim \frac{1}{8\ln(0.5N)} t_{ce}$, where N is the number of stars in the cluster and $t_{ce} \sim (4.8 \times 10^{10}\text{ years})\, v_m^3/[n(m_1+m_2)^2]$ is the timescale for close stellar encounters – the typical time one must wait for a single encounter between stars of mass m_1 and m_2 in a system with mean velocity v_m at density $n(\text{stars pc}^{-3})$ to produce a 90 degree deflection in the direction of travel for one of the stars. The close encounter timescale is of order 17 Gyr using typical globular cluster densities and velocity dispersions – so too long to be relevant except in the densest, central part of the cluster. But the two-body relaxation timescale for a globular cluster is $t_{2body} \sim 10^8$ years in the center, and $\sim 10^9$ years at the core radius, r_c – so that in either density regime the typical globular cluster is dynamially relaxed after a Hubble time.

But dSphs have such a low density (fitted values of $r_c \sim 300$ pc and $r_t \sim 1600$ pc) that they have a two-body relaxation timescale of 10^{13} years in their centers. With such long relaxation timescales it would seem that *there is no good reason that the King profile should work for the low density dSph systems*[5]– and yet King profiles seem to work rather well (at least in the most luminous parts of dSphs) and are often adopted by observers and theorists alike to describe the luminous parts of dSphs. That said, the custom is not universal, and many astronomers, unhappy with the fact that the physics producing King profiles in globular clusters does not work for dSphs, shy from adopting it even as a simple morphological description, preferring, for example, the Sersic function (which is a generalized form of the de Vaucouleurs law: $\ln I(R) = \ln I_e + b[1 - (R/R_e)^{1/n}]$) or Dehnen models (double-power law density models: $\rho(r) = \rho_0/[(r/a)^{\alpha}(1 + r/a)^{\beta-\alpha}]$, where α, β, and ρ_0 are constants). Until the true physics underlying the luminosity profiles of dSphs is known and we can adopt a corresponding, physically motivated analytical form, it doesn't seem necessary to abandon a perfectly good and simple morphological description for which many astronomers have an innate intuition of both shape and scales set by its characteristic parameters (r_c and r_t) as long as one is clear that it is being adopted *only* as an empirical description (after all, the King profile was originally derived as an *empirical* description to begin with). But, to maintain as much clarity as possible, it is preferable to avoid use of the term "tidal radius" as a descriptor for the parameter r_t in the context of dSphs, and use alternative expressions such as "King radius," "limiting radius," or "King limiting radius" for that point in the density profile at which the density rapidly declines.

This is not merely a semantical point, but one rather critical to interpreting and understanding dSph morphological studies. For example, in a study of the distribution of RR Lyrae stars in the Sculptor dSph, Innanen and Papp (1979) found a very nice fit to their data with a King profile, except for the presence of an extended, low density, excess population at large radii, beyond the r_t of their fit to the higher density parts of the Sculptor profile; these authors referred to these excess stars as "extratidal," under the assumption that r_t really corresponds to the tidal radius of the system. Eskridge (1988), using deep photographic counts, similarly found evidence for a Sculptor population beyond the nominal King function at the largest radii. And in their extensive photographic survey of all of the then-known MW dSphs, Irwin and Hatzidimitriou (1995) found evidence for "excesses" of stars beyond the King profile for most of these dwarf systems. Since the King function is not necessarily physically motivated for dSph systems, it is not clear what these excess populations represent, whether a population of bound or unbound stars, so the description of these stars as "extratidal" is appropriately suspect.

However, with more focused efforts using more sensitive CCD photometry and careful filtering to improve their detection against contamination by foreground MW populations, these "excesses" of dSph stars have been observed to extend to much larger radii than seen in the earlier, photographic studies (Kuhn *et al.*, 1996; Palma *et al.*, 2003; Muñoz *et al.*, 2005, 2006; Westfall *et al.*, 2006; Sohn *et al.*, 2007, see Figure 2.8). The density profile of these stars follows a much shallower, typically exponential profile that "breaks" from the steeply declining King profile just interior (Figure 2.7). Spectroscopy of these widely separated stars proves their association with the dSphs and therefore the reality of the extended populations (Majewski *et al.*, 2005; Muñoz *et al.*, 2005, 2006; Sohn *et al.*, 2007), but has not completely resolved their dynamical status (e.g., bound versus unbound) and origin. A possible explanation is that these stars represent an extended, bound "halo" population of the dSphs (e.g. Hayashi *et al.*, 2003; Mashchenko *et al.*, 2005;

[5] King (1966) reasoned that perhaps for dSphs, some other physics other than two-body relaxation (he proposed violent relaxation) could be responsible for their King profile-like morphology.

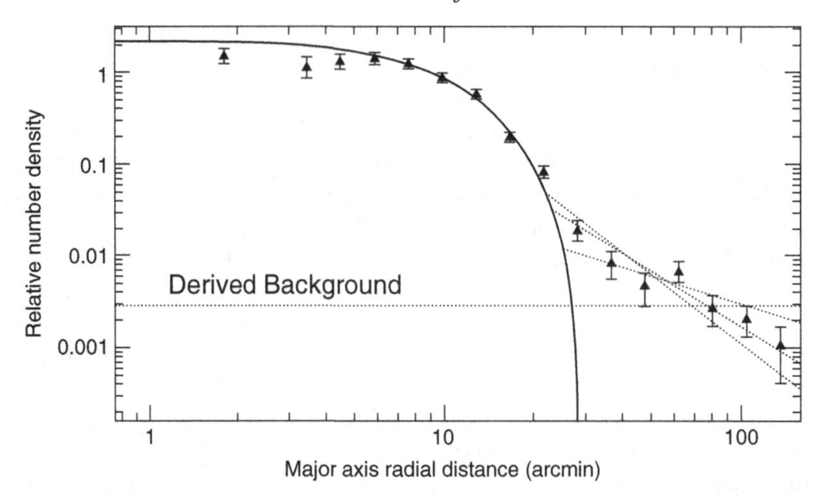

FIG. 2.7. Classic "King+break" density profile observed for the Carina dSph by Muñoz *et al.* (2006) via a specialized search for Carina RGB stars using Washington + *DDO*51 filter photometry, which is effective for removing the overwhelming foreground of MW dwarf stars and improving the signal-to-background contrast of the dSph. The dotted line shows the very low equivalent background density level obtained with the Washington + *DDO*51 technique; this background has been derived from a spectroscopic analysis of stars outside the King limiting radius and subtracted off to produce the final profile shown. All stars have been binned into elliptical annuli and are shown at the corresponding major axis radius, with Poisson errors. The curving solid line shows the King profile derived previously by Irwin and Hatzidimitriou (1995). Outside the King limiting radius, the density values shown come almost entirely from spectroscopically confirmed Carina members. The dashed lines show $r^{-1.5}$, r^{-2}, and $r^{-2.5}$ power laws compared to the excess population "breaking" from the King profile. Figure modified from Muñoz *et al.* (2006).

McConnachie *et al.*, 2007); because the luminous mass of these systems is insufficient to keep these populations bound (e.g., Moore, 1996), the "bound halo of stars" explanation implies very large dark matter contents in the dSphs to keep these extended distributions of stars bound – for example, at least 16,000 times more dark than luminous matter implied in the case of Carina to keep its most displaced, spectroscopically verified member star bound (Muñoz *et al.*, 2006), and implying an actual mass of the system rivaling that of the SMC (or the LMC, in the case of Ursa Minor; Muñoz *et al.*, 2005).

Moreover, actual models of the expected dark matter distributions of dSphs show that they must extend *well beyond* the radii of these widely separated stars to keep them bound (e.g., Peñarrubia *et al.*, 2008a,b, – see Figure 2.9 – and Walker *et al.*, 2007 – see Figure 2.11). It is worth pointing out that the light profiles of dSphs do not match those predicted for dark matter halos created in high resolution, N-body cosmological simulations, which show central cusps (in contrast to the flat central "King profiles" observed) and gently changing, logarithmic slopes (i.e., the Navarro *et al.*, 1996, or "NFW," profile; Figure 2.9). The "observed core" versus "predicted cusp" problem in MW dSph galaxies is actually a more general problem for the CDM model for structure formation on all galaxy-sized scales (Section 2.2.2). But in addition, the general differences between the presumed mass and observed light profiles of galaxies show that they must be generated through independent processes (e.g., Mashchenko *et al.*, 2005). In the case of dSphs, it is commonly held that the baryonic components traced by the surface brightness profiles are so deeply embedded in dark matter halos that the "features in the light profile [are] of limited dynamical relevance" (Hayashi *et al.*, 2003).

Alternatively, the extended break populations (or some part of them) may truly be "extratidal" – a signature of mass loss through *tidal stripping.* The process of tidal stripping has long been well understood from N-body simulations of satellites orbiting in

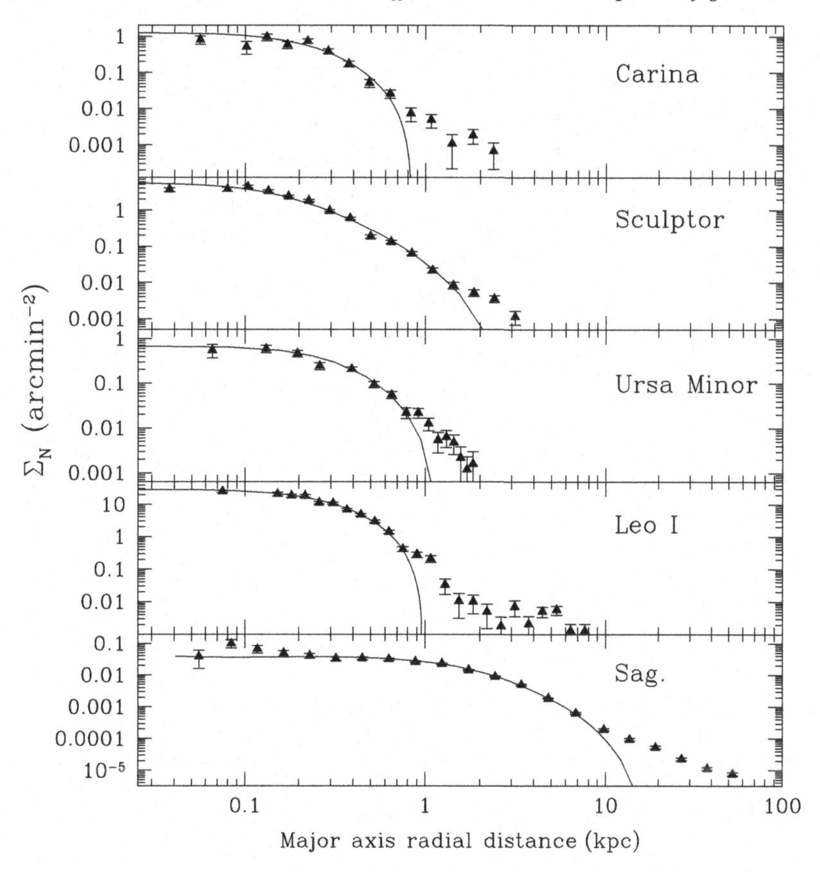

FIG. 2.8. Background-subtracted, radial number density profiles of RGB stars selected using Washington+$DDO51$ photometry for five dSphs. The solid lines show King profiles fitted to the central regions. Error bars represent Poisson errors. Data from the following references: Carina (Muñoz *et al.*, 2006), Sculptor (Westfall *et al.*, 2006), Ursa Minor (Palma *et al.*, 2003), Leo I (Sohn *et al.*, 2007), Sgr (Majewski *et al.*, 2003).

larger (e.g., MW-like) gravitational potentials, with stars pulled off in the direction of the tidal force (i.e., radially) and then forming pairs of *tidal tails* – with the stars pulled outside the tidal radius on the inner side of the satellite having negative orbital energy relative to the satellite core and advancing ahead to form the leading arm, and those escapees on the far side having a higher relative orbital energy and therefore orbiting at larger radius with a slightly longer period and forming a trailing tail (see Figure 2.15). Well-known examples of MW systems undergoing the process – for example, the Palomar 5 globular cluster (Odenkirchen *et al.*, 2001, 2002, 2003, 2009; Grillmair and Dionatos, 2006a, Figure 2.19) and the Sgr dSph (Majewski *et al.*, 2003, 2004; Martínez-Delgado *et al.*, 2007; Casetti-Dinescu *et al.*, 2008) – illustrate this tidal tail formation quite vividly. A motivation for the tidal stripping hypothesis as an explanation for the "break populations" seen around MW dSphs is that the net density profiles (i.e., "King + break") strongly resemble the combinations of bound core + unbound stars seen in N-body simulations of tidally stripped model satellites. Moreover, in one case – the Sgr dSph – it is well established that its "King + break" radial profile is produced as a result of tidally stripped stars; the Sgr system provides a clear demonstration that tidal stripping can produce the observed dSph "King + break" morphology, and the question is whether this process naturally explains the structure of other dSphs sharing this morphology with Sgr, or whether a second mechanism for producing similarly looking break populations exists.

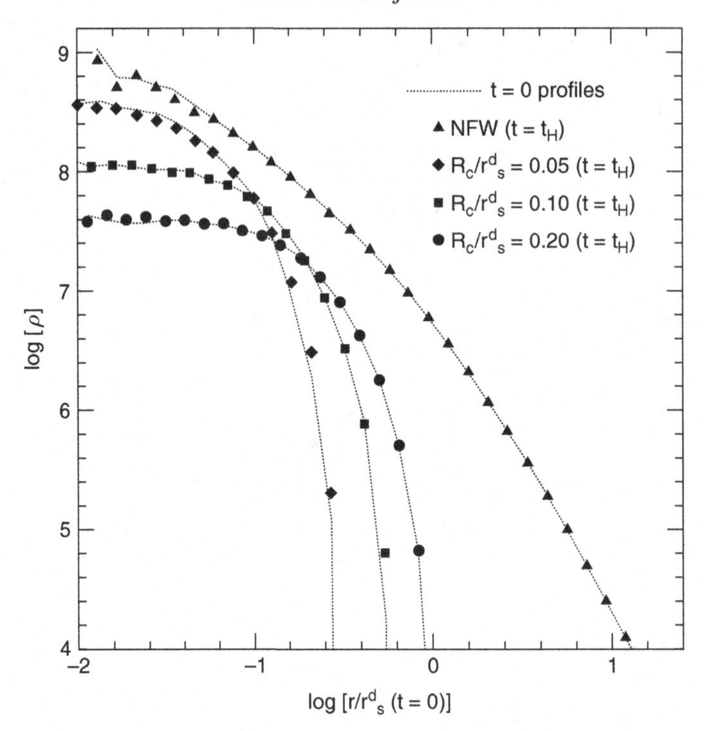

FIG. 2.9. Nominal configuration of a dSph galaxy with a luminous dwarf galaxy following a "cored" King-like profile (several versions shown) embedded in an "cusped" NFW profile (the outer curve). Figure from Peñarrubia *et al.* (2008b).

The following statement by Hayashi *et al.* (2003) aptly describes the importance of resolving the nature of these break populations: "the stars beyond the luminous cutoff detected, for example, by Majewski et al. (2000) may not actually represent a true population of unbound, extratidal stars but rather correspond to a radially extended component bound to Carina. Confirmation of the extratidal nature of such stars through independent means, such as detecting tidal tails or apparent rotation in Carina's extended envelope would thus provide a strong argument against this conclusion." We return to this issue in Sections 2.2.2 and 2.4.

2.1.4 *The ultrafaint dwarf galaxies*

We close this section by discussing briefly the remarkable discovery of a new class of galactic system in the LG, and one whose numbers make them clearly the most common type of galaxy in the LG, and, by inference, the present day universe. By 2005 there were nine known "classical" dSph satellites of the MW (discovered at a pace of about one per decade since Shapley's work), and seven known dSph satellites of M31 (apart from the dIrr and dE satellites around the MW and M31, respectively). The advent of the Sloan Digital Sky Survey resulted in the discovery in only a few years (2005–2008) of 13 "ultra-faint" MW dwarf satellite galaxies (e.g., Willman *et al.*, 2005; Belokurov *et al.*, 2006b; Zucker *et al.*, 2006a,b). These systems predominantly have stellar populations resembling those of dSph systems, and this gives rise to the sometimes-used label "ultrafaint dSphs" for the new class of galaxies. Because the SDSS only covered $\sim 1/4$ of the sky, it is likely that there are at least four times more of these systems around the MW than discovered so far. Many additional examples (along with other, more classical dSphs) were found around M31, mostly through the "PAndAS" survey and its MegaCam precursor (Ibata *et al.*, 2007; Martin *et al.*, 2009). While classical dSphs or dSph/dIrr galaxies have luminosities of $-8 > M_V > -17$ (from 10^5 to 5×10^8 solar luminosities), ultrafaint dSphs

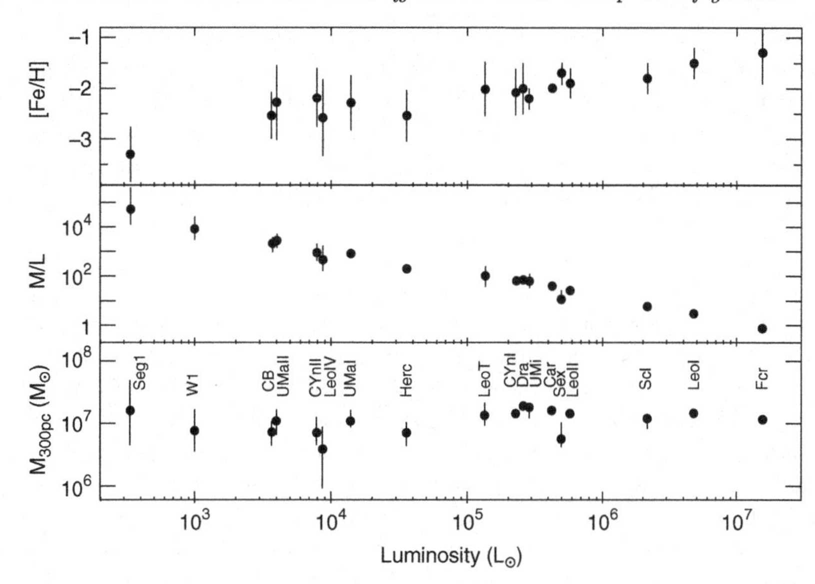

FIG. 2.10. The dependence of metallicity ([Fe/H]), mass-to-light ratio and mass (within 300 pc) as a function of luminosity for MW dSphs. Though their luminosities span almost 5 orders of magnitude, for MW dSphs the mass within 300 pc remains nearly constant at about $10^7\ \mathfrak{M}_\odot$. The derived $\mathfrak{M}/L$ rises dramatically with decreasing luminosity. Figure from Geha *et al.* (2009).

have luminosities as low as only 1000 solar luminosities, or lower (see Figure 2.10)! These systems have exacerbated the debate over the dark matter content versus tidal stripping in satellites, although their velocity dispersions are found to be high, which would imply enormous dark matter fractions if the systems are in equilibrium, these systems also often show very distorted morphologies that suggest tidal stripping (e.g., Belokurov *et al.*, 2007b). On the other hand, the number of stars in these systems is so small that the apparently distorted morphologies could simply be a product of low number statistics (Martin *et al.*, 2008; Muñoz *et al.*, 2008). The question of how readily these systems might be tidally disrupting is important, however, because their large numbers could make them significant contributors of tidally stripped stellar debris in the MW.

2.2 Evidence for dark matter in Local Group dwarf galaxies

2.2.1 *Measuring dwarf galaxy masses*

A useful parameter in the discussion of dark matter is the *mass-to-light ratio*, which is the total mass of a star system, $\mathfrak{M}$, expressed as a ratio to the mass it would have if all of the system's light, L, were emitted from stars of solar luminosity and mass. If $n_\odot = L/L_\odot$ is that number of suns, then the mass-to-light ratio is

$$\frac{\mathfrak{M}}{n_\odot \mathfrak{M}_\odot} = \frac{\mathfrak{M}/\mathfrak{M}_\odot}{L/L_\odot} = 10^{0.4(M-M_\odot)} \frac{\mathfrak{M}}{\mathfrak{M}_\odot} \tag{2.2}$$

where the right-hand quantity is expressed in terms of absolute magnitudes, M. Obviously, systems laden with dark matter have high $\mathfrak{M}/L$. Technically, as expressed here, the mass-to-light ratio is a bolometric quantity, but of course we typically measure luminosities in specific passbands, and it is important to specify that passband explicitly, because the mass-to-light ratio will vary depending on the spectral energy distribution of the system. For example, an older galaxy not actively producing a lot of stars and dominated by red giant starlight will tend to have a lower $\mathfrak{M}/L_I$ (mass-to-light in the I band) than $\mathfrak{M}/L_B$ (mass-to-light in the B band). It is common to use the V filter for mass-to-light ratios. For systems with no dark matter (e.g., star clusters) the $\mathfrak{M}/L_V$ will evolve with the mean age of its stars, from $\mathfrak{M}/L_V \sim 1$ $[\mathfrak{M}/L_V]_\odot$ for a young stellar population dominated by young main sequence stars and supergiants, to larger values as the dominant stars contributing light have lower mean luminosity, and reaching $\mathfrak{M}/L_V \sim 3$–4 $[\mathfrak{M}/L_V]_\odot$ for the oldest, metal-poor systems that we know (from here on, we will assume all mass-to-light ratios are in solar units, and dispense with writing $[\mathfrak{M}/L_V]_\odot$ on the right-hand side).

To estimate the dark matter content of galaxies, one must therefore assess their masses in some way, measure the luminosities, and then compare the derived $\mathfrak{M}/L_V$ (say) estimate to that expected for systems with the same stellar populations but no dark matter. Generally, the actual masses are inferred dynamically via the motions of stars or gas. Fortunately, we are now able to measure radial velocities of individual stars for many resolved stars systems across the LG; for more distant systems it is still common to rely on either radio observations of the 21-cm H I emission, or area-pixelated spectroscopy of the integrated light of stars (e.g., using a long-slit or an integrated field unit to sample the galaxy light). In the case of spiral and dIrr galaxies in the LG, we want to measure their *rotation curve*, from which we can derive mass as a function of radius via $\mathfrak{M}(r) = v^2 r/\mathrm{G}$. Indeed, it is through the measurement of spiral galaxy *rotation curves* that the presence of galactic-scale dark matter was first inferred (e.g., Rubin *et al.*, 1978). In the MW, the presence of dark matter can be inferred from the fact that a census of the light interior to the solar circle suggests a corresponding mass that would lead to a revolutionary speed of the Sun about the Galactic center of around 160 km s^{-1}, whereas we know that the actual speed is more like $\sim$230 km s^{-1} (or more, e.g., Reid *et al.*, 2009). Even for an unresolved system, as long as it is a rotationally supported, disk-like galaxy, the dynamical mass can be obtained from the integrated 21-cm line width as $\mathfrak{M}_{\mathrm{H}}(r) = K\Delta V^2 r_{\mathrm{H}}/\mathrm{G}(\sin i)^2$, where r_{H} is the H I radius, ΔV is the width of the H I velocity profile, $\sin i$ accounts for the inclination of the galaxy, and K is a constant (e.g., Roberts, 1962; Fisher and Tully, 1975; Lo *et al.*, 1993; Young and Lo, 1997). Most relevant to the present discussion, when these various methods are applied to LG dIrr systems we typically find $\mathfrak{M}/L_V \sim$ 3–20, a value suggesting the presence of dark matter in these systems at a fractional level comparable to that of the most modestly dark dSph systems.

To derive the masses of non-rotating systems like dSphs (or dE), which are dynamically supported by the "pressure" of a kinematically "hot" swarm of stars on more or less random orbits, one can use the balance of kinetic and potential energy given by the scalar virial theorem, $2K + W = 2(\mathfrak{M}\langle v^2\rangle/2) + (-\mathrm{G}\mathfrak{M}^2/r_g) = 0$, to derive that $\mathfrak{M} = \langle v^2\rangle r_g/\mathrm{G}$, where $\langle v^2\rangle$ is the mean-squared speed of the stars in the system, and r_g is the "gravitational radius" of the system. The latter is not well defined for a "fuzzy" system like a dwarf galaxy, but it is found that $r_g \sim r_h/0.4$, where r_h is the half-light radius, that radius within which half of the light of the system is contained. Thus, with $\mathfrak{M} = 2.5\langle v^2\rangle r_h/\mathrm{G}$ we have a means to estimate the mass of a system from the *velocity dispersion* of its stars.

Unfortunately, use of the virial theorem to measure mass in this way is only approximate because proper application takes account of the actual structure of the galaxy – specifically, the shape of its radial density profile. King (1966) showed a methodology where W can be calculated (and tabulated) for *isothermal spheres* by integration of the appropriately matching King profile to this system. In this case, one finds (Illingworth, 1976) $\mathfrak{M} = 167 r_c \mu \langle v_r^2\rangle_o$, where r_c is the core radius defined earlier (in parsecs), $\langle v_r^2\rangle_o$ is the *central radial velocity dispersion* of the system (in km s^{-1}), and μ is a tabulated mass function appropriate to a given set of King parameters r_c and r_t. Following this "core-fitting" or "King's method" and the virial theorem, but recognizing that $K = \mathfrak{M}\langle v^2\rangle/2 = (\mathfrak{M}/L)L_{\mathrm{tot}}\langle v^2\rangle/2$ for a system of total luminosity L_{tot} and constant $\mathfrak{M}/L$, Rood *et al.* (1972) (see also King and Minkowski, 1972; Richstone and Tremaine, 1986) derived a useful means for estimating the central $\mathfrak{M}/L$ of a spherical, isotropic, constant $\mathfrak{M}/L$ system from fully observable quantities, arriving at

$$\frac{\mathfrak{M}}{L} = \eta \frac{9\sigma_p^2(0)}{2\pi G I(0) R_{\rm hb}} \qquad (2.3)$$

where $I(0)$ is the central surface brightness, $\sigma_p(0)$ is the observed central velocity dispersion projected on the line-of-sight, which is approximately $(\langle v^2 \rangle/3)^{1/2}$ for an isotropic system, and $R_{\rm hb} = r_c$ is the half-light radius, and where the mass-to-light corresponds to the photometric passband in which the surface brightness and light profile are measured. The constant η is unity in almost all cases of nearly spherical systems. Equation 2.3 becomes less accurate for anisotropic stellar orbits and/or non-spherical systems. Application of the these methods in the case of LG dE systems typically yields very modest central $\mathfrak{M}/L \sim$ 1–5. But it is in the application of these methods to dSph systems that things get very interesting!

Aaronson (1983) first measured the velocity dispersion of a dSph (the Draco system) – from only *three* stars – and got the surprising (at the time) and impressive $\mathfrak{M}/L \sim 31$. As work proceeded on other systems (e.g., Armandroff and Da Costa, 1986; Mateo *et al.*, 1991, 1993), a trend of very large central velocity dispersions – typically 6–9 km s^{-1} for the classical dSphs – and therefore large inferred central $\mathfrak{M}/L$ continued. This work established that dSph galaxies were clearly very different in structure from any other stellar systems known at the time.

But soon an even more intriguing aspect to the dSphs emerged: as demonstrated by, for example, Vogt *et al.* (1995) and Mateo (1998), there is a strong correlation of inferred $\mathfrak{M}/L$ with M_V – stretching to as high as 100 $(\mathfrak{M}/L)_\odot$ for the intrinsically faintest classical dSphs – so that the inferred masses of LG dSphs using the core-fitting method all seem to be around 10^7 $M_\odot$, despite spanning 8 magnitudes of absolute luminosity. One might think that with the addition of the many newly found ultrafaint dSphs this nice coincidence might go away, since they substantially increase the overall dSph sample size, significantly expand the luminosity range of that sample, and have among them a great variety of appearances including some that look very fragile, as if they might not be held together even by dark matter. But the kinematics of the ultrafaint systems suggest that they have *even higher* $\mathfrak{M}/L$, of order 10^2–10^3; yet, despite these even higher $\mathfrak{M}/L$, it seemed at first that the ultrafaints might have lower than the "magical" $\sim 10^7$ $\mathfrak{M}_\odot$ of the classical dSphs when using the core-fitting method (e.g., Simon and Geha, 2007). But using more sophisticated analyses (e.g., Geha *et al.*, 2009), it is found that there *is* in fact a convergence in mass when looking at all dSphs to an equivalent radius (e.g., one set by the radial extent of the ultrafaint dSphs). For example, if you limit the measurement of the mass to a radius of, say, 300 pc, the dSphs once again look "universal" in mass, at $\sim 10^7$ $\mathfrak{M}_\odot$ (Strigari *et al.*, 2008; Geha *et al.*, 2009, Figure 2.10). Is this a remarkable coincidence, or is it telling us something fundamental about the formation and structure of galaxies on small, sub-halo scales?

2.2.2 *dSph galaxies as local laboratories of dark matter physics*

The cosmological implications and possible meaning of the remarkable dSph mass convergence, as well as other cosmological issues related to the general topic of the formation of galaxies within the CDM paradigm were much more fully discussed by other lecturers at the Winter School. We wish here only to make a few comments with regard to scientific and practical implications if dSph systems are really as dark matter rich as implied by the mass determinations discussed previously, and the potential impact that tidal disruption might have on these cosmological issues.

Recent large particle-count numerical simulations of the evolution of the universe in the presence of CDM make a rich variety of predictions about the structure and dynamics of matter from galactic to the largest scales and have had great success in matching observations for the latter. But on galaxy scales these models still have several problems matching observations, including the "missing satellites problem," the "central cusps problem," and several "disk angular momentum" problems (e.g., the inability to make large enough galactic disks). A current focus for advancing dark matter theory is attempting to resolve these problems on small (galaxy) scales. Obviously, understanding and explaining the dynamics of the LG, the MW, and its satellite system are central to making progress in the theory of dark matter, hierarchical formation, and galaxy evolution. But in particular, knowledge of the extent to which MW satellites may be tidally disrupting plays a key role in constraining dark matter physics in several ways.

Missing satellites problem

The present CDM models suggest that the MW of today is very lumpy, surrounded by hundreds or thousands of dark matter sub-halos – far more than the number of known satellites around the MW (Kauffmann *et al.*, 1993; Klypin *et al.*, 1999; Moore *et al.*, 1999). This is the missing satellites problem, and it describes an apparent shortfall of satellites at all relevant mass scales. One explanation for the shortfall might be that the satellites are mostly dark. For example, it might be that only a small fraction of the dark matter lumps ever form stars – perhaps those that happened to have been much larger at former times (e.g., Kravtsov *et al.*, 2004) or those that managed to form stars before the epoch of reionization (e.g., Bullock *et al.*, 2000; Somerville, 2002; Moore *et al.*, 2006, see extensive discussion in Bullock, Chapter 3, this volume).

But, given the recent discovery of the numerous ultrafaint dSphs, one might question whether the problem is with the observers and amounts to a simple matter of accounting. After all, if a dozen ultrafaint dwarfs have been found in the ~1/4 of the sky covered by the Sloan Digital Sky Survey, one might expect the MW to have at least 50 of them. Unfortunately, although this higher number of known dwarf galaxy satellites does help somewhat to "close the gap" between the predicted and observed numbers of satellites, it only does so at the lower end of the sub-halo mass function and doesn't fix the shortfall at all satellite masses; one still needs to invoke mechanisms for suppressing galaxy formation in dark matter sub-halos to accomplish that (e.g., Simon and Geha, 2007).

Of course, another potential reason for the observed satellite deficit is that a significant fraction of the accreted satellites may have succumbed to tidal destruction, and their remains may account for the obvious network of stellar substructure now observed in the MW halo (e.g., Bullock *et al.*, 2001; Somerville, 2002). Because dynamical friction operates more efficiently on more massive systems, the larger systems are more likely to be destroyed by their hosts, which may help alleviate the observed discrepancy for larger mass satellites. In addition, significant tidal stripping can apparently occur before the satellites are accreted by their host halo – and even before that host is formed – through complex and dramatic interactions with other, neighboring merging sub-halos (Kravtsov *et al.*, 2004). Given these predictions coming from numerical simulations of hierarchical galaxy formation, it should not be unexpected to find that tidal effects play a role in the lives of some not insignificant fraction of present day luminous satellites. Cosmologically motivated, CDM-based N-body models (e.g., Bullock and Johnston, 2005) suggest that about 20% to 30% of "classical" MW dSphs should currently be having tidal disruption of their luminous parts – or about 2–3 systems (?) out of the 11 classical MW dwarf satellites. Sgr is clearly one example, and, as we argue elsewhere (Sections 2.1.3 and 2.4), Carina, Leo I, and possibly Ursa Minor may also be cases of tidally disrupting systems. Clearly, determining the true number of MW satellites undergoing luminous disruption can provide a useful test of CDM models in the context of the missing satellites problem.

Cusps versus cores

The microscopic nature of dark matter affects the way it clusters around galaxies and thus can be probed by exploration of the LG, the MW, and its satellites. The prevailing

CDM models predict that galaxy mass profiles should be cusped, but rotation curves of galaxies suggest that they have cored profiles, and the luminous matter profiles are certainly observed to be cored. This is clearly the case for the luminous density profiles of dSph galaxies (e.g., Figures 2.7 and 2.8), but the shape of the *total* mass density profile is still uncertain and controversial. For example, Kleyna *et al.* (2003) argue that the double-peaked stellar structure of the Ursa Minor dSph (e.g., Irwin and Hatzidimitriou, 1995; Palma *et al.*, 2003; Martínez-Delgado *et al.*, 2008) is unexpected if it has a cusped dark matter halo as expected from CDM: the cold kinematics and slow relative velocity of the secondary peak to the main body suggest that this entity – possibly an internally orbiting, disrupted globular cluster – could only have survived a Hubble time if it lived within a host possessing a cored mass profile, whereas this cold clump would have been quickly erased (in less than a Gyr) within a cusped mass distribution. In a similar vein, calculations suggest that if the Formax dSph had a cuspy mass profile, its family of five globular clusters would have sunk to the center by dynamical friction in much less than a Hubble time (Goerdt *et al.*, 2006; Sánchez-Salcedo *et al.*, 2006). Analysis of the surface brightness profiles and velocity dispersions for some dSph galaxies have also led to a preference for lower density, cored mass distributions (e.g., Gilmore *et al.*, 2007; Battaglia *et al.*, 2008). If these implied cored profiles and the inferred low central "phase space densities" (ρ/σ^3, where ρ is the space density and σ is the radial velocity dispersion) are primordial and not the result of some modification in the dynamical life of the satellite, a warm dark matter particle is implied (e.g., gravitinos or light sterile neutrinos), rather than one of the WIMP candidates of CDM (e.g., axions, neutralinos).

Thus, dwarf spheroidal galaxies conveniently probe an interesting regime where, in principle, the dark matter physics can be discriminated. But more sophisticated methods for modeling the velocity distributions – and larger data sets – are needed to arrive at a conclusive definition of the mass profiles. The mass *profile* can be derived using the Jeans equation (spherically symmetric, collisionless Boltzmann equation):

$$\mathfrak{M}(r) = -\frac{r^2}{\mathrm{G}}\left(\frac{1}{\nu(r)}\frac{d\nu\sigma_r^2}{dr} + 2\frac{\beta(r)\sigma_r^2}{r}\right) \tag{2.4}$$

where $\mathfrak{M}(r)$ is the mass profile, $\nu(r)$ is the stellar density distribution, σ_r^2 is the one-dimensional velocity dispersion of those stars and $\beta(r)$ is the velocity anisotropy of their orbits. Fortunately, the data sets of radial velocities for stars in individual dSphs are rapidly growing in size (e.g., Figure 2.11) and radial extent (Figure 2.12), the latter including the difficult work to find the rare dSph stars at the greatest possible radii where their densities are extremely low relative to the enormous contaminating foreground of MW stars.

Although such data sets should make it possible to map the mass profiles over significant dSph scales, unfortunately, having only radial velocities admits degenerate solutions because of uncertainty over the radial variation of the velocity anisotropy, $\beta(r)$. With this uncertainty, even with thousands of radial velocities, one cannot definitively distinguish between cored and cusped dSph halos (Figure 2.13). To break the degeneracies, knowledge of the transverse velocities is needed to constrain the anisotropy parameter. At the distance of interesting, dark matter-dominated systems like Draco, one would need to have microarcsecond-precision astrometry to $V = 17$–20 to obtain sufficiently accurate transverse velocities from the proper motions of the brightest dSph stars. Figure 2.13 shows how this will be a feasible experiment for a sample as large as 200 of the brightest red giants using 100 days of observing on NASA's SIM Lite astrometric satellite – not an unreasonable investment to discriminate definitely between warm and cold dark matter at the 3σ level.

Until such astrometry is feasible, it will not be possible to be certain how to interpret dSph velocity dispersion profiles (Figures 2.11 and 2.12), whether solely in the regime

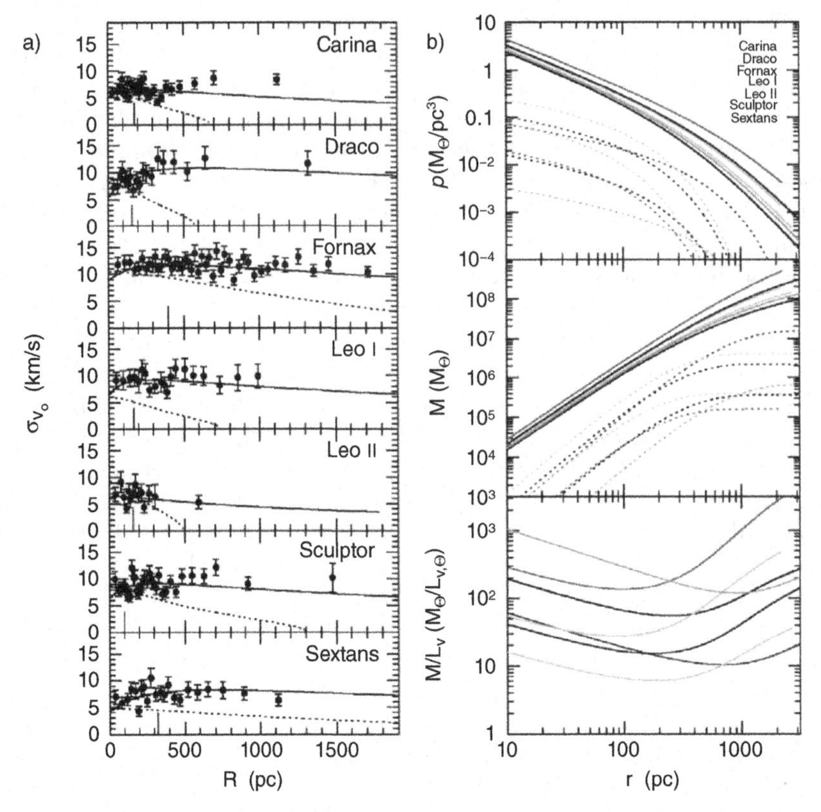

FIG. 2.11. (a): Measured and model velocity dispersion profiles for MW dSphs from Walker *et al.* (2007). MFL profiles from the surface brightness profiles are shown as dashed lines while best-fitting NFW profiles assuming a constant $\beta(r)$ are shown as solid lines. The short, vertical lines mark the luminous core radii from Irwin and Hatzidimitriou (1995). (b): Solid lines show the density, mass, and mass-to-light profiles corresponding for Walker *et al.*'s best-fitting NFW profiles, while dotted lines show the baryonic density and mass profiles assuming exponential density laws with $\mathfrak{M}/L = 1$ and scalelengths from Irwin and Hatzidimitriou (1995). Figure from Walker *et al.* (2007).

of dark matter solutions, where uncertainty in the $\beta(r)$ profile will persist and translate into uncertainty in the derived mass profiles, or admitting the possibility of the influence of tidal disruption. For example, the remarkably flat, and even rising, velocity dispersion profiles observed at the greatest dSph extents probed can be fitted by very extended dark matter profiles with dramatically rising mass-to-light ratios with radius and large total satellite masses, as seen, for example, in the solutions by Walker *et al.* (2007) shown in Figure 2.11 (see also Kleyna *et al.*, 2002; Łokas *et al.*, 2005; Mashchenko *et al.*, 2005; Koch *et al.*, 2007, from among numerous other examples of such work). But models of tidally disrupting satellites *also* create such profiles, where the contributions of tidally stripped stars can account for the continuing flatness/rise of the dispersion profile to large radius (e.g., Piatek and Pryor, 1995; Kroupa, 1997; Kleyna *et al.*, 1999; Mayer *et al.*, 2001; Sohn *et al.*, 2007; Muñoz *et al.*, 2008, – see Section 2.4 and Figure 2.18). In this regard, the velocity dispersion profile of the Sgr dSph (Figure 2.12), where the flatness at large radii is definitively from tidal disruption, is most illustrative. Again we see that determining which of the MW dSphs are tidally disrupting, and to what extent, will play an important role in clarifying dark matter physics, in this case in the potential re-evaluation of the dynamics driving the observed velocity dispersion profiles at large dSph radii.

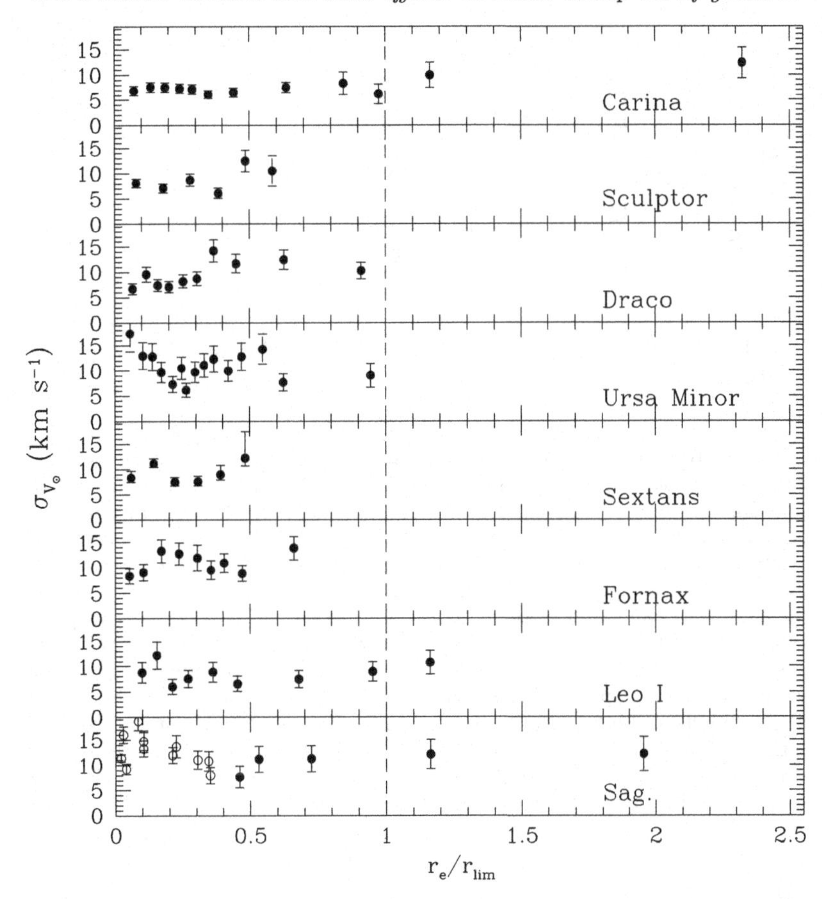

FIG. 2.12. Velocity dispersion profiles for MW dSphs, and including data on stars to parts of the dSphs at extremely low surface brightness – where individual stars that are part of the dSph are greatly swamped by contamination from the MW and require specialized photometric techniques to identify (see description of the methodology in the listed sources). In this representation, the profiles are normalized to the King limiting radius, $r_{\rm lim}$ of the fitted density profiles. The data are from Carina – Muñoz *et al.* (2006), Sculptor – Westfall *et al.* (2006), Draco, and Ursa Minor – Muñoz *et al.* (2005), Sextans – Walker *et al.* (2007), Leo I – Sohn *et al.* (2007), Sgr – Ibata *et al.* (1997) and Majewski *et al.* (in preparation).

Microscopic nature of dark matter

Another way – and a most direct one – where tidally disrupting MW dSphs can have a significant impact on understanding dark matter physics is in aiding experiments designed to detect the dark matter particle. If dSph systems are full of dark matter but also tidally disrupting (Section 2.4), then the tidal debris from such systems will contain both dark and baryonic matter. Depending on the filling fraction of such tidal tails in the Galactic volume, there can be a significant chance that these tidal tails will bring a useful flux of dark matter particles to these detectors on (actually, typically *inside*) the Earth (e.g., Gelmini *et al.*, 2004; Lewis and Freese, 2004; Freese *et al.*, 2005; Gondolo and Gelmini, 2005; Savage *et al.*, 2006).

2.3 Some previously proposed alternatives to dark matter in dSphs and an introduction to tides

In the era when we only knew of the "classical" dSphs, and before CDM theory was well advanced, the notion of such large $\mathfrak{M}/L$ as inferred for those dSphs was not universally

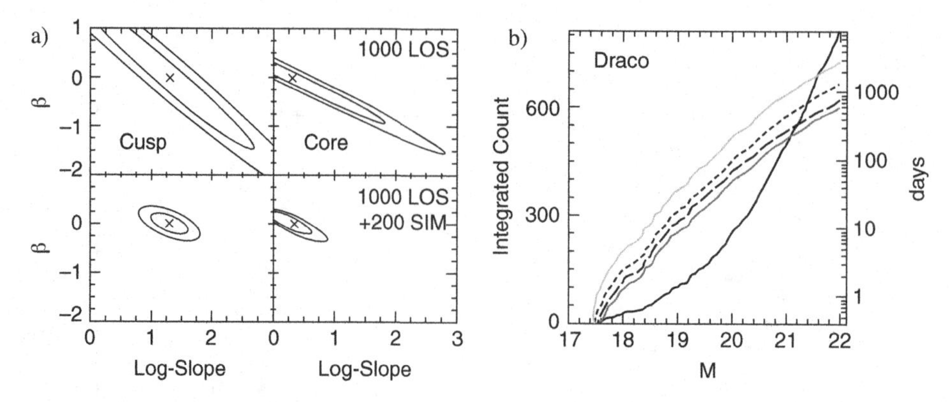

FIG. 2.13. (a): A demonstration of the ability to recover information on the nature of the dark matter profile from observations of dSph stars, from analytical modeling by Strigari *et al.* (2007). Ellipses indicate the 68% and 95% confidence regions for the errors in the measured dark halo density profile slope (measured at twice the King core radius) and velocity anisotropy parameter β in the case where only radial velocities are available for 1,000 stars in a particular dSph (top panels). A significant improvement is derived from the addition of 200 proper motions using NASA's SIM Lite interferometer to obtain 5 km s^{-1} precision transverse velocities (bottom panels). The left/right panels correspond to cusped/cored halo models for dSphs where the small $\times$'s indicate the fiducial, input model values. Note the large degeneracy of solutions when only line of sight velocities are available. (b): With SIM Lite we can obtain proper motions for 200 stars in the Draco dSph to Washington $M \sim V = 19$, as shown by the Washington M-band luminosity function (rightmost black line and left axis; Muñoz *et al.*, 2005). The other lines represent the number of days (right axis) necessary to observe all of the stars to a given magnitude limit with the SIM Lite satellite to a given transverse velocity uncertainty, from top to bottom: 3, 5, 7, and 10 km s^{-1}. From Strigari *et al.* (2009).

accepted. After all, it seemed that Nature had presented a clear trend whereby $\mathfrak{M}/L$ correlated with the sizes of stellar structures – from dark matter-less, $\mathfrak{M}/L \sim 10^0$ in globular clusters to $\mathfrak{M}/L \sim 10^{0-1}$ in single large galaxies, to $\sim 10^{1-2}$ measured for galaxy pairs and $\sim 10^2$ in galaxy groups, and rising to $\sim 10^{2-3}$ for clusters to superclusters; the relatively puny dSphs upset this apparent trend. Clearly, dSphs are quite unusual in this context, with some, like the Ursa Minor and Draco systems, having $\mathfrak{M}/L$ two orders of magnitude larger than either globular clusters (a population of star systems generally fainter than dSphs, but in the MW having some examples with luminosities approaching those of dSphs) or larger galaxies only a few magnitudes brighter. (Of course, this $\mathfrak{M}/L$ difference between dSphs and other systems of similar mass and size scale becomes even more extreme when we include the ultrafaint systems, with their *even higher* inferred dark matter fractions.) Discomfort with the large $\mathfrak{M}/L$'s inferred for dSphs was also increased by our lack of knowledge about what the dark matter is.

As a result, a cottage industry developed to consider alternative explanations for the measured large radial velocity dispersions in dSphs that led to the inference of large $\mathfrak{M}/L$'s. These alternative explanations generally fall into three categories:

(i) Problems with the technical procedures used to measure the true, intrinsic velocity dispersions in dSphs, that lead to their overestimation.

(ii) Problems with our understanding of the physical state of dSphs that lead to incorrect application/interpretation of the observed velocities and the physics of dSph structure.

(iii) Problems with our understanding of gravity itself.

Presently it is generally thought that (i) and (iii) are no longer an issue, whereas aspects of (ii) are still discussed (e.g., see also Kroupa, Chapter 4, this volume), but controversial; nevertheless, in the interests of historical completeness, I will briefly discuss each of these categories, starting with the two alternatives (i and iii) now considered unlikely.

2.3.1 *Potential problems with the velocity dispersions*

Practically, radial velocity surveys of stellar systems do not measure the *intrinsic* velocity dispersions of stars, but the intrinsic dispersions *inflated* by measurement errors as well as other dynamical processes. In the absence of covariances, these effects add in quadrature:

$$\sigma^2_{v,observed} = \sigma^2_{v,intrinsic} + \sigma^2_{v,errors} + \sigma^2_{v,other} \tag{2.5}$$

Thus, one needs to estimate $\sigma_{v,errors}$ and $\sigma_{v,other}$ correctly to retrieve the desired $\sigma_{v,intrinsic}$ from $\sigma_{v,observed}$. Underestimating the velocity errors and other dynamical effects will yield an overestimate of $\sigma_{v,intrinsic}$. Naturally, early concerns were raised that underestimated velocity precisions might be a problem. However, dSph star velocities are now routinely measured to 1–2 km s^{-1} accuracy, which is much smaller than the observed dSph dispersions of 5–9 km s^{-1}, so that errors in estimating the uncertainties are negligible.

What about other dynamical effects that could be contributing to the observed velocity dispersions? One potential source could be motions within the atmospheres of the stars themselves. It is well known that the red giant and asymptotic giant branch stars (the only stars currently accessible to spectroscopy in MW dSphs) exhibit atmospheric "jitter," which must contribute to inflating the observed dispersions. However, recent, high precision ($<$0.1 km s^{-1}) radial velocity monitoring of relatively large samples of metal-poor MW *field* K giants shows that the jitter is at most at the 1 km s^{-1} level in almost all cases, with the magnitude of the effect apparently tapering off among the cooler, metal-poor giants typical of those found in dSphs (Bizyaev *et al.*, 2006).

Another potential inflation of the velocity dispersion could come about if dSphs contain large numbers of binary stars. This is a problem that has been explored in detail by several groups, who, through simulations of the effects of binaries as well as spectroscopic monitoring of dSph stars, have shown that dSphs do not have an unusual binary distribution compared to, say, stars in the solar neighborhood, and that the contribution of binary star motion to inflating the velocity dispersion is at most a 1–2 km s^{-1} effect (Armandroff *et al.*, 1995; Hargreaves *et al.*, 1996; Olszewski *et al.*, 1996). More recently, this was confirmed with data collected having a near-decade span in epochs by Kleyna *et al.* (2002), who repeated observations of 61 stars measured earlier by Armandroff *et al.* (1995). On the other hand, Kleyna *et al.* (2002) also point out that the existence of binary stars *can* produce unusual, non-Gaussian tails in the dSph velocity distributions, and these can increase errors in the measured dispersions and/or distort the statistical significance of (e.g., maximum likelihood) fits to the measured velocity distributions.

2.3.2 *Potential problems in our understanding of gravity*

If gravity is not strictly Newtonian, the non-Newtonian deviations might be more evident in the dynamics of extremely low density objects like dSphs, perhaps to the level of obviating the need for dark matter to explain the observed dynamics. The most famous proposed example of this idea, Modified Newtonian Dynamics (MOND) was proposed by Milgrom (1983a). The basic idea is that in acceleration regimes lower than $a_0 \approx 10^{-8}$ cm s$^{-2} \approx cH_0/6$ (where c is the speed of light and H_0 is the Hubble constant), the effective gravitational attraction approaches $\sqrt{g_n a_0}$, where g_n is the usual Newtonian acceleration.

Though criticized as an ad hoc theory not deriving from any fundamental physical causes, MOND has enjoyed great empirical success reproducing the rotation curves of numerous galaxies (e.g., Sanders and Verheijen, 1998; Sanders and McGaugh, 2002), and the theory has remained remarkably persistent over the years by being able to work on scales ranging from dSph galaxies to superclusters of galaxies. On the basis of their low surface brightnesses, Milgrom (1983b) predicted that the low internal accelerations of dSphs would yield "dynamical masses" 10 times or more than accounted for by their stars. Indeed, using MOND, Sanders and McGaugh (2002) reduce the $\mathfrak{M}/L$ of the classical

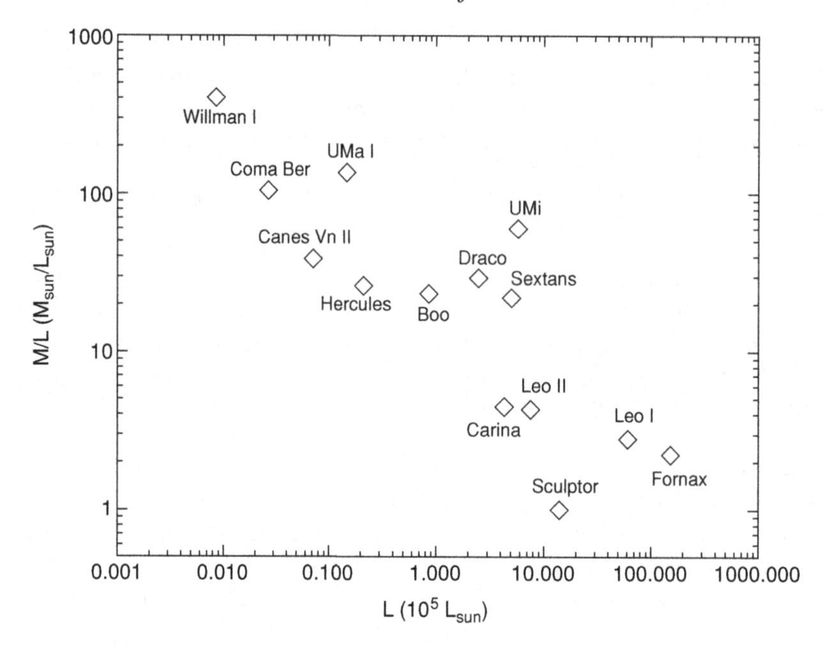

FIG. 2.14. Inferred MOND mass-to-light ratios as a function of their luminosity for those Galactic dSph galaxies lying from 40–260 kpc from the Galactic center and having measured velocity dispersion estimates. While most of the classical dSphs have low inferred $\mathfrak{M}/L$, the ultrafaint systems still require dark matter. Figure from Sánchez-Salcedo and Hernandez (2007).

MW dSphs down to of order only a few in solar units, with the exception of Ursa Minor and Draco at $\mathfrak{M}/L_V \approx 30$–$40$, which, however, are reported with enormous error bars. More impressively, Angus (2008) has recently produced velocity dispersion *profiles* for the classical MW dSphs that are very good matches to extant data to large dSph radii. On the other hand, the newly found ultrafaint dwarfs may represent a new challenge to MOND, since Sánchez-Salcedo and Hernandez (2007) suggest that $\mathfrak{M}/L > 20$ are required for these objects even when working in a MOND gravity paradigm (Figure 2.14).

Nevertheless, it has been difficult to refute MOND as an alternative to dark matter by finding, for example, a galaxy that strictly follows a Newtonian response into the relevant low acceleration regime. Currently, the primary strike against MOND has arisen in quite another context – the recent Clowe *et al.* (2007) results on the Bullet Cluster of galaxies, which seems to give direct evidence for the existence of dark matter. By comparing the location of the baryonic mass in this interacting pair of galaxy clusters (as traced by x-ray emission from the hot gas in the clusters) to the location of the total mass (traced through gravitational lensing), it has been shown, at 8σ statistical significance, that the center of the baryonic mass is considerably displaced from the center of the total mass (though the result may still yet not be 100% definitive in killing MOND – cf. Angus *et al.*, 2006).

2.3.3 *Potential problems with misunderstanding the physical state of dSphs*

A number of ideas have been postulated as potential problems in our understanding of the physical state of dSphs that might caution us regarding interpretation of their observed kinematics and inferences about their true dark matter contents.

For example, could dSphs have very unusual stellar populations – for example, an excess of low mass stars – that make them have extremely skewed, but completely baryonic, $\mathfrak{M}/L$? This idea has been tested on MW dSphs using deep starcount analyses from *Hubble Space Telescope* data, with the result that the luminosity functions of dSphs are not found to be strongly atypical (Feltzing *et al.*, 1999).

Most ideas contemplating problems with our understanding of dSphs involve models where dSphs are not in dynamical equilibrium. After all, an implicit assumption in the use of the virial theorem, core-fitting, or the Jeans's equation for deriving masses of dSphs from their observed velocity dispersions is that the systems are in dynamical equilibrium. Proposed non-equilibrium models for dSphs have spanned a large range in degree:

(i) Models where dSphs are completely, or almost completely unbound and their appearance as bound systems is an allusion, for example, due to projection effects, or a product of "lucky timing" in seeing the dwarfs at a special phase in their dynamical evolution.
(ii) Models where dSphs are bound, but greatly affected by external forces (e.g., inducing internal resonances).
(iii) Models where dSphs are mildly affected by external forces, but just enough to confuse observers (e.g., from tidal mass loss).

These non-equilibrium models were sometimes inspired by the unusual appearances of at least some classical dSphs. In particular, the Ursa Minor system, typically found to have $\mathfrak{M}/L_V \approx 60$ (Mateo, 1998), is both strongly contorted into an S-shape *and* found to contain a double nucleus (Kleyna *et al.*, 2003; Palma *et al.*, 2003; Martínez-Delgado *et al.*, 2008) – a suspiciously unusual, and perhaps suspicious, state. Of course, as mentioned earlier, many of the ultrafaint dwarfs also look very distorted (e.g., Belokurov *et al.*, 2007b) – though, again, this could be due to the small number statistics used to trace their structure.

dSphs as unbound concentrations of tidal stream stars

The idea here is that if dSphs are totally disrupted into streams of stars, then of course, due to the Keplerian nature of the stellar orbits, we would expect to see stars pile up at the apogalacticon point of the stream orbit. This pile-up might form something resembling a dSph, but the "internal" dynamics would be completely non-equilibrium. Greater pileups would be expected for more elliptical orbits. And in a variation on this idea suggested by Kuhn (1993), orbital resonances could help create spatial coherences in the tidal star streams.

It is hard to imagine these types of models explaining most – or even a small fraction – of the known dSphs. First, where such dSphs in a state of complete dissolution, one might expect to see obvious tidal tails extending over large ranges of their orbits, but only the case of the Sgr dSph exhibits such strong tails. Second, the strongest pileups at apogalacticon would occur for the most elliptical orbits; however, such strongly radial orbits for tidally disrupting dwarfs in MW-like potentials are found in N-body simulations to produce large, umbrella-like *shells*, rather than strongly concentrated balls of stars resembling dSphs. But perhaps the most serious problem for these models is that most dSphs have non-zero Galactocentric radial velocities, which indicates that they cannot be at apogalacticon.

dSphs as nearly completely unbound systems

Kroupa (1997) and Klessen and Kroupa (1998) proposed that a completely baryonic dSph severely perturbed by tidal effects could, when near complete dissolution, be so out of virial equilibrium that the observed velocities would be poor reflections of the true $\mathfrak{M}/L$. This near-end state of the system would resemble something looking like a dSph, but with a velocity dispersion inflated due to significant anisotropy. In addition, the effect could be substantially enhanced for systems on eccentric orbits due to the additional overprojection of tidal debris along the line-of-sight.

These models were quickly tested in several ways. Klessen and Zhao (2002) showed that velocity gradients would be expected across dSphs structured this way and Klessen *et al.* (2003) found that no such gradients were seen in the Draco dSph. In addition,

Kroupa-like models with elongated structures along the line of sight should produce obvious signatures in the color-magnitude distributions of stars – for example, vertically smeared horizontal branches. These signatures are also not seen (Klessen *et al.*, 2003).

Bound dSphs significantly affected by tides and resonances

Kuhn and Miller (1989) proposed that resonant interactions between internal dSph stellar orbits and the orbit of the satellite itself could substantially "heat" dSphs (see also Kuhn, 1993; Fleck and Kuhn, 2003). Moreover, as mentioned, the same kinds of resonances can create spatially coherent lumps of unbound stars in tidal streams.

These models have been criticized by Pryor (1996) and Sellwood and Pryor (1998), who pointed out that (1) the original models did not use realistic potentials and when realistic, logarithmic potentials are used, the claimed resonances were not excited; and (2) the tidal remnants created from resonances are ten times larger than observed dSphs. Fleck and Kuhn (2003) revisited the problem with updated, more realistic models and found that whereas dSph velocity dispersions inflated by as much as an order of magnitude could be created, this dynamical state exists only for a short fraction of a satellite's orbital period; thus, this would seem to rule out the model as an explanation for the typical dSph.

Tides and tidal effects

Tidal forces from parent galaxies certainly act on their satellites. If these tides are severe enough, can they inflate the dSph satellite velocity dispersions? Tides, in fact, were first postulated as an explanation for the unusual morphologies of dSphs by Hodge (1961a,b, 1962) and Hodge and Michie (1969). Since then, the net influence of tides on the MW dSphs has been greatly debated.

Of course, tidal disruption of dSph galaxies is now established to have contributed to the buildup of the MW, with the Sgr dSph system a vivid example (e.g., Ibata *et al.*, 2001; Majewski *et al.*, 2003). But other dSphs are likely to have succumbed to tidal shredding to account for at least some of the growing number of discovered MW halo substructures, not all of which are attributable to globular clusters (e.g., Vivas *et al.*, 2001; Newberg *et al.*, 2002; Belokurov *et al.*, 2006a, 2007a; Duffau *et al.*, 2006; Fuchs *et al.*, 2006; Grillmair, 2006, 2009; Grillmair and Dionatos, 2006b). The halo substructures are often seen in the shape of long, coherent tidal streams. These streams are created because stars from the progenitor satellite that find themselves beyond the gravitational grip of the satellite and escape with negative energy relative to the satellite form the leading tidal stream arm, which is strung out along, but just inside, the future orbital path of the satellite, while the escapees with positive relative energy form the trailing stream arm along the past orbital path, but in a slightly higher orbit (Figure 2.15). It is common to refer to a "tidal radius" beyond which a star from the satellite is more bound to the MW and which defines the extent of the bound satellite (as in, e.g., Section 2.1.3 and Figure 2.15), but it should be kept in mind that there is no well-defined "tidal boundary" and whether a star will become stripped from the satellite at a particular radius depends on a number of factors, including the actual orbit of the star within the satellite (with, for example, retrograde-orbiting stars within the satellites being held more tightly than prograde-orbiting stars – Keenan and Innanen, 1975). Nevertheless, the tidal radius of a satellite is often estimated as the Jacobi/Roche limit $r_{tide} \approx R(\mathfrak{M}_{sat}/3\mathfrak{M}_{MW})^{1/3}$, where $\mathfrak{M}_{sat}$ is the satellite mass within r_{tide} and $\mathfrak{M}_{MW}$ is the MW mass within the satellite orbital radius R. It is interesting to rewrite this in terms of mean densities between the satellite and the MW to find that this tidal limit is given approximately when $\mathfrak{M}_{sat}/r_{tide}^3 \approx 3\mathfrak{M}_{MW}/R^3$. Of course, for a satellite on an

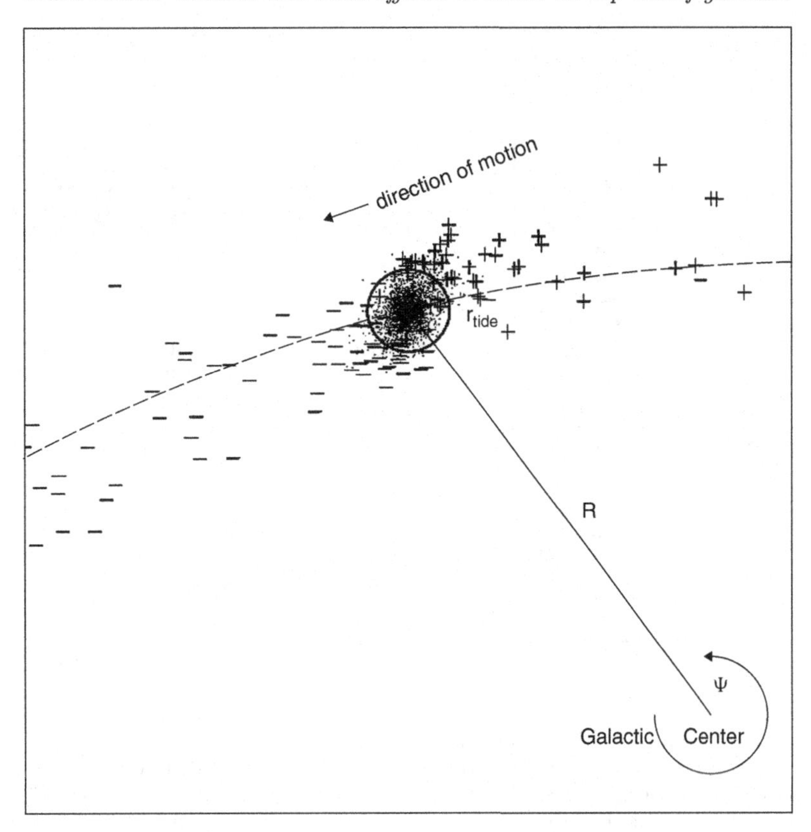

FIG. 2.15. Results of an N-body simulation of the tidal disruption of a MW satellite, as seen projected on the orbital plane of the satellite, with part of that orbit shown by the dashed line. Dots represent particles that are still bound to the satellite, while the "−" and "+" symbols represent unbound particles on orbits with lower and higher energy than that of the satellite orbit, respectively. The circle corresponds in this case to $r_{tide} \approx R(\mathfrak{M}_{sat}/\mathfrak{M}_{MW})^{1/3}$. From Johnston (1998).

elliptical orbit, the tidal radius shrinks toward perigalacticon, and is more accurately given by

$$r_{tide} = ka \left[\frac{\mathfrak{M}_{sat}}{\mathfrak{M}_{MW}(a)}\right]^{1/3} \left[\frac{(1-e)^2}{[(1+e)^2/2e]\ln[(1+e)/(1-e)]+1}\right]^{1/3} \tag{2.6}$$

where e is the orbital eccentricity, a is the semi-major axis radius of the orbit, and k is a proportionality constant (Oh *et al.*, 1992). At this point in its orbit the satellite is most vulnerable to tidal disruption and this is when the most debris is lost.

Mass loss is an obvious, and now established, manifestation of tides on dSphs, and clearly, as the case of the Sgr stream shows, this effect of tides can be dramatic. But can it be large enough to inflate the velocity dispersions of stars in the still bound dSph core, and, therefore, the inferred $\mathfrak{M}/L$? The question has been addressed by several groups, but perhaps the most widely cited papers on the subject are those of Piatek and Pryor (1995), "PP95," and Oh *et al.* (1995), "OLA95." Because of their influence, these papers are worth discussing in more detail. It is worth mentioning that at the time of these papers, there were no known MW stellar tidal streams, and it was not obvious that any of the classical MW dSphs were suffering tidal disruption, except for the just discovered Sgr system, which at least showed signs of tidal distortion in the first analyses of its structure (Ibata *et al.*, 1994).

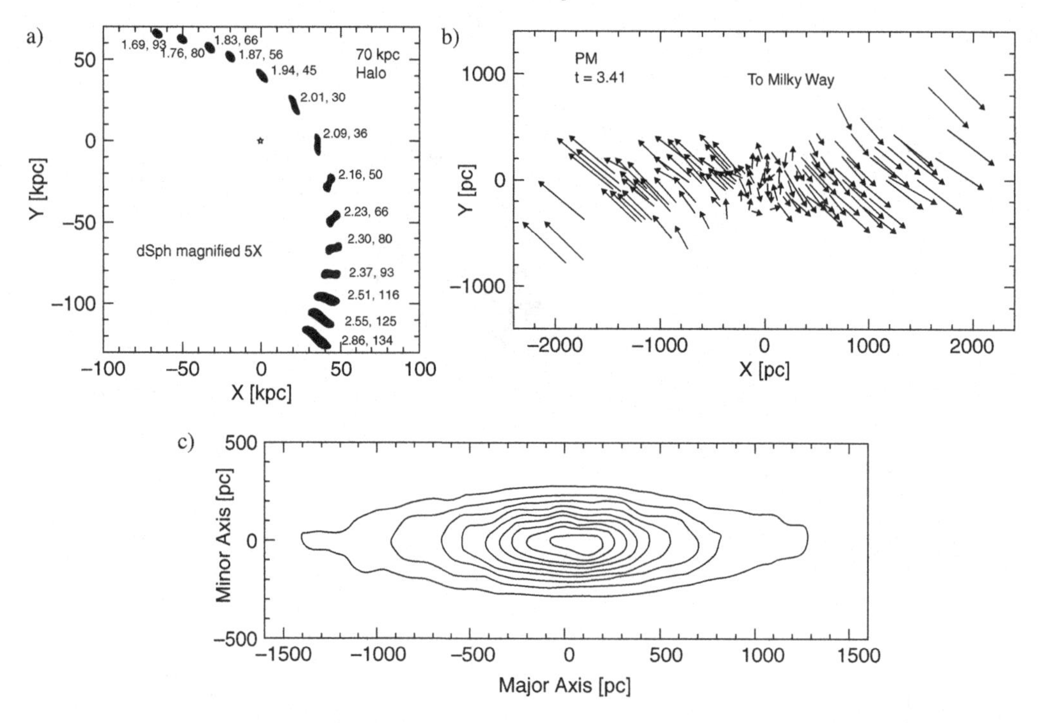

FIG. 2.16. Figures from the very specific modeling by Piatek and Pryor (1995) that have had a lasting influence on the subject of tides and dark matter in dSphs. (a): Low-mass satellite being quickly destroyed on a perigalactic passage. (b): Velocities in the dSph, showing the shearing motion induced by tides and leading to the "S-shape." (c): Surface brightness distribution of the tidally disrupted, low-mass satellite.

OLA95 carried out N-body simulations on low mass ($2 \times 10^6\ \mathfrak{M}_{sun}$) satellites (i.e., with no dark matter) moving in dynamically gentle orbits ($e < 0.5$, perigalacticon >50 kpc) around an MW-like galaxy and making several perigalacticon passages. They found that these satellites flatten in the orbital plane, creating long strands of tidal debris with a power-law density distribution. Even so, OLA95 found that the velocity dispersion of the unbound, but not yet dispersed, stars is similar to that of the original equilibrium value. From this, OLA95 concluded that tidal stripping is not inflating the measured dSph velocity dispersions, and that, even if tides are important in affecting the dSphs, one still needs high mass (i.e., the presence of significant dark matter) to explain their structure.

PP95 also modeled dark matter-less, low mass (3×10^5 and $8 \times 10^5\ \mathfrak{M}_\odot$) satellites, but on very eccentric orbits (perigalacticon = 30 kpc, apogalacticon = 210 kpc; Figure 2.16). Again the satellites were found to form long tails of tidal debris, but significant inflation of $\mathfrak{M}/L$ was rarely observed, and only briefly – for a few 10^8 years, at perigalacticon. Therefore, PP95 concluded that since not all dSphs can be in this special orbital phase, tides must not be causing inflation of dSph velocity dispersions. In addition, PP95 point out that their models give two telltale signs for when a dSph is undergoing tidal disruption: (1) the surface brightness distribution of the dSph will become more elliptical with radius from the satellite center, and (2) the stripped stars just entering the tidal tails will create a velocity gradient (i.e., shearing) across the major axis of the system that will resemble an apparent systemic rotation (Figure 2.16).

As mentioned, these two seminal numerical studies seem to dominate the discussion of tidal effects in dSphs. But it is worth pointing out that whereas these pioneering studies were not technically flawed, OLA95 and PP95 studied *highly specialized* test cases. The OLA95 study focused on low-mass satellites on rarely seen *circular* orbits, whereas PP95 explored satellites of such low $\mathfrak{M}/L$ ($\sim$1) that the fluffy dwarf galaxies don't even survive one orbit (their goal was to explore systems in near-total destruction from one strong episode of tidal interaction). Because of their narrowly focused explorations of parameter space, it is somewhat dangerous to conclude that these models are applicable to the more general case of real dSphs on elliptical orbits. Unfortunately, the results of these two early studies are frequently applied out of their proper context. The OLA95 and PP95 analyses were addressing the specific question of whether tides could inflate velocity dispersions as a way to eliminate the necessity of dark matter; they were not intended to rule out all tidal effects. Nervetheless, based on the PP95 results, it is frequently (incorrectly) assumed that apparent rotation (velocity shearing) is a *necessary* signature of tidal disruption – even though PP95 showed that this phenomenon occurred in only some cases and only at very large dwarf radii. Moreover, we know of at least one example – the Sgr dSph – where tides most certainly *are* affecting a MW satellite galaxy.

Is Sgr an exception, or a paradigm for (at least some) dSphs? As we discuss below, there are other cases of MW dSphs that evidence suggests are being tidally disrupted, and for which N-body models that include *both* tides *and* dark matter can explain their structure and dynamics.

2.4 Modeling the structure and dynamics of dSphs with dark matter and tides

2.4.1 *Successful tidal disruption models of classical dSphs*

Recognizing the value of dSph N-body particle modeling like that done by OLA95 and PP95, but also the shortcomings of the focused applications of these particular studies, now more than a decade old, a more comprehensive assessment of the extent to which Galactic tides might be affecting dSphs is timely. In particular, a new treatment of the problem is needed that includes (1) more realistic satellites, *including* dark matter contributions to the mass, and (2) a much larger, systematic coverage of parameter space, including a variety of satellite masses, sizes, densities, and orbital shapes (eccentricities). We discuss the results of one such study, described in detail in Muñoz *et al.* (2008), hereafter "M08," but also used in Sohn *et al.* (2007), hereafter "S07," that helps lay a broader foundation to the interpretation, and reassessment, of the possible role of tides in dSph galaxies. To make the analysis both as realistic and directly relevant as possible, M08's extensive probe of parameter space adopted firm observational constraints, with a focus on making dSph satellite models match two specific empirical test cases presently having the most extensive photometric and dynamical data to large radii (Figures 2.7, 2.8, and 2.12) – the Carina (Muñoz *et al.*, 2006) and Leo I (S07) galaxies. In both cases, the studies cited have shown that these two classical dSphs show strong evidence of tidal disruption, in the form of stripped stars.

At the onset, it must be admitted that while the M08 analysis does seek to model dSph satellites containing dark matter, this is done in a simplified way, with the assumption that both the baryonic and dark mass track one another (and, therefore, the light profile of the system). While such one-component, "mass-follows-light" (MFL) models can be criticized as overly simplistic and unrealistic representations of the relative baryonic and dark matter distributions of the *precursors* of dSphs (which may resemble the relative contributions of baryonic and dark matter shown in Figure 2.9), it is not clear that MFL models have, in fact, been ruled out as viable descriptions of *present* MW dSphs, *if they are tidally disrupting.* Previous simulations of dwarf satellites using multi-component precursors consistent with the predictions of CDM cosmogonies have shown that when undergoing significant tidal stripping, these satellites actually evolve into systems that resemble MFL models (Mayer *et al.*, 2001; Klimentowski *et al.*, 2007, Figure 2.17), and when the extended dark matter halos are stripped to the point that stars are also being

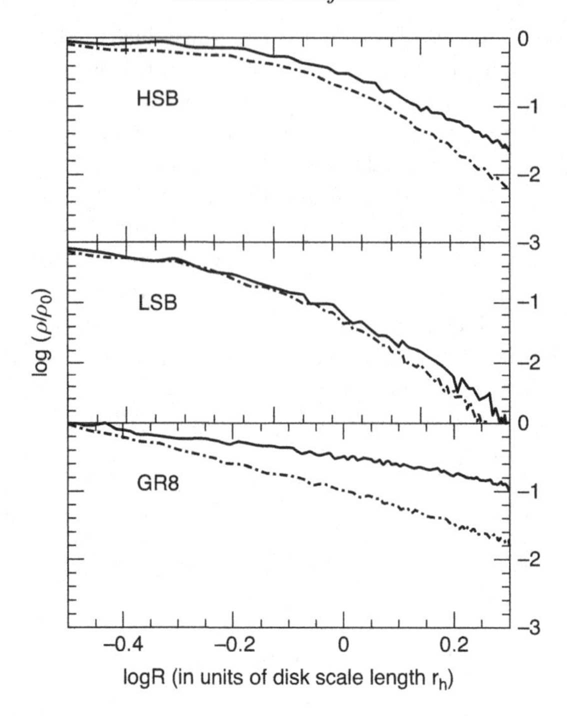

FIG. 2.17. A comparison between the final total mass (*solid lines*) and the stellar mass profiles (*dot-dashed lines*) for N-body, smoothed particle hydrodynamical simulations of initially rotationally supported, two-component (i.e., luminous matter embedded in an extended dark matter halo) dwarf galaxies that become tidally stirred and stripped while orbiting in a MW-like potential, by Mayer *et al.* (2001). The models include a low surface brightness (LSB) dwarf with disk scalelength 4.4 kpc that eventually turns into a dSph, a high surface brightness (HSB) dwarf with a 2 kpc disk scalelength that eventually turns into a dwarf elliptical, and a very faint dIrr, like the LG system GR8, with disk scalelength 76 pc and a very high central dark matter density, that evolves into an extremely dark matter dominated dSph, like Ursa Minor and Draco. Note that only in the last case does the dark matter still greatly "cocoon" the baryonic matter, whereas, particularly in the LSB/dSph case, the end state is essentially a mass-follows-light system. Compare with Figures 2.9 and 2.11. Figure from Mayer *et al.* (2001).

lost, then the system behaves basically as MFL (Bullock and Johnston, 2005). Since the goal of M08 is to attempt to describe the structure of present day dSphs with tidal stripping to see if a combination of tides and dark matter can explain their current properties, it is useful to consider the simplified MFL form from the start; the alternative of building the satellite precursors with multi-component models (i.e., where the baryonic and dark matter components follow independent distributions), though more sophisticated, does have the problem that it can admit degenerate solutions between which present observations cannot definitively discriminate (e.g., see Figure 2.13). Nevertheless, there has been some interesting recent work using multi-component dSph modeling (e.g., Koch *et al.*, 2007; McConnachie *et al.*, 2007; Walker *et al.*, 2007; Peñarrubia *et al.*, 2008a,b) that relies on the extended dark matter halos to explain the relatively flat or rising dSph velocity dispersion profiles at large radii (see, e.g., Figure 2.11 as well as some earlier references already cited in Section 2.2.2).

The goal of the present discussion (and that of M08) is different: to see whether tidal disruption can explain the flat/rising velocity dispersion profiles. Indeed, as was found by the M08 and S07 studies, quite good matches to all extant data for the Carina and Leo I systems, including their velocity dispersion profiles, can be found with tidally disrupting MFL models (e.g., Figure 2.18) selected from large libraries of N-body simulations created using 10^5 particles initially arranged in a Plummer distribution – a configuration where the three-dimensional density as a function of radius is given by

$$\rho(r) = \left(\frac{3\mathfrak{M}}{4\pi a^3}\right)\left(1 + \frac{r^2}{a^2}\right)^{-5/2} \tag{2.7}$$

where $\mathfrak{M}$ is the total mass and a is the scale radius – and orbiting within a static MW model.[6] The simulations explored a wide range of satellite parameters, with masses (dark + baryonic) spanning 1×10^6 to 3×10^8 $\mathfrak{M}_\odot$, Plummer scalelengths of 90, 194, 280, and 480 pc, and orbits shaped from nearly circular to an eccentricity, $e = (r_a - r_p)/(r_a + r_p)$, of 0.9, where r_a and r_p are the apo- and perigalactica, respectively. The evolution of the model satellites was followed for five radial orbits or 10 Gyr, whichever occurs first, but the orbits were always designed to end with the model satellite having the current radial velocity and position observed for the real satellites. In the end, 250 separate simulations were run to model the Carina dSph and 100 were run for Leo I.

Figure 2.18 shows models that match particularly well the light profile, velocity, and velocity dispersion profiles measured for the Carina dSph. For both the Leo I and Carina cases, the best model matches to the data were found for satellites on very radial orbits (with $r_p/r_a = 13/400$ kpc in the case of Leo I and 15/100 kpc for Carina). Such radial orbits are actually expected in the Leo I case, which has a very large observed radial velocity (200 km s^{-1} in the Galactic Standard of Rest) despite its enormous, 270 kpc distance from the Galactic center, so that a greatly plunging orbit is obviously indicated. In the case of Carina, the model that best matches the photometric and spectroscopic data on the core has an orbit that also matches the derived orbit for the system by Piatek *et al.* (2003), who measured the proper motion of the system using Hubble Space Telescope data.

Orbits with this eccentric and with such close perogalactica greatly subject the satellites to tidal shocking and result in the release of stellar mass. In the case of the Carina model, the satellite loses mass at a rate of about 10% per orbit, from an initial mass of $\sim 3.8 \times 10^7$ $\mathfrak{M}_\odot$ about 8 Gyr ago to a final bound mass now of $\sim 1.8 \times 10^7$ $\mathfrak{M}_\odot$. The mass loss rate for Leo I is less, about 2% per Gyr, but this is due to the very long, ~6.5 Gyr orbital period for this system on an orbit with an apogalacticon projected to be as high as 450 kpc. This mass loss explains and fits the "break" populations seen in the extended light profiles of these dSphs as unbound stars forming tidal tails. Muñoz et al. (2006) has mapped these tidal tails over many degrees for Carina, while in the Leo I system these unbound stars are shown to lead to an asymmetry in the velocity distribution of the stars observed over a smaller area by S07; the asymmetry results because of an imbalanced projection of leading versus trailing tidal debris along the line of sight.

Most significantly, despite the fact that in these two test cases the dSph data are matched by models where the satellite is experiencing significant mass loss, the models also show that the satellites have $\mathfrak{M}/L_V \sim 40$ and ~5 for Carina and Leo I, respectively, *which is exactly what one obtains for the $\mathfrak{M}/L_V$ of these systems when one applies the core-fitting technique* (Section 2.2.1). Extensive testing of the models shows that the measured $\mathfrak{M}/L$ remains stable for most of the lifetime of the satellite and that core-fitting always gives an accurate satellite mass. Whereas this is consistent with the findings of PP95 and OLA95 – both suggested that if a satellite retains a significantly bound core, then the central velocity dispersion does indeed reflect the instantaneous system mass – the new models, in addition, prove that *(1) classical MW dSph satellites can be dark-matter-filled and also experiencing luminous, baryonic mass loss, and (2) that this mass loss is a natural explanation for the extended break populations seen in the luminosity profiles of these systems.* Moreover, the M08/S07 studies

[6] The Galactic potential used was the prescription given by Law *et al.* (2005), which incorporates a Miyamoto and Nagai (1975) disk, a Hernquist and Ostriker (1992) spheroid, and a logarithmic halo.

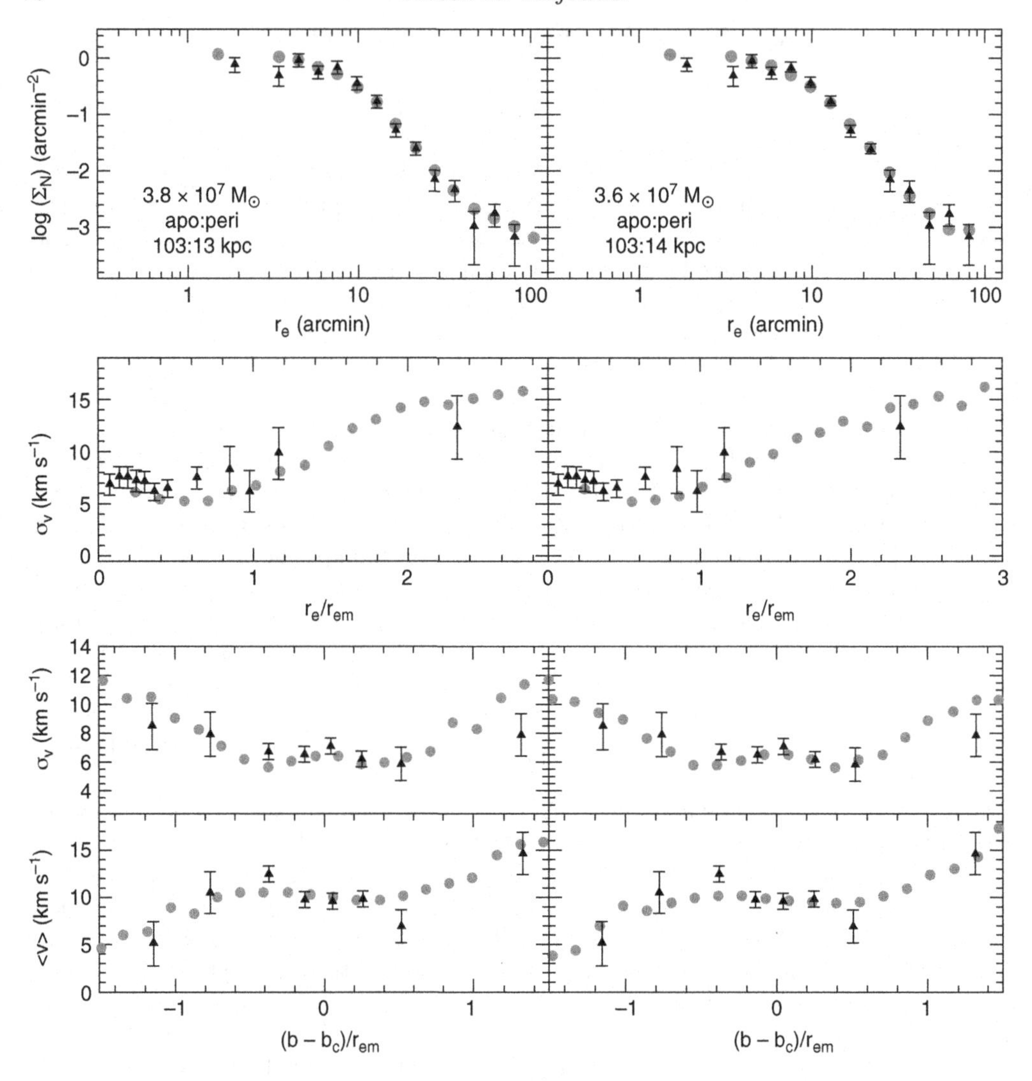

FIG. 2.18. Matches of MFL model satellites containing dark matter and experiencing tidal disruption (*circles*) to data on the Carina dSph from Muñoz *et al.* (2006), shown as filled triangles with error bars. The panels, from top to bottom, show the number density profile of Carina as a function of radius, the observed velocity dispersion as a function of radius (normalized to the King limiting radius), velocity dispersion as a function of position projected along the major axis, and the mean trend of radial velocity as a function of position projected along the major axis. Figure from M08.

also illustrate how *MFL modeling* can satisfactorily explain the present appearance and dynamics of dSphs; specifically, such models can *simultaneously* match the observed density laws, velocity profiles, velocity dispersion profiles, and the inferred large central $\mathfrak{M}/L$ of dSphs, as well as account for the clear presence of a tidal tail in Carina and the asymmetric velocity distribution for Leo I – and do this with fewer free parameters than required for multi-component models. For the MFL models discussed here, only the satellite mass, scale, and orbit shape needed to be varied to find rather good matches of the models to real dSphs. Thus, tidal disruption does seem a viable alternative explanation to extended dark matter halos to explain the properties of at least some MW dSphs.

An interesting question is whether it is a coincidence that tidal disruption models work well for Carina and Leo I, which were only selected by M08/S07 for this modeling because they are the dSphs with the most extensive observations to large radius (Figures 2.7, 2.8, and 2.12) – apart from the obvious tidally disrupting Sgr system – and therefore can well constrain such models. Clearly, it would be helpful to continue gathering data on all of the MW dSphs to ever larger radii to see whether examples could be found where the extended dark matter versus tidal disruption models can clearly be discriminated.

2.4.2 *Six common misperceptions regarding tides and dSph galaxies*

In addition to allowing viable MFL models of tidally disrupting satellites to be found for specific dSphs, the extensive libraries of N-body simulations generated by M08 and S07 enable testing of a number of commonly held notions about the effects of tides on dSphs. Many of these notions are driven by the significant influence that the earlier OLA95 and PP95 papers had on the field. What the new models show is that the appearances of dSphs can often be deceiving with regard to ongoing tidal disruption, and that many misperceptions about tides and dSphs abound:

- *Misperception 1 – Lack of a "rotation" signature means that tidal disruption is not happening:* OLA95 and PP95 showed that escaping debris in their model satellites creates a "rotation-like," shearing signature (Figure 2.16) due to the fact that tidal forces tend to pull stars off in the Galactic radial direction (before these stars then drift into the leading and trailing tidal arms; see Figure 2.15), which often has a significant projection on the line of sight. Thus, from the leading to trailing side of the satellite along its orbit a variation in the mean radial velocity of dSph stars is expected (see, e.g., the inflection visible in the Carina data shown in Figure 2.18). Thus, a common test that is applied to velocity data on dSphs is a search for this "S-shaped" inflection of velocities as a test of tidal stripping.

  However, a large fraction of models created by M08, including many having enormous mass loss rates, show *no* significant degree of velocity variation across the face of the dSph over angular spans of the sky typically probed by radial velocity studies of these systems. In many cases this dynamical signature of tidal disruption does not become evident until much larger spatial scales are probed. Thus, the search for the characteristic velocity inflection should not be considered a definitive test of whether a system is experiencing tidal disruption.
- *Misperception 2 – Tidally disrupting satellites have pronounced ellipticities:* In PP95, we see illustrations of very elongated satellites, stretched out by tides (Figure 2.16). But this classical appearance applies in their treatment to a satellite at the point of near-total destruction in a point-mass Galactic potential! Among the M08 models, similarly elongated dSphs can be seen, but these happen when the systems are nearly completely destroyed.

  On the other hand, many M08 systems (which were initially set up to be spheres) that still contain significant amounts of mass – though vigorously disrupting – are seen to remain *completely spherical* for most of their lives.
- *Misperception 3 – No detected tidal tails means no tidal disruption:* A common complaint against the idea that Galactic dSphs may be having luminous mass loss is that, except in the very obvious case of the Sgr dSph, no visible tidal tails be seen.

  But what the M08 models show is that tidal tails form over much larger spatial scales than the typical King-limiting radii of these systems. This point was demonstrated previously by Grillmair *et al.* (1995) and is obvious in the light profile of the tidally disrupting Palomar 5 globular cluster, where one needs to go several King-limiting radii before any detectable asymmetry is seen (Figure 2.19). And even if the tidal tails are formed across the sky area surveyed for a dSph, they may be too faint to be detectable. Indeed, in the best-matching Carina model shown in Figure 2.19, any obvious elongation due to the tidal tails does not appear until one goes down to the extremely faint isophotal surface brightness of 34 mag arcsec^{-2}.

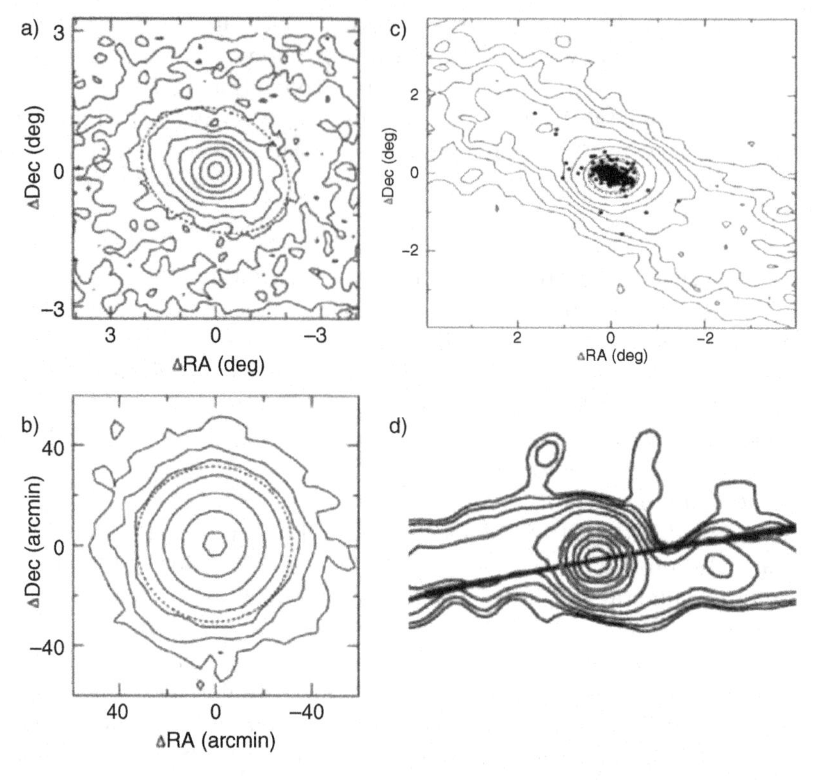

FIG. 2.19. (a): Isodensity contours for one of M08's model satellites right before it becomes completely unbound (5% of initial mass still bound). Only when the satellite is on the verge of being destroyed does the bound component of the satellite enlarge and elongate. (b): Isodensity contours for one of M08's best-fitting Carina models (the left one in Figure 2.18). Only particles within two King-limiting radii (which is marked by the dotted ellipse) are used in this representation so as to mimic the surface brightness limitations of current photometric surveys. The contour levels have been selected to be roughly evenly spaced in radius. The density contours show a remarkably regular structure, with no tail-like structures visible despite the satellite having already lost nearly half its initial mass. (c): Isodensity contours for the same Carina model shown in the bottom left panel, but showing the densities of particles over a 4° radius. Again, the contours have been plotted roughly evenly spaced in radius along the minor axis at surface brightness levels 35.5, 34.9, 34.6, 34.2, 33.8, 33.3, 32.0, and 30.4 mag arcsec^{-1}. This time, the development of tidal tails is clearly observed, but over scales much larger than the King-limiting radius. To illustrate how far out one needs to go to detect clear tidal tails, Carina RV members identified by Muñoz *et al.* (2006) have been overplotted (as filled circles) for comparison. The search for widely displaced dSph members has been most vigorously pursued for Carina, with the results only now showing clear evidence of the Carina tidal tails. (d): Isodensity contours of the tidally disrupting MW globular cluster Palomar 5, showing how tidal tails form over larger spatial scales than the typical King limiting radius (shown as the dashed circle). All figures from M08, except lower right figure modified from Odenkirchen *et al.* (2009).

- *Misperception 4 – No S-shaped morphology means no tidal disruption:* The same dynamics that lead to the expectation of a velocity inflection in Galactic satellites (tidal mass loss along the radial vector to the Galactic center) will also lead to S-shaped morphologies if the satellite is viewed with the correct perspective. An obvious example of this morphology is given by the disrupting globular cluster Pal 5 (e.g., Odenkirchen *et al.*, 2003, Figure 2.19).

  However, this S-shaped distortion takes place in the orbital plane of the satellite, and for many distant dSphs the difference between our perspective and that from the Galactic center (a vantage point at which the distortion would be collapsed perpendicular to the plane of the sky and therefore not manifest in the projected morphology) is negligible. As an example, M08 show that their best-fitting Carina model shows no S-distortion at its great distance from us.

- *Misperception 5 – Flat or rising velocity dispersion profiles REQUIRE extended dark matter halos:* As discussed earlier, the observed flat or rising velocity dispersion profiles in dSphs are commonly interpreted in the context of extended dark matter profiles. If extended distributions of dSph stars are proven to exist, either photometrically as break populations or via a census by radial velocity membership, a typical reaction seems to be that we need to extend the dark matter halos of these dSphs to keep these stars bound, and if the velocity dispersions of these stars continues to be high to larger radius, the dark matter halo must greatly expand if all of these stars are assumed to be bound.

  However, as a number of studies have found (PP95, Mayer *et al.*, 2001, M08), tidal disruption can also produce velocity dispersion profiles that are flat or rising to large radii (e.g., Figure 2.18). Indeed, as has been discussed by, for example, Łokas *et al.* (2005) and Klimentowski *et al.* (2007), one must be very careful how one analyzes the velocity distributions at large radii, because the presence of even some unbound stars can inflate the velocity dispersion profiles and give the illusion of extended dark matter distributions around the system.
- *Misperception 6 – The combined observations of dSphs show that they are not being tidally disrupted:* Perhaps the primary point we want to make in this section is that, in fact, N-body simulations of orbiting satellites that include *both* (relatively) modest amounts of dark matter (as, e.g., reflected in the masses derived by core-fitting) *and* tidal disruption can present satellites with appearances that are deceivingly simple. Indeed, it is easy to find model satellites losing baryonic mass and showing little evidence of "extratidal stars" in their density profile, no rotation signature in their kinematics, no S-shape or elliptical morphology, and flat velocity dispersion profiles that might also be explained by invoking an extended dark matter halo. The lesson is that one should not dismiss the possibility of tidal disruption occurring even if one observes a satellite without any of the commonly held "telltale" signatures.

2.5 A special case: the role of tides in the origin of the Magellanic Stream

So far our focus has been on evidence of tidal effects in LG dwarf galaxies based on observations of stars. But we would be remiss to neglect the first strong evidence discovered for disruption of a MW dwarf satellite – the Magellanic Stream (MS; Figure 2.20). Discovered originally via 21-cm radio observations as a narrow, arcing, "180°-long" hydrogen filament apparently emanating from and trailing behind an extensive H I envelope around the Magellanic Clouds (the "Inter-Cloud Region") by Mathewson *et al.* (1974) (see earlier discoveries of sections by van Kuilenburg, 1972; Wannier and Wrixon, 1972), the MS is still only manifest as a pure H I feature, despite a number of searches for associated stars. Like the similarly elongated (stellar) Sgr stream, the MS is (1) a vivid example of ongoing hierarchical galaxy formation in our own Galactic backyard, (2) presents a special opportunity to explore merger interactions between satellites and parent galaxies, and (3) is an exceedingly important tool for not only precisely determining the orbital dynamics of the disrupting satellite but also for constraining the shape and strength of the Galactic potential as well. However, because it is apparently purely gaseous, the MS is clearly different from other known MW streams, which are purely stellar and clearly have an origin in tidal stripping. Since its discovery, the role of tides in the origin of the MS has been greatly debated.

To date, no single theory has been settled upon to explain the origin of the MS. An early idea that the gas was primordial (Mathewson, 1976; Mathewson and Schwarz, 1976)

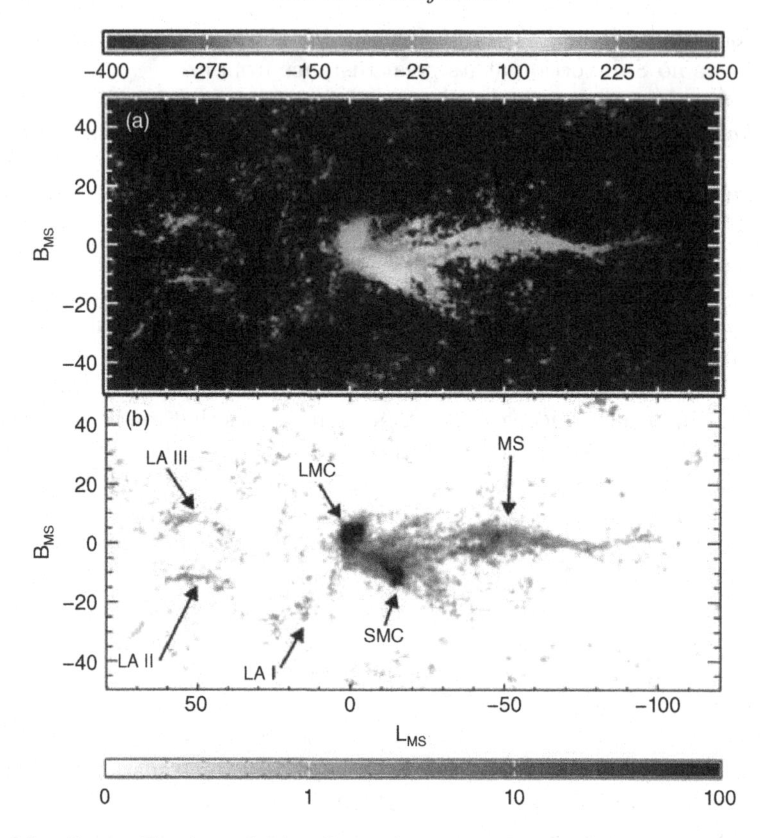

FIG. 2.20. Magellanic Clouds and Magellanic Stream H I gas as distributed on the sky as mapped by N08, shown in the Magellanic coordinate system: (a) Map where color indicates Local Standard of Rest velocity (V_{LSR}), and intensity indicates column density. (b) Gray-scale map of H I column density (10^{19} atoms cm^{-2}) and indicating the various pieces of the gaseous Magellanic system, including the three Leading Arm (LA) complexes I–III (including the three "clumps" of LA I clearly seen in panel a), the Large Magellanic Cloud (LMC), Small Magellanic Cloud (SMC), and the trailing Magellanic Stream (MS). Figure from N08.

was refuted by abundance measures of MS gas (e.g., Lu *et al.*, 1994). Strong early support had been lent to the idea of a ram-pressure stripping origin (Mathewson *et al.*, 1974, 1977, 1987; Meurer *et al.*, 1985; Sofue, 1994) because, at first, only that prominent part of the MS that trails the Magellanic Clouds was known, whereas one expects both trailing and leading features if the gas were tidal in origin. Moreover, the density contours of the H I gas on the leading side of the Clouds are very steep, suggesting a compressional front. The source of the ram pressure is presumed to be a halo of hot coronal gas (or, in the case of some of the earlier models, an intergalactic wind).[7] Moore and Davis (1994) argued that the hot corona was needed to explain the present position of the MS and argued against the tidal model because their N-body simulations of tidal disruption produced widely dispersed debris that looked nothing like the rather well collimated *gaseous* stream.

Nevertheless, a number of other tidal models were created to explain the presence and nature of the MS with various degrees of success (Mirabel and Turner 1973 and similar work by Clutton-Brock, 1972; Wright, 1972; Davies and Wright, 1977; Lin and Lynden-Bell, 1977, 1982; Kunkel, 1979; Murai and Fujimoto, 1980, 1986; Tanaka, 1981; Shuter, 1992; Gardiner and Noguchi, 1996; Fujimoto *et al.*, 1999). To achieve workable solutions, however, these early models often required some important assumptions

[7] Oort (1970), as referenced in Mathewson *et al.* (1974), invoked "snowplowing" through an intergalactic wind to explain the radial velocity distribution of the gas.

and/or "fine tuning," such as specific gravitational potentials or particular three body interactions and timings between the Large Magellanic Cloud (LMC), Small Magellanic Cloud (SMC), and the MW (Murai and Fujimoto, 1980, 1986; Lin and Lynden-Bell, 1982; Wayte, 1991; Gardiner and Noguchi, 1996; Fujimoto *et al.*, 1999), and even M31 (Shuter, 1992), though arguably these were often no more contrived than conditions utilized in some of the ram-pressure models.

However, subsequent studies showing evidence that a leading arm counterpart to the MS actually exists (Lu *et al.*, 1998; Putman *et al.*, 1998, see Figure 2.20) gives substantial credence to the tidal origin picture. In particular, Putman *et al.* (1998) used large-scale, high resolution H I data from the Parkes telescope to show the existence of filamentary, but kinked H I structures "pseudo-continuously" connected (kinematically and spatially) in front of the LMC for about 50 degrees toward the Galactic plane. It is very difficult for a leading arm to be created by ram pressure forces. On the other hand, as found by Gardiner (1999) – using a model of SMC tidal destruction from an SMC-LMC encounter 1.5 Gyr ago at perigalactic passage of both MCs – one can apparently reproduce some, but not all, of the details of these gaseous structures unless one introduces *some* ram pressure drag for "fine tuning." Mastropietro *et al.* (2005) later performed a large ram pressure simulation of the MS (coming out of the LMC) that also included tidal forces and were able to reproduce the general features of the MS, including its extend, shape, column density gradient, and velocity gradient – but they could not, however, reproduce the striking spatial bifurcation of the MS discovered by Putman *et al.* (2003), nor the leading arm feature. N-body tidal simulations by Connors *et al.* (2004, 2006) seem to give the closest reproduction of the MS to date, including the spatial bifurcation of the stream, the leading arm (and its bent shape), and velocity distribution. In these models most of the particles are stripped from the SMC during a close encounter with the LMC+MW $\sim$1.5 Gyr ago. However, like most tidal models, the Connors *et al.* (2004, 2006) simulations still have trouble reproducing the column density gradient along the MS, whereas ram pressure models are typically able to match this feature of the observations much better.

But in the end, perhaps the most troubling problem for the tidal models is that gravity should act upon both gas *and* stars, yet *no stellar component to the MS has ever been detected.* And now the problems for both the tidal *and* ram pressure models become compounded if, as the recent proper motion measurements of the Magellanic Clouds suggest, they are on their first passage by the MW (Kallivayalil *et al.*, 2006a,b; Besla *et al.*, 2007). There simply isn't enough time for either the ram pressure or tidal mechanisms, either working alone or together, to produce an MS as long as is observed if the Clouds are just now reaching the MW environment for the first time. Since neither mechanism can successfully explain all aspects of the Magellanic system, either alone or working together, clearly additional physics is at play.

A proposal for that additional physics has recently been made by Nidever *et al.* (2008), "N08" hereafter, who explored the MS with the new Leiden-Argentine-Bonn (LAB) all-sky H I survey (Kalberla *et al.*, 2005). Using the high velocity resolution (1.3 km s^{-1}) afforded by this survey, N08 was able to take advantage of the velocity coherence of the MS filaments to trace one of them back to their origin in the intensely star-forming 30 Doradus region of the LMC, and to show, in addition, that a large portion of the leading arm counterpart to the MS must also derive from there (Figure 2.21). That at least half of the MS and most of the leading arm originates in the LMC is contrary to numerous previous claims that these H I features originate in the SMC or the Inter-Cloud Region.

The two filaments of the MS show strong periodic patterns in position and a sinusoidal pattern in velocity (see parts of this pattern in Figure 2.21), and N08 hypothesize that one reason for these "oscillatory" patterns is that it is an imprint of the LMC rotation, which gyrates the off-centered 30 Doradus region as the LMC moves along its trajectory around the MW. Adopting this hypothesis and the known LMC rotation curve, N08 estimate the drift rate of the MS gas away from the MCs and, from the length of the

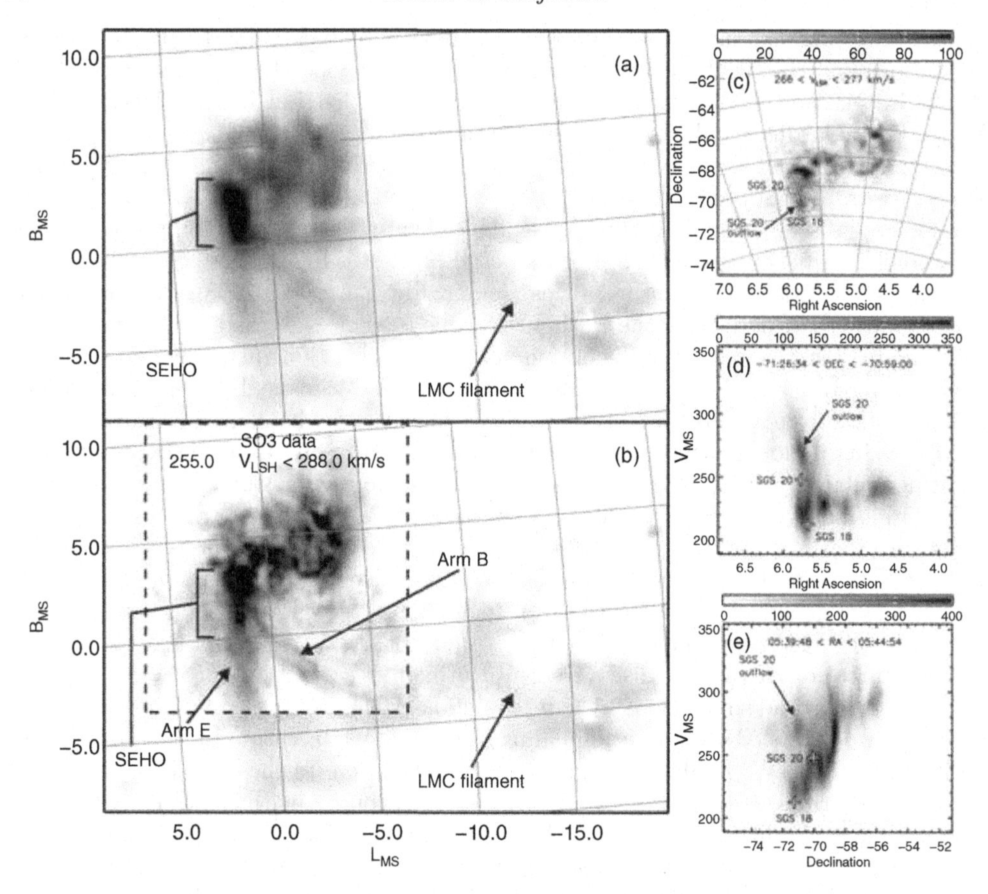

FIG. 2.21. *Left panels*: Integrated H I intensity maps of the LMC and MS (after velocity filtering – see N08) showing the "LMC filament" of MS emanating from the "Southeast H I Overdensity (SEHO)" associated with 30 Doradus. Panel (a) shows the LAB data, while (b) shows the higher spatial resolution data from Staveley-Smith *et al.* (2003) substituted in the region outlined by the dashed box. *Right panels*: Integrated intensity of LMC H I gas showing the outflow from LMC Supergiant Shell (SGS) 20: (c) Column density (10^{19} atoms cm^{-2}; integrated over the indicated range of V_{LSR}), (d) V_{LSR} vs. right ascension (integrated over the indicated declination range; gray scale in Kelvins), and (e) V_{LSR} vs. declination (integrated over indicated right ascension range, in Kelvins). The locations of SGS 20 and 18 are shown in all three panels. Even though the outflow is spatially more aligned with SGS 18 than SGS 20 (panel c), panels (d) and (e) show this to be a projection effect. The gas outflow is connected in position and velocity to a gas bubble surrounding SGS 20 but has moved farther south and is spatially projected onto SGS 18. These figures show that SGS 20 is blowing out a large amount of gas from the LMC. Figures from N08.

MS, determine its age to be at least ~1.7 Gyr. This is longer than the current MW interaction timescale for a first-pass LMC scenario and, at first, would seem to bring no further clarification as to how the MS could have had time to get to its currently observed length. However, through analysis of the high-spatial resolution Parkes data of the LMC (Staveley-Smith *et al.*, 2003), N08 were able to identify large gas outflows from supergiant shells in the 30 Doradus region that are creating the LMC filament of the Magellanic Stream and its leading arm counterpart (Figure 2.21).

These outflows make evident a new formation mechanism for the MS that can plausibly account for those observed features that have long been a problem for the ram pressure and tidal models, as well as the new complications imposed by the first-pass scenario. In the new picture proposed by N08 (1) the intense star formation in the 30 Doradus region has created superbubbles and supergiant shells in the LMC's interstellar gas; (2) some of the supergiant shells are sufficiently energetic that they can provide a significant boost to the kinetic energy of the gas enabling it to more easily be stripped from the system; and (3) once the gas is far enough away from the LMC, tidal forces, and ram pressure pull/push the gas to form the MS and the leading arm structure. In this scenario, the ram pressure and tidal forces aren't responsible for all of the work of *stripping* the gas, and they primarily serve to *shape it* into the stream features observed. This blowout hypothesis explains the long-standing mystery of why there are no stars in the stream, since the supergiant shells only push out gas, not stars. Thus, a gas-only stream is a natural consequence of the blowout scenario, even in the presence of tidal forces. Moreover, this process provides perhaps the best explanation yet for the long tidal stream in the face of a first MW passage scenario, since great proximity to the MW is no longer needed for strong tidal/ram pressure forces to strip out gas, and the stream-like coherence of the lost gas can organize in a much weaker tidal field at greater distance from the MW, and maybe even from the overall gravitational potential of the LG itself as the MCs fell in from Mpc scale distances. However, only detailed and complex hydrodynamical modeling of this proposed coordination of vigorous star formation, ram pressure forces and tidal interactions between the infalling LMC and the MW, as well as between the LMC and the LG generally, will be able to verify this promising hypothesis for the origin of the MS system. First steps toward this goal have already been taken by Olano (2004), who has demonstrated that blowout from the MCs can create the large-scale features of the MS, as well as the high velocity H I cloud system of the MW. Olano's model explored blowout from the MCs that results from an interaction between the two Clouds 570 Myr ago. The N08 analysis identifies a more specific site (the 30 Doradus region of the LMC) where the process is active, as well as an approximate timescale (~1.7 Gyr) over which the process might have taken place, and provides much more specific observational data to help guide future models of this first known, and perhaps still most dramatic, example of a satellite galaxy presently being substantially influenced by, and losing mass to, the MW.

REFERENCES

Aaronson, M. 1983. Accurate radial velocities for carbon stars in Draco and Ursa Minor – The first hint of a dwarf spheroidal mass-to-light ratio. *ApJ*, **266**(Mar.), L11–L15.

Abazajian, K. N., and 203 colleagues 2009. The seventh data release of the Sloan Digital Sky Survey. *ApJS*, **182**(June), 543–558.

Angus, G. W. 2008. Dwarf spheroidals in MOND. *MNRAS*, **387**(July), 1481–1488.

Angus, G. W., Famaey, B., and Zhao, H. S. 2006. Can MOND take a bullet? Analytical comparisons of three versions of MOND beyond spherical symmetry. *MNRAS*, **371**(Sept.), 138–146.

Armandroff, T. E. and Da Costa, G. S. 1986. The radial velocity, velocity dispersion, and mass-to-light ratio of the Sculptor dwarf galaxy. *AJ*, **92**(Oct.), 777–786.

Armandroff, T. E., Olszewski, E. W., and Pryor, C. 1995. The mass-to-light ratios of the Draco and Ursa Minor Dwarf Spheroidal Galaxies. I. Radial velocities from multifiber spectroscopy. *AJ*, **110**(Nov.), 2131–2165.

Baade, W. 1944. The resolution of Messier 32, NGC 205, and the central region of the Andromeda Nebula. *ApJ*, **100**(Sept.), 137–146.

Baade, W. and Hubble, E. 1939. The new stellar systems in Sculptor and Fornax. *PASP*, **51**(Feb.), 40–44.

Battaglia, G., Helmi, A., Tolstoy, E., Irwin, M., Hill, V., and Jablonka, P. 2008. The kinematic status and mass content of the Sculptor dwarf spheroidal galaxy. *ApJ*, **681**(July), L13–L16.

Beaton, R. L., and 8 colleagues. 2007. Unveiling the boxy bulge and bar of the Andromeda Spiral Galaxy. *ApJ*, **658**(Apr.), L91–L94.

Belokurov, V., and 20 colleagues. 2006a. The field of streams: Sagittarius and its siblings. *ApJ*, **642**(May), L137–L140.

Belokurov, V., and 32 colleagues. 2006b. A faint new Milky Way satellite in Bootes. *ApJ*, **647**(Aug.), L111–L114.

Belokurov, V., and 25 colleagues. 2007a. The Hercules-Aquila Cloud. *ApJ*, **657**(Mar.), L89–L92.

Belokurov, V., and 33 colleagues. 2007b. Cats and dogs, hair and a hero: a quintet of new Milky Way companions. *ApJ*, **654**(Jan.), 897–906.

Besla, G., Kallivayalil, N., Hernquist, L., Robertson, B., Cox, T. J., van der Marel, R. P., and Alcock, C. 2007. Are the Magellanic Clouds on their first passage about the Milky Way? *ApJ*, **668**(Oct.), 949–967.

Bizyaev, D., and 9 colleagues 2006. The Space Interferometry Mission Astrometric Grid Giant Star Survey. I. Stellar parameters and radial velocity variability. *AJ*, **131**(Mar.), 1784–1796.

Brown, T. M., Ferguson, H. C., Smith, E., Kimble, R. A., Sweigart, A. V., Renzini, A., Rich, R. M., and VandenBerg, D. A. 2003. Evidence of a significant intermediate-age population in the M31 halo from main-sequence photometry. *ApJ*, **592**(July), L17–L20.

Bullock, J. S. and Johnston, K. V. 2005. Tracing galaxy formation with stellar halos. I. Methods. *ApJ*, **635**(Dec.), 931–949.

Bullock, J. S., Kravtsov, A. V., and Weinberg, D. H. 2000. Reionization and the abundance of galactic satellites. *ApJ*, **539**(Aug.), 517–521.

Bullock, J. S., Kravtsov, A. V., and Weinberg, D. H. 2001. Hierarchical galaxy formation and substructure in the Galaxy's stellar halo. *ApJ*, **548**(Feb.), 33–46.

Cannon, R. D., Hawarden, T. G., and Tritton, S. B. 1977. A new Sculptor-type dwarf elliptical galaxy in Carina. *MNRAS*, **180**(Sept.), 81P–82P.

Casetti-Dinescu, D. I., Carlin, J. L., Girard, T. M., Majewski, S. R., Peñarrubia, J., and Patterson, R. J. 2008. Kinematics of stars in Kapteyn Selected Area 71: sampling the Monoceros and Sagittarius tidal streams. *AJ*, **135**(June), 2013–2023.

Chapman, S. C., and 11 colleagues. 2007. Strangers in the night: discovery of a dwarf spheroidal galaxy on its first Local Group infall. *ApJ*, **662**(June), L79–L82.

Chilingarian, I., Cayatte, V., Revaz, Y., Dodonov, S., Durand, D., Durret, F., Micol, A., and Slezak, E. 2009. A population of compact elliptical galaxies detected with the Virtual Observatory. *Science*, **326**(Dec.), 1379–1382.

Chou, M.-Y., and 11 colleagues. 2007. A 2MASS all-sky view of the Sagittarius Dwarf Galaxy. V. Variation of the metallicity distribution function along the Sagittarius stream. *ApJ*, **670**(Nov.), 346–362.

Clowe, D., Randall, S. W., and Markevitch, M. 2007. Catching a bullet: direct evidence for the existence of dark matter. *Nuclear Physics B Proceedings Supplements*, **173**(Nov.), 28–31.

Clutton-Brock, M. 1972. How are intergalactic filaments made? *Ap&SS*, **17**(Aug.), 292–324.

Colless, M., and 28 colleagues 2001. The 2dF Galaxy Redshift Survey: spectra and redshifts. *MNRAS*, **328**(Dec.), 1039–1063.

Connors, T. W., Kawata, D., and Gibson, B. K. 2006. N-body simulations of the Magellanic stream. *MNRAS*, **371**(Sept.), 108–120.

Connors, T. W., Kawata, D., Maddison, S. T., and Gibson, B. K. 2004. High-resolution N-body simulations of galactic cannibalism: the Magellanic stream. *PASA*, **21**, 222–227.

Dalcanton, J. J., and 26 colleagues. 2009. The ACS Nearby Galaxy Survey Treasury. *ApJS*, **183**(July), 67–108.

Davies, R. D. and Wright, A. E. 1977. A tidal origin for the Magellanic Stream. *MNRAS*, **180**(July), 71–88.

Davis, M., Huchra, J., Latham, D. W., and Tonry, J. 1982. A survey of galaxy redshifts. II – The large scale space distribution. *ApJ*, **253**(Feb.), 423–445.

Dolphin, A. E., and 8 colleagues. 2003. Deep Hubble Space Telescope imaging of Sextans A. III. The star formation history. *AJ*, **126**(July), 187–196.

Duffau, S., Zinn, R., Vivas, A. K., Carraro, G., Méndez, R. A., Winnick, R., and Gallart, C. 2006. Spectroscopy of QUEST RR Lyrae Variables: The New Virgo Stellar Stream. *ApJ*, **636**(Jan.), L97–L100.

Elson, R. A. W. 1999. Stellar dynamics in globular clusters. *Globular Clusters*, 209–248.

Eskridge, P. B. 1988. The structure of the sculptor dwarf elliptical galaxy. I – The radial profile. *AJ*, **95**(June), 1706–1716.

Evans, N. W. and Wilkinson, M. I. 2000. The mass of the Andromeda galaxy. *MNRAS*, **316**(Aug.), 929–942.

Feltzing, S., Gilmore, G., and Wyse, R. F. G. 1999. The Faint Optical Stellar Luminosity function in the Ursa Minor Dwarf Spheroidal Galaxy. *ApJ*, **516**(May), L17–L20.

Fisher, J. R. and Tully, R. B. 1975. Neutral hydrogen observations of DDO dwarf galaxies. *A&A*, **44**(Nov.), 151–171.

Fleck, J.-J. and Kuhn, J. R. 2003. Parametric dwarf spheroidal tidal interaction. *ApJ*, **592**(July), 147–160.

Freese, K., Gondolo, P., and Newberg, H. J. 2005. Detectability of weakly interacting massive particles in the Sagittarius dwarf tidal stream. *Phys. Rev. D*, **71**(Feb.), 043516–043531.

Fuchs, B., Phleps, S., and Meisenheimer, K. 2006. CADIS has seen the Virgo overdensity and parts of the Monoceros and "Orphan" streams in retrospect. *A&A*, **457**(Oct.), 541–543.

Fujimoto, M., Sawa, T., and Kumai, Y. 1999. The Magellanic Stream and the Magellanic Cloud System. *IAU Symposium*, **186**, 31–38.

Gardiner, L. T. 1999. N-body simulations of the Magellanic Stream. *Stromlo Workshop on High-Velocity Clouds*, **166**, 292–301.

Gardiner, L. T. and Noguchi, M. 1996. N-body simulations of the Small Magellanic Cloud and the Magellanic Stream. *MNRAS*, **278**(Jan.), 191–208.

Geha, M., Willman, B., Simon, J. D., Strigari, L. E., Kirby, E. N., Law, D. R., and Strader, J. 2009. The least-luminous galaxy: spectroscopy of the Milky Way satellite Segue 1. *ApJ*, **692**(Feb.), 1464–1475.

Geller, M. J. and Huchra, J. P. 1989. Mapping the Universe. *Science*, **246**(Nov.), 897–903.

Gelmini, G., Gondolo, P., and Soldatenko, A. 2004. Detectability of the Sgr dwarf leading tidal stream with Auger, EUSO, or OWL. *Phys. Rev. D*, **70**(July), 023010–023015.

Gerhard, O. 2002. The Galactic Bar. *The dynamics, structure and history of galaxies: a workshop in honour of Professor Ken Freeman*, **273**, 73–83.

Gilmore, G., Wilkinson, M. I., Wyse, R. F. G., Kleyna, J. T., Koch, A., Evans, N. W., and Grebel, E. K. 2007. The observed properties of dark matter on small spatial scales. *ApJ*, **663**(July), 948–959.

Goerdt, T., Moore, B., Read, J. I., Stadel, J., and Zemp, M. 2006. Does the Fornax dwarf spheroidal have a central cusp or core? *MNRAS*, **368**(May), 1073–1077.

Gondolo, P. and Gelmini, G. 2005. Compatibility of DAMA dark matter detection with other searches. *Phys. Rev. D*, **71**(June), 123520–123530.

Gottesman, S. T., Hunter, J. H., and Boonyasait, V. 2002. On the mass of M31. *MNRAS*, **337**(Nov.), 34–40.

Graham, A. W. 2002. Evidence for an outer disk in the prototype "compact elliptical" Galaxy M32. *ApJ*, **568**(Mar.), L13–L17.

Grebel, E. K. 2004. The evolutionary history of Local Group irregular galaxies. *Origin and Evolution of the Elements*, 237–257.

Grebel, E. K., Gallagher, J. S., III, and Harbeck, D. 2003. The progenitors of dwarf spheroidal galaxies. *AJ*, **125**(Apr.), 1926–1939.

Grillmair, C. J. 2006. Detection of a 60°;-long dwarf galaxy debris stream. *ApJ*, **645**(July), L37–L40.

Grillmair, C. J. 2009. Four new stellar debris streams in the Galactic Halo. *ApJ*, **693**(Mar.), 1118–1127.

Grillmair, C. J. and Dionatos, O. 2006a. A 22° tidal tail for Palomar 5. *ApJ*, **641**(Apr.), L37–L39.

Grillmair, C. J. and Dionatos, O. 2006b. Detection of a 63° cold stellar stream in the Sloan Digital Sky Survey. *ApJ*, **643**(May), L17–L20.

Grillmair, C. J., Freeman, K. C., Irwin, M., and Quinn, P. J. 1995. Globular clusters with tidal tails: deep two-color star counts. *AJ*, **109**(June), 2553–2585.

Hayashi, E., Navarro, J. F., Taylor, J. E., Stadel, J., and Quinn, T. 2003. The structural evolution of substructure. *ApJ*, **584**(Feb.), 541–558.

Hargreaves, J. C., Gilmore, G., and Annan, J. D. 1996. The influence of binary stars on dwarf spheroidal galaxy kinematics. *MNRAS*, **279**(Mar.), 108–120.

Harrington, R. G. and Wilson, A. G. 1950. Two new stellar systems in Leo. *PASP*, **62**(June), 118–120.

Hernquist, L. and Ostriker, J. P. 1992. A self-consistent field method for galactic dynamics. *ApJ*, **386**(Feb.), 375–397.

Hodge, P. W. 1961a. The Fornax dwarf galaxy. II. The distribution of stars. *AJ*, **66**(Aug.), 249–257.

Hodge, P. W. 1961b. The distribution of stars in the Sculptor dwarf galaxy. *AJ*, **66**(Oct.), 384–389.

Hodge, P. W. 1962. Distribution of stars in the Leo II Dwarf Galaxy. *AJ*, **67**(Mar.), 125–129.

Hodge, P. W. 1971. Dwarf Galaxies. *ARA&A*, **9**, 35–66.

Hodge, P. W. and Michie, R. W. 1969. The structure of dwarf elliptical galaxies of the Local Group. *AJ*, **74**(June), 587–596.

Howley, K. M., Geha, M., Guhathakurta, P., Montgomery, R. M., Laughlin, G., and Johnston, K. V. 2008. Darwin tames an Andromeda Dwarf: unraveling the orbit of NGC 205 using a genetic algorithm. *ApJ*, **683**(Aug.), 722–749.

Howley, K., Guhathakurta, P., Geha, M., Kalirai, J., van der Marel, R., Yniguez, B., Cuillandre, J., and Gilbert, K. 2011. The Splash Survey: internal stellar kinematics of the nearby compact elliptical M32. *Bull. Am. Astron. Soc.*, **43**(Jan.), #207.04.

Ibata, R., Lewis, G. F., Irwin, M., Totten, E., and Quinn, T. 2001. Great circle tidal streams: evidence for a nearly spherical massive dark halo around the Milky Way. *ApJ*, **551**(Apr.), 294–311.

Ibata, R., Gilmore, G., and Irwin, M. J. 1994. A dwarf satellite galaxy in Sagittarius. *Nature*, **370**(July), 194–196.

Ibata, R., Martin, N. F., Irwin, M., Chapman, S., Ferguson, A. M. N., Lewis, G. F., and McConnachie, A. W. 2007. The haunted halos of Andromeda and Triangulum: a panorama of galaxy formation in action. *ApJ*, **671**(Dec.), 1591–1623.

Ibata, R., Wyse, R. F. G., Gilmore, G., Irwin, M. J., and Suntzeff, N. B. 1997. The kinematics, orbit, and survival of the Sagittarius Dwarf Spheroidal Galaxy. *AJ*, **113**(Feb.), 634–655.

Illingworth, G. 1976. The masses of globular clusters. II – Velocity dispersions and mass-to-light ratios. *ApJ*, **204**(Feb.), 73–93.

Innanen, K. A. and Papp, K. A. 1979. Extratidal variables and the dynamics of the Sculptor dwarf galaxy. *AJ*, **84**(May), 601–603.

Irwin, M. and Hatzidimitriou, D. 1995. Structural parameters for the Galactic dwarf spheroidals. *MNRAS*, **277**(Dec.), 1354–1378.

Irwin, M. J., Bunclark, P. S., Bridgeland, M. T., and McMahon, R. G. 1990. A new satellite galaxy of the Milky Way in the constellation of Sextans. *MNRAS*, **244**(May), 16P–19P.

Johnston, K. V. 1998. A prescription for building the Milky Way's halo from disrupted satellites. *ApJ*, **495**(Mar.), 297–308.

Jones, D. H., and 32 colleagues. 2009. The 6dF Galaxy Survey: final redshift release (DR3) and southern large-scale structures. *MNRAS*, **399**(Oct.), 683–698.

Kalberla, P. M. W., Burton, W. B., Hartmann, D., Arnal, E. M., Bajaja, E., Morras, R., and Pöppel, W. G. L. 2005. The Leiden/Argentine/Bonn (LAB) Survey of Galactic HI. Final data release of the combined LDS and IAR surveys with improved stray-radiation corrections. *A&A*, **440**(Sept.), 775–782.

Kallivayalil, N., van der Marel, R. P., Alcock, C., Axelrod, T., Cook, K. H., Drake, A. J., and Geha, M. 2006a. The proper motion of the Large Magellanic Cloud using HST. *ApJ*, **638**(Feb.), 772–785.

Kallivayalil, N., van der Marel, R. P., and Alcock, C. 2006b. Is the SMC bound to the LMC? The Hubble Space Telescope proper motion of the SMC. *ApJ*, **652**(Dec.), 1213–1229.

Karachentsev, I. D., Kashibadze, O. G., Makarov, D. I., and Tully, R. B. 2009. The Hubble flow around the Local Group. *MNRAS*, **393**(Mar.), 1265–1274.

Kauffmann, G., White, S. D. M., and Guiderdoni, B. 1993. The formation and evolution of galaxies within merging dark matter haloes. *MNRAS*, **264**(Sept.), 201–218.

Keenan, D. W. and Innanen, K. A. 1975. Numerical investigation of galactic tidal effects on spherical stellar systems. *AJ*, **80**(Apr.), 290–302.

King, I. 1962. The structure of star clusters. I. an empirical density law. *AJ*, **67**(Oct.), 471–485.

King, I. R. 1966. The structure of star clusters. III. Some simple dynamical models. *AJ*, **71**(Feb.), 64–75.

King, I. R. and Minkowski, R. 1972. *External Galaxies and Quasi-Stellar Objects*, IAU Symposium: Evans, D. S. and Wills, D. and Wills, B. J.

Klessen, R. S. and Kroupa, P. 1998. Dwarf spheroidal satellite galaxies without dark matter: results from two different numerical techniques. *ApJ*, **498**(May), 143–155.

Klessen, R. S. and Zhao, H. 2002. Are dwarf spheroidal galaxies dark matter dominated or remnants of disrupted larger satellite galaxies? A possible test. *ApJ*, **566**(Feb.), 838–844.

Klessen, R. S., Grebel, E. K., and Harbeck, D. 2003. Draco: a failure of the tidal model. *ApJ*, **589**(June), 798–809.

Kleyna, J., Geller, M., Kenyon, S., and Kurtz, M. 1999. Measuring the dark matter scale of Local Group dwarf spheroidals. *AJ*, **117**(Mar.), 1275–1284.

Kleyna, J., Wilkinson, M. I., Evans, N. W., Gilmore, G., and Frayn, C. 2002. Dark matter in dwarf spheroidals – II. Observations and modelling of Draco. *MNRAS*, **330**(Mar.), 792–806.

Kleyna, J. T., Wilkinson, M. I., Gilmore, G., and Evans, N. W. 2003. A dynamical fossil in the Ursa Minor Dwarf Spheroidal Galaxy. *ApJ*, **588**(May), L21–L24.

Klimentowski, J., Łokas, E. L., Kazantzidis, S., Prada, F., Mayer, L., and Mamon, G. A. 2007. Mass modelling of dwarf spheroidal galaxies: the effect of unbound stars from tidal tails and the Milky Way. *MNRAS*, **378**(June), 353–368.

Klypin, A., Kravtsov, A. V., Valenzuela, O., and Prada, F. 1999. Where are the missing galactic satellites? *ApJ*, **522**(Sept.), 82–92.

Klypin, A., Zhao, H., and Somerville, R. S. 2002. ΛCDM-based models for the Milky Way and M31. I. Dynamical models. *ApJ*, **573**(July), 597–613.

Koch, A., Kleyna, J. T., Wilkinson, M. I., Grebel, E. K., Gilmore, G. F., Evans, N. W., Wyse, R. F. G., and Harbeck, D. R. 2007. Stellar kinematics in the remote Leo II Dwarf Spheroidal Galaxy–another brick in the wall. *AJ*, **134**(Aug.), 566–578.

Kravtsov, A. V., Gnedin, O. Y., and Klypin, A. A. 2004. The tumultuous lives of Galactic Dwarfs and the missing satellites problem. *ApJ*, **609**(July), 482–497.

Kroupa, P. 1997. Dwarf spheroidal satellite galaxies without dark matter. *New A*, **2**(July), 139–164.

Kuhn, J. R. 1993. Unbound dwarf spheroidal galaxies and the mass of the Milky Way. *ApJ*, **409**(May), L13–L16.

Kuhn, J. R. and Miller, R. H. 1989. Dwarf spheroidal galaxies and resonant orbital coupling. *ApJ*, **341**(June), L41–L45.

Kuhn, J. R., Smith, H. A., and Hawley, S. L. 1996. Tidal disruption and tails from the Carina Dwarf Spheroidal Galaxy. *ApJ*, **469**(Oct.), L93–L96.

Kunkel, W. E. 1979. On the origin and dynamics of the Magellanic Stream. *ApJ*, **228**(Mar.), 718–733.

Law, D. R. and Majewski, S. R. 2010. The Sagittarius Dwarf Galaxy: a model for evolution in a triaxial Milky Way halo. *ApJ*, **714**(May), 229–254.

Law, D. R., Johnston, K. V., and Majewski, S. R. 2005. A Two Micron All-Sky Survey view of the Sagittarius Dwarf Galaxy. IV. Modeling the Sagittarius tidal tails. *ApJ*, **619**(Feb.), 807–823.

Lewis, M. J. and Freese, K. 2004. Phase of the annual modulation as a tool for determining the mass of the weakly interacting massive particle. *Phys. Rev. D*, **70**(Aug.), 043501–043508.

Lin, D. N. C. and Lynden-Bell, D. 1977. Simulation of the Magellanic Stream to estimate the total mass of the Milky Way. *MNRAS*, **181**(Oct.), 59–81.

Lin, D. N. C. and Lynden-Bell, D. 1982. On the proper motion of the Magellanic Clouds and the halo mass of our galaxy. *MNRAS*, **198**(Feb.), 707–721.

Lo, K. Y., Sargent, W. L. W., and Young, K. 1993. The H I structure of nine intrinsically faint dwarf galaxies. *AJ*, **106**(Aug.), 507–529.

Łokas, E. L., Kazantzidis, S., Majewski, S. R., Law, D. R., Mayer, L., and Frinchaboy, P. M. 2010. The Inner structure and kinematics of the Sagittarius Dwarf Galaxy as a product of tidal stirring. *ApJ*, **725**(Dec.), 1516–1527.

Łokas, E. L., Mamon, G. A., and Prada, F. 2005. Dark matter distribution in the Draco dwarf from velocity moments. *MNRAS*, **363**(Nov.), 918–928.

Lu, L., Sargent, W. L. W., Savage, B. D., Wakker, B. P., Sembach, K. R., and Oosterloo, T. A. 1998. The metallicity and dust content of HVC 287.5+22.5+240 – evidence for a Magellanic Clouds origin. *AJ*, **115**(Jan.), 162–167.

Lu, L., Savage, B. D., and Sembach, K. R. 1994. Probing the galactic disk and halo: metal abundances in the Magellanic Stream. *ApJ*, **437**(Dec.), L119–L122.

Majewski, S. R., and 8 colleagues. 2005. Exploring halo substructure with giant stars. VI. Extended distributions of giant stars around the Carina Dwarf Spheroidal Galaxy: how reliable are they? *AJ*, **130**(Dec.), 2677–2700.

Majewski, S. R., and 11 colleagues. 2007. Discovery of Andromeda XIV: a dwarf spheroidal dynamical rogue in the Local Group? *ApJ*, **670**(Nov.), L9–L12.

Majewski, S. R., and 10 colleagues. 2002. *Modes of Star Formation and the Origin of Field Populations*, Astronomical Society of the Pacific Conference Series: Grebel, E. K. and Brandner, W. eds.

Majewski, S. R., and 12 colleagues. 2004. A Two Micron All Sky Survey View of the Sagittarius Dwarf Galaxy. II. Swope Telescope spectroscopy of M giant stars in the dynamically cold Sagittarius Tidal Stream. *AJ*, **128**(July), 245–259.

Majewski, S. R., *et al.* In preparation.

Majewski, S. R., Ostheimer, J. C., Patterson, R. J., Kunkel, W. E., Johnston, K. V., and Geisler, D. 2000. Exploring halo substructure with giant stars. II. Mapping the extended structure of the Carina Dwarf Spheroidal Galaxy. *AJ*, **119**(Feb.), 760–776.

Majewski, S. R., Skrutskie, M. F., Weinberg, M. D., and Ostheimer, J. C. 2003. A Two Micron All Sky Survey View of the Sagittarius Dwarf Galaxy. I. Morphology of the Sagittarius core and tidal arms. *ApJ*, **599**(Dec.), 1082–1115.

Martin, N. F., and 12 colleagues 2009. PAndAS' CUBS: discovery of two new dwarf galaxies in the surroundings of the Andromeda and Triangulum Galaxies. *ApJ*, **705**(Nov.), 758–765.

Martin, N. F., de Jong, J. T. A., and Rix, H.-W. 2008. A comprehensive maximum likelihood analysis of the structural properties of faint Milky Way satellites. *ApJ*, **684**(Sept.), 1075–1092.

Martin, N. F., Ibata, R. A., Irwin, M. J., Chapman, S., Lewis, G. F., Ferguson, A. M. N., Tanvir, N., and McConnachie, A. W. 2006. Discovery and analysis of three faint dwarf galaxies and a globular cluster in the outer halo of the Andromeda galaxy. *MNRAS*, **371**(Oct.), 1983–1991.

Martínez-Delgado, D., Peñarrubia, J., Gabany, R. J., Trujillo, I., Majewski, S. R., and Pohlen, M. 2008. The ghost of a dwarf galaxy: fossils of the hierarchical formation of the nearby spiral galaxy NGC 5907. *ApJ*, **689**(Dec.), 184–193.

Martínez-Delgado, D., Peñarrubia, J., Jurić, M., Alfaro, E. J., and Ivezić, Z. 2007. The Virgo stellar overdensity: mapping the infall of the Sagittarius tidal stream onto the Milky Way disk. *ApJ*, **660**(May), 1264–1272.

Mashchenko, S., Couchman, H. M. P., and Sills, A. 2005. Modeling star formation in dwarf spheroidal galaxies: a case for extended dark matter halos. *ApJ*, **624**(May), 726–741.

Mastropietro, C., Moore, B., Mayer, L., Wadsley, J., and Stadel, J. 2005. The gravitational and hydrodynamical interaction between the Large Magellanic Cloud and the Galaxy. *MNRAS*, **363**(Oct.), 509–520.

Mathewson, D. S. 1976. The Magellanic Stream and other gas concentrations in the Local Group (invited review). *The Galaxy and the Local Group.* Royal Greenwich Observatory Bulletins: Dickens, R. J. and Perry, J. E. and Smith, F. G. and King, I. R. eds.

Mathewson, D. S. and Schwarz, M. P. 1976. The origin of the Magellanic Stream. *MNRAS*, **176**(Aug.), 47P–51P.

Mathewson, D. S., Cleary, M. N., and Murray, J. D. 1974. The Magellanic stream. *ApJ*, **190**(June), 291–296.

Mathewson, D. S., Schwarz, M. P., and Murray, J. D. 1977. The Magellanic stream – The turbulent wake of the Magellanic clouds in the halo of the Galaxy. *ApJ*, **217**(Oct.), L5–L8.

Mathewson, D. S., Wayte, S. R., Ford, V. L., and Ruan, K. 1987. The 'high velocity cloud' origin of the Magellanic system. *Astronomical Society of Australia, Proceedings*, vol. 7, no. 1, 19–25.

Mateo, M. L. 1998. Dwarf galaxies of the Local Group. *ARA&A*, **36**, 435–506.

Mateo, M., Olszewski, E., Pryor, C., Welch, D. L., and Fischer, P. 1993. The Carina dwarf spheroidal galaxy – How dark is it? *AJ*, **105**(Feb.), 510–526.

Mateo, M., Olszewski, E., Welch, D. L., Fischer, P., and Kunkel, W. 1991. A kinematic study of the Fornax dwarf spheroidal galaxy. *AJ*, **102**(Sept.), 914–926.

Mayer, L., Governato, F., Colpi, M., Moore, B., Quinn, T., Wadsley, J., Stadel, J., and Lake, G. 2001. The Metamorphosis of Tidally Stirred Dwarf Galaxies. *ApJ*, **559**(Oct.), 754–784.

McConnachie, A. W., Peñarrubia, J., and Navarro, J. F. 2007. Multiple dynamical components in Local Group dwarf spheroidals. *MNRAS*, **380**(Sept.), L75–L79.

Merrifield, M. R. 2004. The Galactic Bar. *Milky Way surveys: the structure and evolution of our Galaxy*, **317**(Dec.), 289–302.

Meurer, G. R., Bicknell, G. V., and Gingold, R. A. 1985. A drag dominated model of the Magellanic stream. *Astronomical Society of Australia, Proceedings*, vol. 6, no. 2, 195–198.

Milgrom, M. 1983a. A modification of the Newtonian dynamics as a possible alternative to the hidden mass hypothesis. *ApJ*, **270**(July), 365–370.

Milgrom, M. 1983b. A modification of the Newtonian dynamics – implications for galaxies. *ApJ*, **270**(July), 371–389.

Mirabel, I. F. and Turner, K. C. 1973. A search for neutral hydrogen remnants of strong tidal disruption of the Small Magellanic Cloud. *A&A*, **22**(Feb.), 437–440.

Miyamoto, M. and Nagai, R. 1975. Three-dimensional models for the distribution of mass in galaxies. *PASJ*, **27**, 533–543.

Moore, B. 1996. Constraints on the global mass-to-light ratios and on the extent of dark matter halos in globular clusters and dwarf spheroidals. *ApJ*, **461**(Apr.), L13–L16.

Moore, B. and Davis, M. 1994. The origin of the Magellanic stream. *MNRAS*, **270**(Sept.), 209–221.

Moore, B., Diemand, J., Madau, P., Zemp, M., and Stadel, J. 2006. Globular clusters, satellite galaxies and stellar haloes from early dark matter peaks. *MNRAS*, **368**(May), 563–570.

Moore, B., Ghigna, S., Governato, F., Lake, G., Quinn, T., Stadel, J., and Tozzi, P. 1999. Dark matter substructure within galactic halos. *ApJ*, **524**(Oct.), L19–L22.

Muñoz, R. R., and 8 colleagues. 2005. Exploring halo substructure with giant stars: the velocity dispersion profiles of the Ursa Minor and Draco Dwarf spheroidal galaxies at large angular separations. *ApJ*, **631**(Oct.), L137–L141.

Muñoz, R. R., and 12 colleagues. 2006. Exploring halo substructure with giant stars. XI. The tidal tails of the Carina Dwarf Spheroidal Galaxy and the discovery of Magellanic Cloud stars in the Carina foreground. *ApJ*, **649**(Sept.), 201–223.

Muñoz, R. R., Majewski, S. R., and Johnston, K. V. 2008. Modeling the structure and dynamics of dwarf spheroidal galaxies with dark matter and tides. *ApJ*, **679**(May), 346–372.

Murai, T. and Fujimoto, M. 1980. The Magellanic Stream and the Galaxy with a massive halo. *PASJ*, **32**, 581–604.

Murai, T. and Fujimoto, M. 1986. Dynamics of the Magellanic system and the galaxy – present status of theoretical understanding. *Ap&SS*, **119**(Feb.), 169–171.

Navarro, J. F., Frenk, C. S., and White, S. D. M. 1996. The structure of cold dark matter halos. *ApJ*, **462**(May), 563–575.

Newberg, H. J., and 18 colleagues. 2002. The ghost of Sagittarius and lumps in the halo of the Milky Way. *ApJ*, **569**(Apr.), 245–274.

Nidever, D. L., Majewski, S. R., and Burton, W. B. 2008. The origin of the Magellanic Stream and its leading arm. *ApJ*, **679**(May), 432–459.

Odenkirchen, M., and 9 colleagues 2003. The extended tails of Palomar 5: a 10° arc of globular cluster tidal debris. *AJ*, **126**(Nov.), 2385–2407.

Odenkirchen, M., and 19 colleagues 2001. Detection of massive tidal tails around the globular cluster Palomar 5 with Sloan Digital Sky Survey commissioning data. *ApJ*, **548**(Feb.), L165–L169.

Odenkirchen, M., Grebel, E. K., Dehnen, W., Rix, H.-W., and Cudworth, K. M. 2002. Kinematic study of the disrupting globular cluster Palomar 5 using VLT spectra. *AJ*, **124**(Sept.), 1497–1510.

Odenkirchen, M., Grebel, E. K., Kayser, A., Rix, H.-W., and Dehnen, W. 2009. Kinematics of the tidal debris of the globular cluster Palomar 5. *AJ*, **137**(Feb.), 3378–3387.

Oh, K. S., Lin, D. N. C., and Aarseth, S. J. 1992. Tidal evolution of globular clusters. I – method. *ApJ*, **386**(Feb.), 506–518.

Oh, K. S., Lin, D. N. C., and Aarseth, S. J. 1995. On the tidal disruption of dwarf spheroidal galaxies around the galaxy. *ApJ*, **442**(Mar.), 142–158.

Olano, C. A. 2004. The high-velocity clouds and the Magellanic Clouds. *A&A*, **423**(Sept.), 895–907.

Olszewski, E. W., Pryor, C., and Armandroff, T. E. 1996. The mass-to-light ratios of the Draco and Ursa Minor Dwarf Spheroidal Galaxies. II. The binary population and its effects on the measured velocity dispersions of dwarf spheroidals. *AJ*, **111**(Feb.), 750–767.

Oort, J. H. 1970. The formation of galaxies and the origin of the high-velocity hydrogen. *A&A*, **7**(Sept.), 381–404.

Palma, C., Majewski, S. R., Siegel, M. H., Patterson, R. J., Ostheimer, J. C., and Link, R. 2003. Exploring halo substructure with giant stars. IV. The extended structure of the Ursa Minor Dwarf Spheroidal Galaxy. *AJ*, **125**(Mar.), 1352–1372.

Peñarrubia, J., McConnachie, A. W., and Navarro, J. F. 2008a. The cold dark matter halos of Local Group dwarf spheroidals. *ApJ*, **672**(Jan.), 904–913.

Peñarrubia, J., Navarro, J. F., and McConnachie, A. W. 2008b. The tidal evolution of Local Group dwarf spheroidals. *ApJ*, **673**(Jan.), 226–240.

Piatek, S. and Pryor, C. 1995. The effect of galactic tides on the apparent mass-to-light ratios in dwarf spheroidal galaxies. *AJ*, **109**(Mar.), 1071–1085.

Piatek, S., Pryor, C., Olszewski, E. W., Harris, H. C., Mateo, M., Minniti, D., and Tinney, C. G. 2003. Proper motions of dwarf spheroidal galaxies from Hubble Space Telescope imaging. II. Measurement for Carina. *AJ*, **126**(Nov.), 2346–2361.

Pietrzyński, G., and 10 colleagues. 2008. The Araucaria Project: the distance to the Sculptor Dwarf Spheroidal Galaxy from infrared photometry of RR Lyrae stars. *AJ*, **135**(June), 1993–1997.

Pryor, C. 1996. Models of dwarf galaxy destruction. *Formation of the Galactic Halo... Inside and Out*, **92**(Apr.), 424–433.

Putman, M. E., and 25 colleagues 1998. Tidal disruption of the Magellanic Clouds by the Milky Way. *Nature*, **394**(Aug.), 752–754.

Putman, M. E., Staveley-Smith, L., Freeman, K. C., Gibson, B. K., and Barnes, D. G. 2003. The Magellanic Stream, high-velocity clouds, and the Sculptor Group. *ApJ*, **586**(Mar.), 170–194.

Reid, M. J., and 13 colleagues 2009. Trigonometric parallaxes of massive star-forming regions. VI. Galactic structure, fundamental parameters, and noncircular motions. *ApJ*, **700**(July), 137–148.

Richstone, D. O. and Tremaine, S. 1986. Measuring mass-to-light ratios of spherical stellar systems by core fitting. *AJ*, **92**(July), 72–74.

Roberts, M. S. 1962. A study of neutral hydrogen in IC 10.. *AJ*, **67**, 431–436.

Rood, H. J., Page, T. L., Kintner, E. C., and King, I. R. 1972. The structure of the Coma Cluster of Galaxies. *ApJ*, **175**(Aug.), 627–648.

Rubin, V. C., Thonnard, N., and Ford, W. K., Jr. 1978. Extended rotation curves of high-luminosity spiral galaxies. IV – Systematic dynamical properties, SA through SC. *ApJ*, **225**(Nov.), L107–L111.

Sánchez-Salcedo, F. J. and Hernandez, X. 2007. Masses, tidal radii, and escape speeds in dwarf spheroidal galaxies under MOND and dark halos compared. *ApJ*, **667**(Oct.), 878–890.

Sánchez-Salcedo, F. J., Reyes-Iturbide, J., and Hernández, X. 2006. An extensive study of dynamical friction in dwarf galaxies: the role of stars, dark matter, halo profiles and MOND. *MNRAS*, **370**(Aug.), 1829–1840.

Sandage, A. 1986. The redshift-distance relation. IX – Perturbation of the very nearby velocity field by the mass of the Local Group. *ApJ*, **307**(Aug.), 1–19.

Sanders, R. H. and McGaugh, S. S. 2002. Modified Newtonian dynamics as an alternative to dark matter. *ARA&A*, **40**, 263–317.

Sanders, R. H. and Verheijen, M. A. W. 1998. Rotation curves of Ursa Major Galaxies in the context of modified Newtonian dynamics. *ApJ*, **503**(Aug.), 97–108.

Savage, C., Freese, K., and Gondolo, P. 2006. Annual modulation of dark matter in the presence of streams. *Phys. Rev. D*, **74**(Aug.), 043531–043539.

Saviane, I., Monaco, L., and Hallas, T. 2010. Morphological transformation of NGC 205? *IAU Symposium*, **262**(Apr.), 426–427. Eds. G. Bruzual and S. Charlot.

Seigar, M. S., Barth, A. J., and Bullock, J. S. 2008. A revised ΛCDM mass model for the Andromeda Galaxy. *MNRAS*, **389**(Oct.), 1911–1923.

Sellwood, J. A. and Pryor, C. 1998. Pulsation modes of spherical stellar systems. *Highlights of Astronomy*, **11**, 638–648.

Shapley, H. 1938. A stellar system of a new type. *Harvard College Observatory Bulletin*, **908**(Mar.), 1–11.

Shapley, H. 1939. Galactic and extragalactic studies, II. Notes on the peculiar stellar systems in Sculptor and Fornax. *Proceedings of the National Academy of Science*, **25**(Nov.), 565–569.

Shectman, S. A., Landy, S. D., Oemler, A., Tucker, D. L., Lin, H., Kirshner, R. P., and Schechter, P. L. 1996. The Las Campanas Redshift Survey. *ApJ*, **470**(Oct.), 172–188.

Shuter, W. L. H. 1992. A new tidal model of the Magellanic Stream. *ApJ*, **386**(Feb.), 101–105.

Simon, J. D. and Geha, M. 2007. The kinematics of the ultra-faint Milky Way satellites: solving the missing satellite problem. *ApJ*, **670**(Nov.), 313–331.

Skillman, E. D., Côté, S., and Miller, B. W. 2003. Star formation in Sculptor group dwarf irregular galaxies and the nature of "transition" galaxies. *AJ*, **125**(Feb.), 593–609.

Sofue, Y. 1994. Fate of the Magellanic Stream. *PASJ*, **46**(Aug.), 431–440.

Sohn, S. T., and 9 colleagues. 2007. Exploring halo substructure with giant stars. X. Extended dark matter or tidal disruption?: The case for the Leo I Dwarf Spheroidal Galaxy. *ApJ*, **663**(July), 960–989.

Somerville, R. S. 2002. Can photoionization squelching resolve the substructure crisis? *ApJ*, **572**(June), L23–L26.

Sparke, L. S. and Gallagher, J. S., III 2000. Galaxies in the universe : an introduction. *Galaxies in the Universe*, Cambridge, UK: Cambridge University Press

Staveley-Smith, L., Kim, S., Calabretta, M. R., Haynes, R. F., and Kesteven, M. J. 2003. A new look at the large-scale HI structure of the Large Magellanic Cloud. *MNRAS*, **339**(Feb.), 87–104.

Strauss, M. A., and 35 colleagues. 2002. Spectroscopic target selection in the Sloan Digital Sky Survey: the main galaxy sample. *AJ*, **124**(Sept.), 1810–1824.

Strigari, L. E., Bullock, J. S., and Kaplinghat, M. 2007. Determining the nature of dark matter with astrometry. *ApJ*, **657**(Mar.), L1–L4.

Strigari, L. E., Bullock, J. S., and Kaplinghat, M. Determining the nature of dark matter with astrometry. *in preparation*.

Strigari, L. E., Bullock, J. S., Kaplinghat, M., Simon, J. D., Geha, M., Willman, B., and Walker, M. G. 2008. A common mass scale for satellite galaxies of the Milky Way. *Nature*, **454**(Aug.), 1096–1097.

Tanaka, K. I. 1981. The Magellanic Stream and the interacting galaxies. *PASJ*, **33**, 247–255.

van der Marel, R. P., Cretton, N., de Zeeuw, P. T., and Rix, H.-W. 1998. Improved evidence for a black hole in M32 from HST/FOS spectra. II. Axisymmetric dynamical models. *ApJ*, **493**(Jan.), 613–620.

van Kuilenburg, J. 1972. A systematic search for high-velocity hydrogen outside the Galactic plane II. *A&A*, **16**(Jan.), 276–281.

Vivas, A. K., and 35 colleagues. 2001. The QUEST RR Lyrae Survey: Confirmation of the Clump at 50 Kiloparsecs and Other Overdensities in the Outer Halo. *ApJ*, **554**(June), L33–L36.

Vogt, S. S., Mateo, M., Olszewski, E. W., and Keane, M. J. 1995. Internal kinematics of the Leo II dwarf spherodial galaxy. *AJ*, **109**(Jan.), 151–163.

Walker, M. G., Mateo, M., Olszewski, E. W., Gnedin, O. Y., Wang, X., Sen, B., and Woodroofe, M. 2007. Velocity dispersion profiles of seven dwarf spheroidal galaxies. *ApJ*, **667**(Sept.), L53–L56.

Wannier, P. and Wrixon, G. T. 1972. An unusual high-velocity hydrogen feature. *ApJ*, **173**(May), L119–L123.

Wayte, S. R. 1991. Review: the interacting Magellanic System. *The Magellanic Clouds*, **148**, 447–452.

Westfall, K. B., Majewski, S. R., Ostheimer, J. C., Frinchaboy, P. M., Kunkel, W. E., Patterson, R. J., and Link, R. 2006. Exploring halo substructure with giant stars. VIII. The extended structure of the Sculptor dwarf spheroidal galaxy. *AJ*, **131**(Jan.), 375–406.

Willman, B., and 14 colleagues. 2005. A new Milky Way dwarf galaxy in Ursa Major. *ApJ*, **626**(June), L85–L88.

Wilson, A. G. 1955. Sculptor-type systems in the Local Group of galaxies. *PASP*, **67**(Feb.), 27–29.

Wright, A. E. 1972. Computational models of gravitationally interacting galaxies. *MNRAS*, **157**, 309–315.

York, D. G., and 144 colleagues. 2000. The Sloan Digital Sky Survey: technical summary. *AJ*, **120**(Sept.), 1579–1587.

Young, L. M. and Lo, K. Y. 1997. The neutral interstellar medium in nearby dwarf galaxies. III. Sagittarius DIG, LGS 3, and PHOENIX. *ApJ*, **490**(Dec.), 710–718.

Zucker, D. B., and 31 colleagues 2006a. A curious Milky Way satellite in Ursa Major. *ApJ*, **650**(Oct.), L41–L44.

Zucker, D. B., and 32 colleagues. 2006b. A new Milky Way Dwarf satellite in Canes Venatici. *ApJ*, **643**(June), L103–L106.

3. Notes on the missing satellites problem

JAMES BULLOCK

The Missing Satellites Problem (MSP) broadly refers to the overabundance of predicted Cold Dark Matter (CDM) sub-halos compared to satellite galaxies known to exist in the Local Group. The most popular interpretation of the MSP is that the smallest dark matter halos in the universe are extremely inefficient at forming stars. The question from that standpoint is to identify the feedback source that makes small halos dark and to identify any obvious mass scale where the truncation in the efficiency of galaxy formation occurs.

Among the most exciting developments in near-field cosmology in recent years is the discovery of a new population satellite galaxies orbiting the Milky Way and M31. Wide field, resolved star surveys have more than doubled the dwarf satellite count in less than a decade, revealing a population of ultrafaint galaxies that are less luminous that some star clusters. For the first time, there are *empirical* reasons to believe that there really are more than 100 missing satellite galaxies in the Local Group, lurking just beyond our ability to detect them, or simply inhabiting a region of the sky that has yet to be surveyed.

Remarkably, both kinematic studies and completeness-correction studies seem to point to a characteristic potential well depth for satellite sub-halos that is quite close to the mass scale where photoionization and atomic cooling should limit galaxy formation. Among the more pressing problems associated with this interpretation is to understand the selection biases that limit our ability to detect the lowest mass galaxies. The least massive satellite halos are likely to host stealth galaxies with very low surface brightness and this may be an important limitation in the hunt for low-mass fossils from the epoch of reionization.

3.1 Perspective and historical context

This chapter focuses on a potential problem for the standard model of structure formation, but before delving into this issue in depth, it is worth providing a bit of perspective. Cosmological observations ranging from the cosmic microwave background (Komatsu *et al.*, 2011) to large-scale galaxy clustering (Reid *et al.*, 2010) to the Lyman-α forest (Viel *et al.*, 2008) all point to a concordance cosmological model built around Dark Energy + Cold Dark Matter (ΛCDM). Specifically, direct constraints on the power spectrum of density fluctuations suggest that structure formation in the universe is hierarchical (Press and Schechter, 1974; White and Rees, 1978), at least down to the scale of large galaxies. This is a remarkable fact and, for those of us interested in galaxy formation, it provides some confidence that we are not wasting our time attempting to build galaxies within the scaffolding provided by a CDM-based cosmology (Blumenthal *et al.*, 1984; Davis *et al.*, 1985). On still smaller scales, stellar streams, and clumps appear to fill the stellar halos of the Milky Way and M31 (Ivezić *et al.*, 2000; Ferguson *et al.*, 2002; Newberg *et al.*, 2002; Majewski *et al.*, 2003; Zucker *et al.*, 2004; Guhathakurta *et al.*, 2006; Ibata *et al.*, 2007; Bell *et al.*, 2008; McConnachie *et al.*, 2009; Watkins *et al.*, 2009) in accordance with expectations from ΛCDM-based models of dwarf galaxy accretion and disruption (Bullock *et al.*, 2001; Bullock and Johnston, 2005; Cooper *et al.*, 2010). Even if the underlying makeup of the universe is somehow different from this, it must look very much like ΛCDM from the scale of the horizon down to $\sim$50 kpc or so. It is within this context that we consider the Missing Satellites Problem in CDM.

A defining characteristic of CDM-based hierarchical structure formation is that dark matter halos are assembled via mergers with smaller systems. In the standard framework of CDM, density inhomogeneities in the early universe are small compared to unity and characterized by a Gaussian distribution with variance σ. If we consider density fluctuations within spheres of radius R and associated mass $M \propto R^3$, we can characterize the linear fluctuation amplitude as

$$\sigma(M, a) \propto D(a)\, F(M) \tag{3.1}$$

where $a = (1+z)^{-1}$ is the expansion factor and $D(a)$ quantifies the gravitational amplification of fluctuations, which grow as $D(a) \propto a$ at early times when the universe is near critical density. CDM predicts that $F(M)$ bends as $F \propto M^{-\alpha}$ with $\alpha = 2/3$ on large scales to $\alpha \to 0$ (logarithmically) on the smallest mass scales (Blumenthal *et al.*, 1984). At any given redshift, the typical mass-scale that is collapsing, $M_*(a)$, can be estimated by determining the non-linear scale via $\sigma(M_*, a) \sim 1$, which implies $M_*(z) \sim D(a)^{1/\alpha}$. This means low-mass scale fluctuations eventually break off from the expansion first. More massive objects are assembled later (at least in part) by many smaller-scale pre-collaps for example, ed regions (Press and Schechter, 1974). The normalization of the power spectrum is often quantified by $\sigma_8 = \sigma(M = M_8, a = 1)$, where M_8 is the mass enclosed within a sphere of radius $8h^{-1}$ Mpc. Structure of a given mass tends to collapse earlier for a higher value of σ_8.

One implication of this hierarchical scenario is that small dark matter halos collapse at high redshift, when the universe is very dense, and as a result have high density concentrations. When these halos merge into larger hosts, their high core densities allow them to resist the strong tidal forces that would otherwise act to destroy them. While gravitational interactions act to strip mass from merged progenitors, a significant fraction of these small halos survive as bound substructures (see e.g, Zentner and Bullock, 2003).

Kauffmann *et al.* (1993) were the first to show using a self-consistent treatment that substructure would likely survive the merging process with abundance in CDM halos. They used a semi-analytic model to show that sub-halos within cluster-size dark matter halos will have approximately the correct number and mass spectrum to account for cluster *galaxies*. On the other hand, the same calculation demonstrated that Milky Way size halos should host a large number of satellite sub-halos, with over ~100 objects potentially massive enough to host observable satellite galaxies (with $L > 10^6\ L_\odot$), even when allowing for fairly substantial supernovae feedback. Given that there are only ~10 satellites brighter than this around the Milky Way, the authors concluded that dwarf galaxy formation would need to be suppressed by some large factor in order to explain the discrepancy. Nevertheless, this dwarf satellite issue did not gain significant attention in the community until these initial semi-analytic estimates were (more or less) verified by direct numerical calculations some ~5 years later (Klypin *et al.*, 1999a; Moore *et al.*, 1999). It was only then that the Missing Satellites Problem (MSP) gained urgency.

Implicit in the title Missing Satellites is the notion of prediction – the satellites are expected by theory but are not seen. It is not surprising then that the problem was named by theorists.[1] Klypin *et al.* (1999a) titled their seminal paper on the topic "Where are the Missing Galactic Satellites?" The notion is that CDM is a successful theory and that the substructure prediction is robust and to be taken seriously. A similar sentiment was echoed in the nearly coeval paper by Moore *et al.* (1999): This is a robust prediction of CDM, so why don't we see more dwarfs? It is indeed the robustness of the Klypin *et al.* (1999a) and Moore *et al.* (1999) calculations that demanded attention to the problem that was sketched by Kauffmann *et al.* (1993). More recent simulations, now with a factor of ~10^3 more particles, have effectively confirmed these first calculations

[1] A strict empiricist may have called it "The Overabundant Substructure Problem," but this is not what the MSP is now called even though many may interpret it this way.

(Diemand *et al.*, 2008; Springel *et al.*, 2008). The mass function of substructure is predicted to rise steeply to the smallest masses, while the luminosity function of observed dwarf satellites is fairly flat. In this chapter, I touch on some of the physical mechanisms that may act to suppress galaxy formation within the smallest halos but focus primarily on the best way to compare predictions with observations. Specifically, the zeroth order concern for any ΛCDM (or variant) model that wishes to match the Local Group satellite population is to correctly reproduce the mass-luminosity relationship obeyed by faint galaxies. This necessarily relies on mass determinations. Once we have robust mass determinations for the dwarfs, we can begin to test specific models for their formation and evolution. Section 3.4 provides an overview of recent efforts to constrain the mass-luminosity relation for dwarf galaxies.

The most important observational development in MSP studies in the last decade has been the discovery of a new population of faint satellite galaxies. Approximately 25 new dwarf galaxy companions of the Milky Way and M31 have been discovered since 2004, more than doubling the known satellite population in the Local Group in five years (Willman *et al.*, 2005; Grillmair, 2006, 2009; Zucker *et al.*, 2006; Belokurov *et al.*, 2007, 2009; Majewski *et al.*, 2007; Martin *et al.*, 2009; Willman, 2010). The majority of these newly discovered dwarfs are less luminous than any galaxy previously known. The most extreme of these, the ultrafaint Milky Way dwarfs, have luminosities[2] smaller than an average globular cluster $L \simeq 10^2$–10^4 $L_\odot$ and were discovered by searches for stellar overdensities in the wide-field maps of the Sloan Digital Sky Survey (SDSS) and the Sloan Extension for Galactic Understanding and Exploration (SEGUE). Follow-up kinematic observations showed that these tiny galaxies have surprisingly high stellar velocity dispersions for their luminosities and sizes ($\sigma_* \sim 5$ km s^{-1}; Martin *et al.*, 2007; Simon and Geha, 2007; Simon *et al.*, 2011) and subsequent mass modeling has shown that they are the most dark matter dominated galaxies known (Wolf *et al.*, 2010; Martinez *et al.*, 2011). Remarkably, these extreme systems are not only the faintest, most dark matter dominated galaxies in the universe but they are also the most metal-poor stellar systems yet studied (Kirby *et al.*, 2008; Geha *et al.*, 2009). All of the new dwarfs, with the exception of Leo T at a distance of more than 400 kpc, have negligible atomic hydrogen fractions (Grcevich and Putman, 2009). Perhaps the most exciting aspect of these recent discoveries is that completeness corrections point to a much larger population of undiscovered dwarfs (Koposov *et al.*, 2008; Tollerud *et al.*, 2008; Walsh *et al.*, 2009; Bullock *et al.*, 2010). For the first time we have empirical evidence to suggest that there really are missing satellite galaxies in the halo of the Milky Way and this is the subject of Section 3.5. These missing objects are likely extremely dark matter dominated, but too dim and/or too diffuse to have been discovered (yet).

3.2 Counting and defining dark matter halos in CDM

3.2.1 *Galaxy halos*

The standard paradigm of galaxy formation today posits that all (or at least an overwhelming majority) of galaxies reside near the centers of extended dark matter halos. The outer edges of the halos are somewhat arbitrarily defined, but a typical convention is that halos *in the field* are characterized by a radius $R_{\rm h}$ and mass $M_{\rm h}$ that obey $M_{\rm h} = \Delta_{\rm h}\, \rho_m\, (4\pi/3)\, R_{\rm h}^3$, such that average density within the halo exceeds the background ρ_m by a factor $\Delta_{\rm h} \sim 200$, which is motivated by the virialized overdensity obtained in approximate spherical collapse models (see, e.g., Bryan and Norman, 1998).[3]

[2] We will assume V-band luminosities throughout.

[3] For our purposes the precise value of $\Delta_{\rm h}$ does not matter, but in principle one needs to be careful about the definition when doing comparisons, especially when considering halos more massive than $M_{\rm h} \sim M_*$.

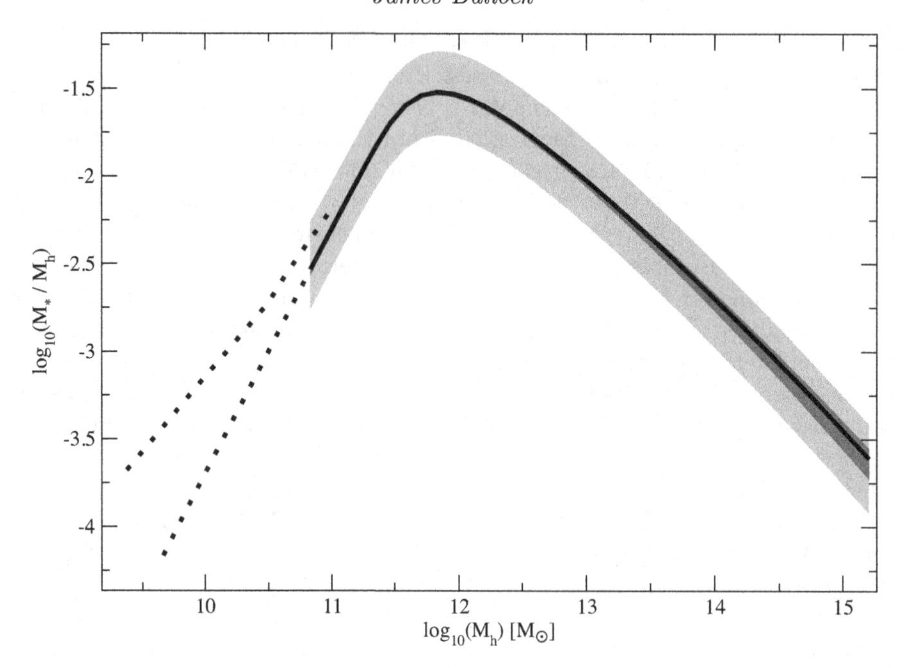

FIG. 3.1. Relationship between stellar mass M_* and halo mass M_h as quantified by the abundance matching analysis of Behroozi *et al.* (2010), which includes uncertainties associated with observed stellar mass functions and possible scatter in the relationship between halo mass and stellar mass (shaded band). The lower dashed line is the extrapolated relationship that results from assuming the stellar mass function continues as a power law for $M_* < 10^{8.5}$ $M_\odot$, using data from the full SDSS (Li *et al.*, 2009). The upper dashed line is extrapolated under the assumption that the slope of the stellar mass function has an upturn at low masses to $\alpha = -1.8$, as found by (Baldry *et al.*, 2008).

With mass defined in this way, N-body simulations reveal that the halo mass function is remarkably well characterized by (see Tinker *et al.*, 2008, and references therein)

$$\frac{dn}{dM_h} = f_h(\sigma) \frac{\rho_m}{M_h^2} \frac{d \ln \sigma^{-1}}{d \ln M_h} \tag{3.2}$$

where $f_h(\sigma)$ is effectively a fitting function that approaches a constant at low mass ($M_d \ll M_*$ or $\sigma \gg 1$) and becomes exponentially suppressed for halos that are too large to have collapsed ($M_d \gg M_*$ or $\sigma \ll 1$).[4] One important observation (which has been recognized at least since White and Rees, 1978) is that the mass function rises steadily toward smaller masses $dn/dM \sim M^{-2}$, while the luminosity function or baryonic mass function of galaxies in the real universe does not rise as steeply. Clearly a one-to-one correspondence between mass and luminosity must break down at small masses.

A more precise way of characterizing the relationship between galaxies and their halos comes from the technique known as abundance matching. One simply asks the question: what mass-luminosity relationship must I impose in order to reproduce the observed luminosity function of galaxies? Remarkably, two-point clustering statistics of bright galaxies can be explained fairly well under the simple assumption that galaxy luminosity L or stellar mass M_* maps to dark matter halo mass M_h in a nearly monotonic way (see Kravtsov *et al.*, 2004a; Conroy and Wechsler, 2009; Behroozi *et al.*, 2010; Moster *et al.*, 2010). Figure 3.1 illustrates the required mapping by showing the ratio of galaxy stellar

[4] Note that the *linear* mass variable M in $\sigma(M)$ has been somewhat magically replaced by the virial mass variable $M \to M_h$ in this equation. It is not outrageous to think that the two should be related, but an exact correspondence would be bizarre given the simplicity of spherical collapse. The function $f_h(\sigma)$ can be regarded as a fudge factor that makes up for this difference.

mass to halo virial mass ($M_*/M_{\rm h}$) as a function of halo mass (Behroozi, 2010, private communication). Dark gray shading indicates statistical and sample variance errors and the light gray shading includes systematic errors. The shaded band is truncated at small stellar masses where the mass functions become incomplete in the Sloan Digital Sky Survey ($M_* \simeq 10^{8.5}$ $M_\odot$). It is clear from this figure that the efficiency of galaxy formation must peak in halos with masses $\sim 10^{12}$ $M_\odot$, and to become increasingly inefficient toward smaller and larger halo masses. An understanding of this behavior – how and why it happens – remains a fundamental goal in galaxy formation.

The two dashed lines in Figure 1 bracket reasonable low-mass extrapolations of the relationship defined at larger masses, with the upper line defined by a power-law extrapolation of the observed stellar mass function from Li *et al.* (2009) and the lower dashed corresponding to a case where the stellar mass function has an upturn, as advocated by Baldry *et al.* (2008). At small masses, the implied range of scalings is $M_* \propto M_h^p$ with $p \simeq$ 2.5–1.8. For example, at the edge of the figure, the power-law extrapolation suggests that $M_{\rm h} = 10^9$ $M_\odot$ halos should host galaxies with stellar masses between $M_* \simeq 5 \times 10^4$ $M_\odot$ and $M_* \simeq 3 \times 10^5$ $M_\odot$. As we shall see, halo masses of $M_{\rm h} \sim 10^9$ $M_\odot$ ($V_{\rm max} \simeq 30$ km s^{-1}) are approximately those required to match the kinematics of $L \sim 10^5$ $L_\odot$ dwarf galaxies (Strigari *et al.*, 2008). Of course, there is no real physics in this extrapolation, but it is encouraging that when one imposes this simple, empirically motivated scaling, we get numbers that are not far off from those required to match dwarf satellite observations (Busha *et al.*, 2010; Kravtsov, 2010). A more general comparison of abundance-matching masses and kinematically derived masses is presented in Tollerud *et al.* (2011).

3.2.2 *Sub-halos*

Galaxies that are as faint as $L \sim 10^5$ $L_\odot$ are very difficult to detect, and, as we discuss in Section 3.4.2, we are only reasonably confident in our census of these objects to distances comparable to the expected virial radius of the Milky Way's halo, and within small, deeply surveyed regions within the virial radius of M31 (Ibata *et al.*, 2007; Majewski *et al.*, 2007; Martin *et al.*, 2009; McConnachie *et al.*, 2009). In addition, these galaxies appear to be embedded within dark matter halos (Mateo, 1998; Simon and Geha, 2007; Walker *et al.*, 2009b; Collins *et al.*, 2010; Kalirai *et al.*, 2010; Wolf *et al.*, 2010). The implication is that in order to confront the faintest galaxies theoretically in the context of ΛCDM, we must discuss them as embedded within sub-halos – dark matter halos that are bound and within $R_{\rm h}$ of a larger host halo.

Sub-halos (unlike field halos) are not naturally characterized by a virial mass, $M_{\rm h}$, because they tend to be truncated by tidal forces at radii smaller than their over-density-defined virial radii (e.g., Klypin *et al.*, 1999b; Kazantzidis *et al.*, 2004; Peñarrubia *et al.*, 2008). The most common way one characterizes a sub-halo is to use $V_{\rm max}$, the maximum value of the circular velocity curve $V_c(r) = (G\,M(r)\,/r)^{1/2}$, which itself is measured in spheres, stepping out from the sub-halo center of mass. The radius $r_{\rm max}$ where $V_{\rm max}$ occurs is typically set by tidal stripping, which more readily unbinds material from the outer parts of galaxy halos than the inner parts (Bullock and Johnston, 2007; Kazantzidis *et al.*, 2011). The choice of $V_{\rm max}$ as a quantifier is particularly nice because it is fairly robust to the details of the halo-finding algorithm used. Another choice is to assign sub-halos a mass directly. Unfortunately, *total* mass assignment to sub-halos is somewhat subjective and can result in differences from halo finder to halo finder, but reasonable choices involve (1) defining the sub-halo boundary at the radius where its density profile approximately reaches the mean background density of the host halo; (2) estimating a local tidal radius iteratively; or (3) removing unbound particles from the sub-halo iteratively in such a way that a boundary emerges fairly naturally (see, e.g., Springel *et al.*, 2008, Figure 14 and related discussion).

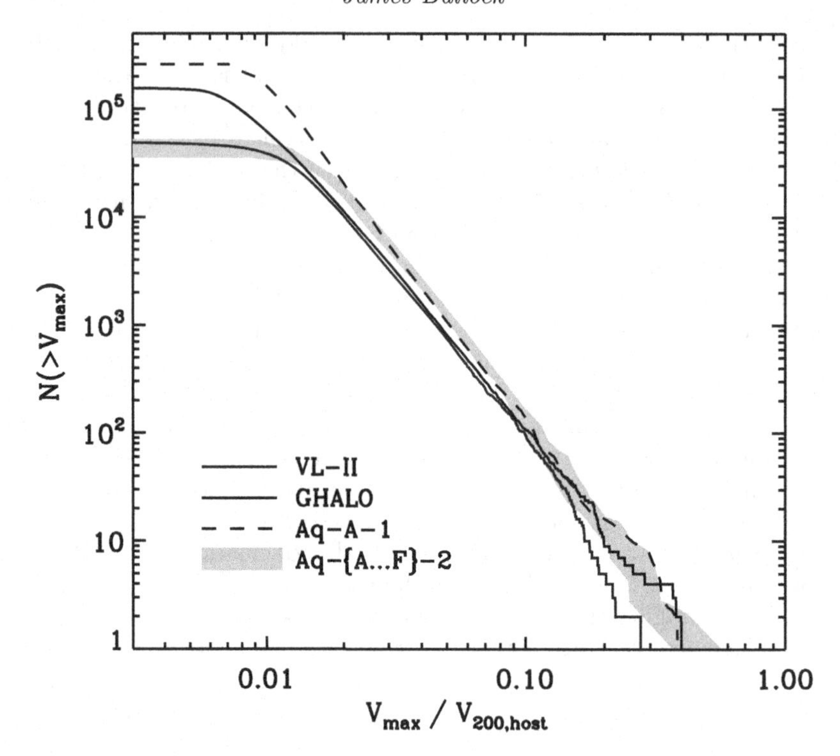

FIG. 3.2. Cumulative sub-halo $V_{\max}$ function within $R_h \simeq 400$ kpc for the three highest resolution simulations of Milky Way size halos from (Kuhlen, 2010, private communication), shown here relative to the circular velocity of each host at its outer radius $V_{\mathrm{h}} \simeq 140$ km s^{-1}. The solid gray line shows VL II sub-halos from Diemand *et al.* (2008), the dashed black line shows Aq-A sub-halos from Springel *et al.* (2008) and the solid dark line shows GHALO sub-halos from Stadel *et al.* (2009). The shaded band provides an estimate of halo-to-halo scatter from a series of lower resolution halos (Aq suite, Springel *et al.*, 2008). The small normalization difference between the GHALO/VL II and Aq halos is likely a result of their having different power spectrum normalizations (Zentner and Bullock, 2003). The most recent determination of the normalization WMAP-7 (Komatsu *et al.*, 2011) is intermediate between the adopted normalizations of the simulations shown here.

Figure 3.2 summarizes cumulative sub-halo $V_{\max}$ functions from the three highest resolution ΛCDM simulations of Milky Way size halos that have yet been run (VL II – Diemand *et al.*, 2008; Aq-A – Springel *et al.*, 2008; and GHALO – Stadel *et al.*, 2009; Kuhlen *et al.*, 2010). Each simulation contains $N_p \simeq 10^9$ particles within the halo radius $R_{\mathrm{h}} = R_{200}$[5] and the sub-halos plotted are those within R_{h}. Because the three halos have slightly different masses ($M_{\mathrm{h}} = 1.3$, 1.9, and 2.5×10^{12} $\mathrm{M}_{\odot}$, for GHALO, VL II, and Aq-A, respectively), the $V_{\max}$ values in Figure 3.2 have been scaled to $V_{\mathrm{h}} = V_{200} = (GM_{\mathrm{h}}/R_h)^{1/2}$ for each system, where $R_{\mathrm{h}} = 347$, 402, 433 kpc and $V_{\mathrm{h}} \simeq 127$, 142, and 157 km s^{-1}, for the three hosts. The shaded band summarizes the results of a series of lower resolution simulations ($N_p \simeq 2 \times 10^8$) presented in Springel *et al.* (2008). The flattening seen in the cumulative $V_{\max}$ functions are entirely a result of resolution limits, as can be seen by comparing the shaded band of lower resolution Aq halos to the dashed line, which is the highest resolution Aq halo.

The most striking feature of the comparison shown in Figure 3.2 is that the simulation lines basically agree. The small normalization shift toward higher number counts for the Aq halos is likely due to the fact that the Aq runs use a higher normalization ($\sigma_8 = 0.9$) than the VL-II and GHALO runs ($\sigma_8 = 0.74$). The difference seen is in line with the

[5] Defined via $\Delta = 200$ over the *background* density.

analytic expectations of Zentner and Bullock (2003), who explored how changes to the power spectrum should affect substructure counts. Note that the most recent WMAP-7 preferred value for the normalization is $\sigma_8 = 0.81$, which is intermediate between the two sets of simulations shown here. Overall, the $V_{\rm max}$ functions are well-fit by a power-law:

$$N(> V_{\rm max}) \simeq 0.15 \left(\frac{V_{\rm max}}{V_{\rm h}} \right)^{-2.94} \tag{3.3}$$

where the normalization has been chosen to be intermediate between the two sets of simulations shown. For a Milky Way size host ($V_{\rm h} \simeq 140$ km s^{-1}; Xue *et al.*, 2008) we expect $N \simeq 350$ sub-halos with $V_{\rm max} > 10$ km s^{-1} within $R_{\rm d} \simeq 400$ kpc of the Milky Way. This number matches well with the predictions from the first MSP papers more than a decade ago (Klypin *et al.*, 1999a; Moore *et al.*, 1999). What we also see is that the count is expected to rise for smaller halos and there is no sign of a physical break. The highest resolution simulation shown in Figure 3.2 has already resolved more than 10^5 sub-halos, and there is no reason to expect the numbers to stop rising in the absence of another scale in the problem.

The only other obvious scale in CDM is the filtering mass in the power spectrum, where fluctuations are suppressed on the scale of the comoving horizon at the time of kinetic decoupling of the dark matter in the early universe (Loeb and Zaldarriaga, 2005; Bertschinger, 2006):

$$M_{\rm min} = 10^{-4} \left(\frac{T_d}{10\,{\rm MeV}} \right)^{-3} {\rm M}_{\odot}. \tag{3.4}$$

The kinetic decoupling temperature T_d depends on the scattering interactions of dark matter with standard model fermions, and is therefore sensitive to the microphysical dark matter model itself (Profumo *et al.*, 2006). Martinez *et al.* (2009) showed that the popular Constrained Minimal Supersymmetric Standard Model (CMSSM) with neutralino dark matter allows $M_{\rm min} \simeq 10^{-9}$–$10^{-6}$ M$_{\odot}$, after accounting for the entire presently viable parameter space of the CMSSM, including low-energy observables, the relic abundance (ρ_m), and direct constraints on standard model quantities. The implication is that there is good reason to suspect that the mass spectrum of substructure in the Milky Way halo continues to rise down to the scale of an Earth mass, and even smaller.

Figure 3.3 provides a cartoon-style estimate of the total substructure abundance within the Milky Way's halo down to the scale of the minimum mass. Plotted is the $N(> V_{\rm max})$ relation from Equation 3.3 for a Milky Way size host. The line becomes dashed where we are extrapolating the power-law beyond the resolving power of state-of-the-art numerical simulations. The upper axis provides a mass scale using bound-mass to $V_{\rm max}$ relation for sub-halos found by Springel *et al.* (2008): $M_{\rm sub} = 3.4 \times 10^7 {\rm M}_{\odot} (V_{\rm max}/10\ {\rm km\,s}^{-1})^{3.5}$. Figure 3.3 is divided into four vertical bands. The first vertical band provides an indication of where we expect this power-law to break. Here, I have taken the $M_{\rm min}$ range from Martinez *et al.* (2009) and expanded it slightly to allow for uncertainties associated with going from the linear power spectrum to sub-halos, and to qualitatively allow for more freedom in the dark matter model (e.g., Profumo *et al.*, 2006). For comparison, the third band shows the mass (or velocity) scale that has been directly resolved in simulations and the last vertical band on the far right indicates the current range of $V_{\rm max}$ values that are consistent with kinematic constraints from observed Milky Way satellite galaxies (as discussed later and in Peñarrubia *et al.*, 2008).

Based on this extrapolated power-law and the range of minimum sub-halo masses, we expect a total of $\sim 10^{11}$ to 10^{17} bound sub-halos to exist within 400 kpc of the Sun! Compared to the total number of satellite galaxies known ($\sim$20; lower dashed line) or even an observational completeness corrected value ($\sim$400; upper dashed line from

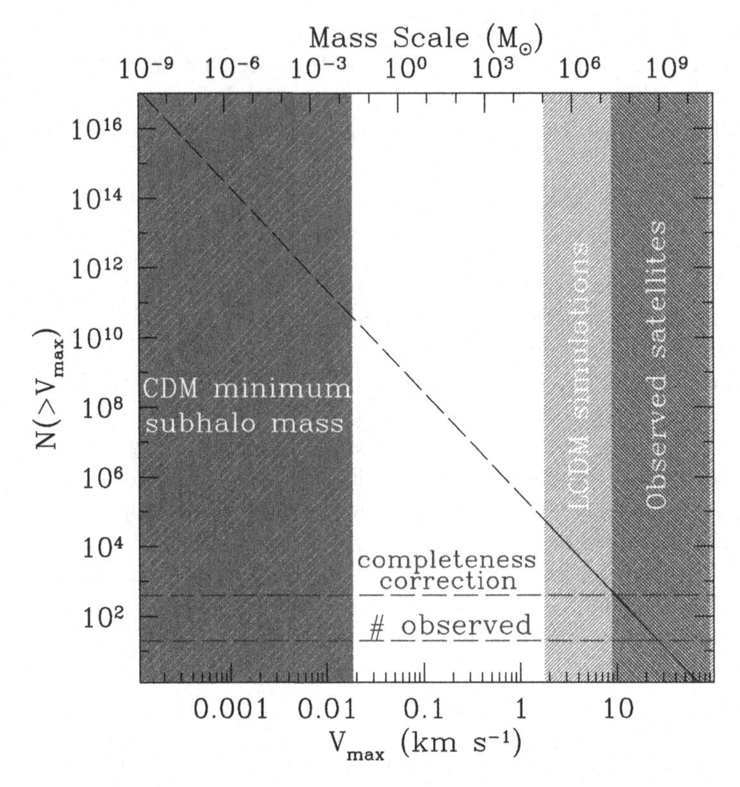

FIG. 3.3. Illustrative sketch of the expected cumulative CDM substructure abundance within the Milky Way's halo. The line is Equation 3.3 for a host halo of circular velocity $V_{\rm h} = 140$ km s^{-1} at $R_{\rm h} = 400$ kpc (for which we would expect a maximum circular velocity of about 200 km s^{-1}). The line becomes dashed where we are extrapolating the power-law beyond the resolving power of state-of-the-art numerical simulations. The first vertical band provides an indication of where we expect this power-law to break for popular CDM dark matter candidates. The lower horizontal dashed line shows the number of Milky Way satellite galaxies known while the upper dashed line is an estimate of the total number of satellite galaxies that exist within 400 kpc of the Milky Way corrected for luminosity-bias and sky coverage limitations of current surveys (Tollerud *et al.*, 2008). The fact that the upper horizontal line intersects the edge of the third vertical band at about the location of the CDM prediction is quite encouraging for the theory.

Tollerud *et al.*, 2008), we are very far from discovering all of the sub-halos that are expected in CDM.

The most extreme way one might characterize the MSP is to say that CDM predicts $>10^{11}$ sub-halos whereas we observe only $\sim 10^2$ satellite galaxies. Although this statement is true, it should not worry anyone because the vast majority of those CDM sub-halos are less massive than the Sun. No one expects a satellite galaxy to exist within a dark matter halo that is less massive than a single star. We it know that galaxy formation must truncate in dark matter structures smaller than some mass.[6] The question is what is that mass? Is there a sharp threshold mass, below which galaxy formation ceases? Can we use the Local Group satellite population as a laboratory for galaxy formation on small scales?

Physical processes like supernova feedback ($V_{\rm max} \sim 100$ km s^{-1}), heating from photoionization ($V_{\rm max} \sim 30$ km s^{-1}), or the ability for gas to cool ($V_{\rm max} \sim 15$ km s^{-1} for Lyman-alpha and ~ 5 km s^{-1} for H_2) each impose a different mass scale of relevance (Dekel and Silk, 1986; Efstathiou, 1992; Tegmark *et al.*, 1997; Bullock *et al.*, 2000;

[6] A conservative requirement would be $M_{\rm h} > (1/f_b){\rm M}_\odot \sim 5\ {\rm M}_\odot$ in order to host a "galaxy" containing one solar mass star.

Ricotti *et al.*, 2001). If, for example, we found evidence for very low-mass dwarf galaxies $V_{\rm max} \sim 5$ km s^{-1} then these would be excellent candidates for primordial H_2 cooling "fossils" of reionization in the halo (Madau *et al.*, 2008; Ricotti, 2010). We would like to measure the mass scale of the satellites themselves and determine whether the number counts of observed satellites are consistent with what is expected given their masses. Unfortunately, it turns out that this comparison is not as straightforward as one might naively expect.

3.3 Defining the problem

3.3.1 *Maximum circular velocity* $V_{\rm max}$

From the perspective of numerical simulations, a $V_{\rm max}$ function such as that shown in Figure 3.2 provides a natural and robust way to characterize the substructure content of a dark matter halo. Ideally, we would like to put the satellite galaxies of the Milky Way and M31 on this plot to make a direct comparison between theory and data. Unfortunately, $V_{\rm max}$ is not directly observable for the dwarf satellites. Instead, the stellar velocity dispersion, σ_*, is the most commonly derived kinematic tracer for dwarf spheroidal (dSph) satellite galaxies in the Local Group.[7] dSph velocity dispersion profiles tend to be flat with radius (Walker *et al.*, 2009a), but for concreteness, I will use σ_* to denote the luminosity-weighted line-of-sight velocity dispersion of the system.

The first MSP papers estimated $V_{\rm max}$ values for the dwarfs by assuming that their stellar velocity dispersion was equal to the dark matter velocity dispersion within the central regions of their host halos: $\sigma_* = \sigma_{\rm dm}^{\rm SIS} = V_{\rm max}/\sqrt{2}$ (Moore *et al.*, 1999) and $\sigma_* = \sigma_{\rm dm}^{\rm NFW} \simeq V_{\rm max}/\sqrt{3}$ (Klypin *et al.*, 1999a). This was a reasonable start, especially given that the main point was to emphasize an order-of-magnitude discrepancy in the overall count of objects. Nevertheless, for detailed comparisons it is important to realize $\sigma_* \neq \sigma_{\rm dm}$.[8] Specifically, since the stars are bound to the *central* regions of dark matter halos, while the dark matter particles can orbit to much larger radii, we generally expect $\sigma_* \lesssim \sigma_{\rm dm}$. This fact motivated the idea that the dwarf galaxies could be more massive than originally supposed and perhaps populate only the most massive sub-halos of the Milky Way (Stoehr *et al.*, 2002; Hayashi *et al.*, 2003; Kravtsov *et al.*, 2004b, who discussed a more nuanced model associated with mass prior to accretion). Below I demonstrate explicitly that the only relationship that can be made without imposing a theoretical model for sub-halo structure is $V_{\rm max} \geq \sqrt{3}\sigma_*$ (Wolf *et al.*, 2010). Cases where $V_{\rm max} \simeq \sqrt{3}\sigma_*$ correspond to those where the edge of the stellar distribution is very close to $r_{\rm max}$. Because $r_{\rm max}$ is typically set by tidal truncation in CDM halos, this last equivalence is expected to hold in cases where the stars have be affected or are beginning to be affected by tidal stripping/stirring in the Milky Way (as shown numerically by Kazantzidis *et al.*, 2011).

One of the problems with placing dSph data on a $V_{\rm max}$ function plot is that the mapping between σ_* and $V_{\rm max}$ depends sensitively on what one assumes about the density profile structure of the sub-halos. The stellar velocity dispersion data probe only the potential within the stellar extent, and one must extrapolate beyond that point in order to estimate $V_{\rm max}$. For the current population of well-studied Milky Way dSph galaxies (see Table 1

[7] The best studied dwarfs have individual velocity measurements for more than a thousand stars (Walker *et al.*, 2009a) and this allows for more nuanced studies of their kinematics involving velocity dispersion profiles, higher-order moments of the velocity distribution, and explorations of the underlying distribution function shape.

[8] This simple fact has caused a lot of confusion in the literature, especially when it relates to attempts to "measure" phase space densities of dark matter in dwarfs, which cannot be done without appealing to theoretical expectations for how and how far the dark matter halos extend beyond the stellar extent of the dwarfs.

in Wolf *et al.*, 2010), the median (3d) half-light radius is $r_{1/2} \simeq 300$ pc, while rotation curves of ΛCDM sub-halos with $V_{\rm max} > 15\ {\rm km\,s^{-1}}$ can peak at radii ranging from ∼1000 to 6000 pc depending on the value of σ_8 (Zentner and Bullock, 2003, Figure 21). For a typical dwarf, this amounts to a factor of ∼3 to ∼20 extrapolation in the assumed mass profile from the point where it is constrained by data (and it can be much worse for the smallest dwarfs). If we restrict ourselves to ΛCDM models with $\sigma_8 = 0.8$ we expect that *median* rotation curves will peak at radii[9]

$$r_{\rm max} \simeq 1100\,{\rm pc}\left(\frac{V_{\rm max}}{15\ {\rm km\,s^{-1}}}\right)^{1.35} . \tag{3.5}$$

Even with fixed σ_8, the extrapolations are large and sensitive to halo-to-halo scatter in profile parameters (the 68% scatter is about a factor of ∼2 in $r_{\rm max}$ at fixed $V_{\rm max}$ according to Springel *et al.*, 2008). For the most extreme cases of ultrafaint dwarf galaxies, the stellar extent is so small ($r_{1/2} \sim$ 30–40 pc) that we must rely on a factor of ∼35 extrapolation in r to reach a typical $V_{\rm max}$ value at $r_{\rm max} \simeq 1100$ pc. One begins to wonder whether there might be a better option than $V_{\rm max}$ for comparing theory to data.

3.3.2 *Kinematic mass determinations*

What quantity is best constrained by stellar velocity dispersion data? Consider a spherically symmetric galaxy that is in equilibrium with stellar density distribution $\rho_*(r)$ and radial velocity profile $\sigma_r(r)$ that is embedded within total mass profile $M(r)$. The Jeans equation is conveniently written as

$$M(r) = \frac{r\,\sigma_r^2}{G}\left(\gamma_* + \gamma_\sigma - 2\beta\right) \tag{3.6}$$

where $\beta(r) \equiv 1 - \sigma_t^2/\sigma_r^2$ characterizes the tangential velocity dispersion and $\gamma_* \equiv -{\rm d}\ln\rho_*/{\rm d}\ln r$ and $\gamma_\sigma \equiv -{\rm d}\ln\sigma_r^2/{\rm d}\ln r$. In principle, $\sigma_r(r)$ can be inferred from the observed line-of-sight velocity dispersion profile $\sigma_{\rm los}(R)$, but this geometric deconvolution depends on the unknown function $\beta(r)$. At this point one may begin to get worried: uncertainties in β will affect both the mapping from observed $\sigma_{\rm los}$ to σ_r *and* the derived relationship between $M(r)$ and σ_r in Equation 3.6. Given that the dSph galaxies are well outside the regime where we can legitimately assume that mass follows light, are we forced to assume that $\beta = 0$ in order to derive any meaningful mass constraint? Thankfully, no.

It turns out that there is a single radius where the degeneracy between stellar velocity dispersion anisotropy and mass is minimized. As shown analytically and numerically by Wolf *et al.* (2010), this radius is close to the 3D deprojected half light radius, $r_{1/2}$, for most galaxies, and it is the mass with this radius, $M_{1/2} = M(< r_{1/2})$, that is best constrained by kinematic data. Figure 3.4 demonstrates this by showing explicitly mass model constraints for the Carina galaxy. By exploring models with variable β values, several authors had seen similar results prior to the Wolf *et al.* (2010) derivation, but with more restrictive conditions and without a full exploration of the implications (van der Marel *et al.*, 2000; Strigari *et al.*, 2007a; Walker *et al.*, 2009b). Indeed, rather than extrapolate mass profiles to estimate a $V_{\rm max}$ value for each dwarf halo, Strigari *et al.* (2007b, 2008) advocated use of the integrated mass within a pre-defined radius that was similar in magnitude to the median $r_{1/2}$ value for the dwarf *population* (either 600 pc for large dwarfs, or 300 pc for smaller dwarfs) as a means of comparing data to simulations. I discuss these comparisons later, but first let me turn to a brief explanation of why the mass is best constrained within $r_{1/2}$.

Qualitatively, one might expect that the degeneracy between the integrated mass and the assumed anisotropy parameter will be minimized at some intermediate radius within

[9] Estimated from the results of Springel *et al.* (2008) and Diemand *et al.* (2008) by chosing an intermediate normalization.

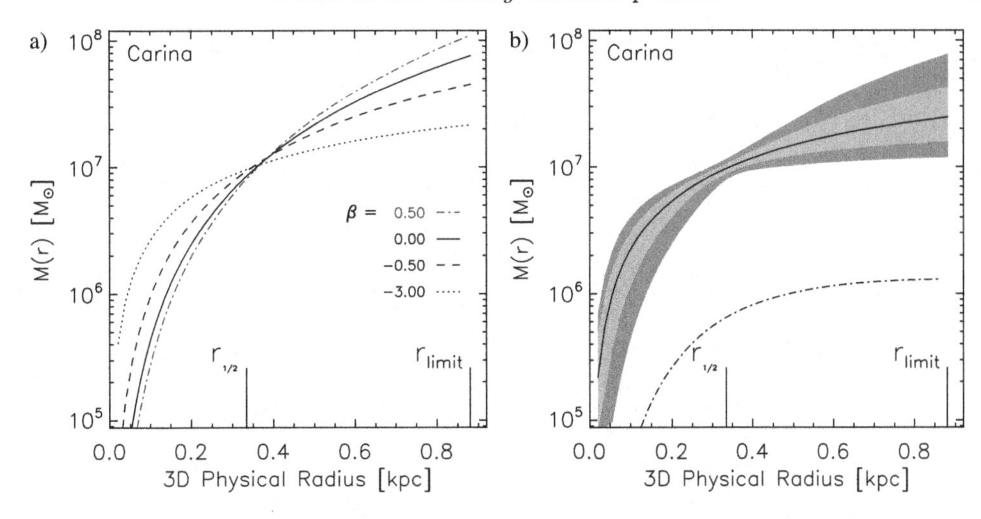

FIG. 3.4. (a) The cumulative mass profile generated by analyzing the Carina dSph using four different constant velocity dispersion anisotropies. The lines represent the median cumulative mass value from the likelihood as a function of physical radius. (b) The cumulative mass profile of the same galaxy, where the black line represents the median mass from our full mass likelihood (which allows for a radially varying anisotropy). The different shades represent the inner two confidence intervals (68% and 95%). The dot-dashed line represents the contribution of mass from the stars, assuming a stellar V-band mass-to-light ratio of 3 $M_\odot/L_\odot$. This figure is from Wolf *et al.* (2010).

the stellar distribution. Think about the line-of-sight velocity dispersion measured along the projected center ($R = 0$) and then at the far edge $R = r_{\rm edge}$ of a spherical, dispersion supported galaxy that has stars extending to a 3D radius of $r_{\rm edge}$. At the center, line-of-sight observations will project onto the radial component with $\sigma_{\rm los} \sim \sigma_r$, while at the edge of the same galaxy, line-of-sight velocities project onto the tangential component with $\sigma_{\rm los} \sim \sigma_t$. Consider a galaxy that is intrinsically isotropic ($\beta = 0$). If this system is analyzed using line-of-sight velocities under the false assumption that $\sigma_r > \sigma_t$ ($\beta > 0$), then the total velocity dispersion at $r \sim 0$ would be underestimated while the total velocity dispersion at $r \sim r_{\rm edge}$ would be overestimated. Conversely, if one were to analyze the same galaxy under the assumption that $\sigma_r < \sigma_t$ ($\beta < 0$), then the total velocity dispersion would be overestimated near the center and underestimated near the galaxy edge. It is plausible then that there is some intermediate radius where attempting to infer the enclosed mass from only line-of-sight kinematics is minimally affected by the unknown value of β.

A more quantitative understanding of the $r_{1/2}$ mass constraint (but less rigorous than that provided in (Wolf *et al.*, 2010) can be gained by rewriting the Jeans equation such that the $\beta(r)$ dependence is absorbed into the definition of $\sigma^2_{\rm tot} = (3 - 2\beta)\sigma_r^2$:

$$GM(r)r^{-1} = \sigma^2_{\rm tot}(r) + \sigma_r^2(r)\,(\gamma_* + \gamma_\sigma - 3)\,. \tag{3.7}$$

It turns out that to very good approximation, the log slope of the stellar light profile at $r_{1/2}$ for a wide range of commonly used stellar profiles is close to -3, such that $\gamma_*(r_{1/2}) = 3$ (Wolf *et al.*, 2010). Under the assumption that the velocity dispersion profile is flat (implying $\gamma_\sigma \ll 3$) we see that at $r = r_{1/2}$ the mass depends only on $\sigma_{\rm tot}$:

$$M(r_{1/2}) \simeq G^{-1}\sigma^2_{\rm tot}(r_{1/2})\,r_{1/2} \simeq 3\,G^{-1}\sigma_*^2\,r_{1/2}\,. \tag{3.8}$$

In this chain of arguments I have used the fact that $\sigma_* = \langle\sigma^2_{\rm tot}\rangle$ and the approximation $\langle\sigma^2_{\rm tot}\rangle \simeq \sigma^2_{\rm tot}(r_{1/2})$. This second approximation works because the stellar-weighted velocity dispersion gets its primary contribution from the radius where $\gamma_* = 3$ (Wolf *et al.*,

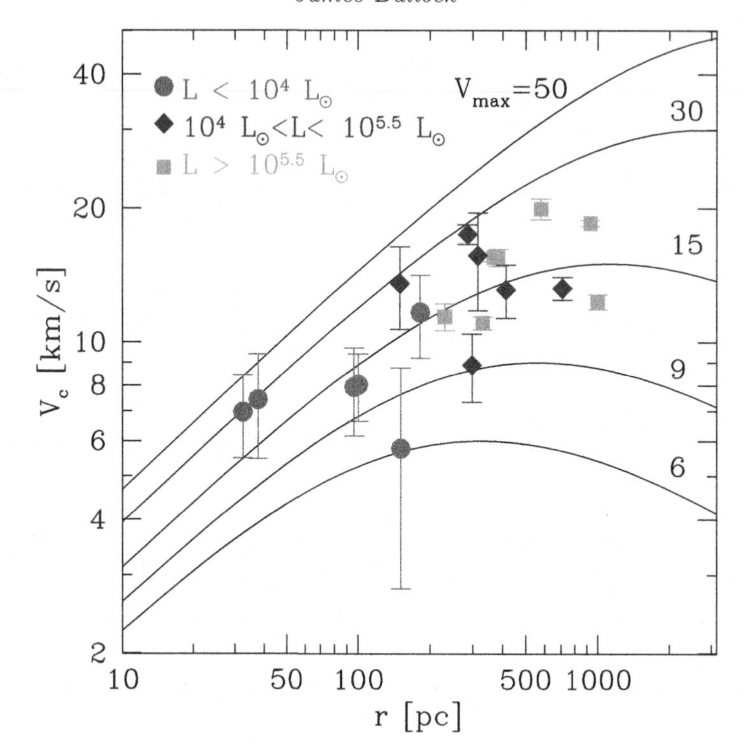

FIG. 3.5. Observed circular velocities $V_c(r_{1/2})$ plotted versus $r_{1/2}$ for each of the Milky Way dSph galaxies discussed in Wolf *et al.* (2010). The circular velocity curve values for the data were determined using Equation 3.9. For reference, we plot $V(r)$ rotation curves for NFW sub-halos obeying a *median* $V_{\rm max}$–$r_{\rm max}$ relationship given by Equation 3.5. Each curve is labeled by its $V_{\rm max}$ value (assumed to be in $\rm km\,s^{-1}$). Dwarf galaxy points are coded by their luminosities (see legend). Notice that the least luminous dwarfs seem to fall along similar $V(r)$ curves as the most luminous dwarfs.

2010). A cleaner way to write this mass estimator is

$$V_c(r_{1/2}) = \sqrt{3}\,\sigma_* \,. \tag{3.9}$$

Since by definition $V_{\rm max} \geq V_c(r_{1/2})$ we can conclude that

$$V_{\rm max} \geq \sqrt{3}\,\sigma_* . \tag{3.10}$$

As mentioned, a popular assumption in the literature has been to assume that dwarf galaxy sub-halos obey $V_{\rm max} = \sqrt{3}\,\sigma_*$ (Klypin *et al.*, 1999a; Bullock *et al.*, 2000; Simon and Geha, 2007). It is likely that this common assumption underestimated the $V_{\rm max}$ values of MW and M31 satellite galaxies significantly. Figure 3.5 plots $V_c(r_{1/2})$ vs. $r_{1/2}$ for the sample of Milky Way dwarf galaxies discussed in Wolf *et al.* (2010). For comparison, the lines show $V(r)$ rotation curves for NFW sub-halos (Navarro *et al.*, 1997) obeying the $V_{\rm max}$–$r_{\rm max}$ relationship given by Equation 3.5. Each curve is labeled by its $V_{\rm max}$ value (assumed to be in $\rm km\,s^{-1}$). In many cases, the rotation curves continue to rise well beyond the $r_{1/2}$ value associated with each point. Note that all of these galaxies are consistent with sitting in halos larger than $V_{\rm max} = 10\ \rm km\,s^{-1}$, but that in many cases the implied extrapolation is significant.

3.3.3 *Mass within a fixed radius: M_{600} and M_{300}*

Strigari *et al.* (2007b) suggested that a more direct way to compare satellite galaxy kinematics to predicted sub-halo counts was to consider their integrated masses within

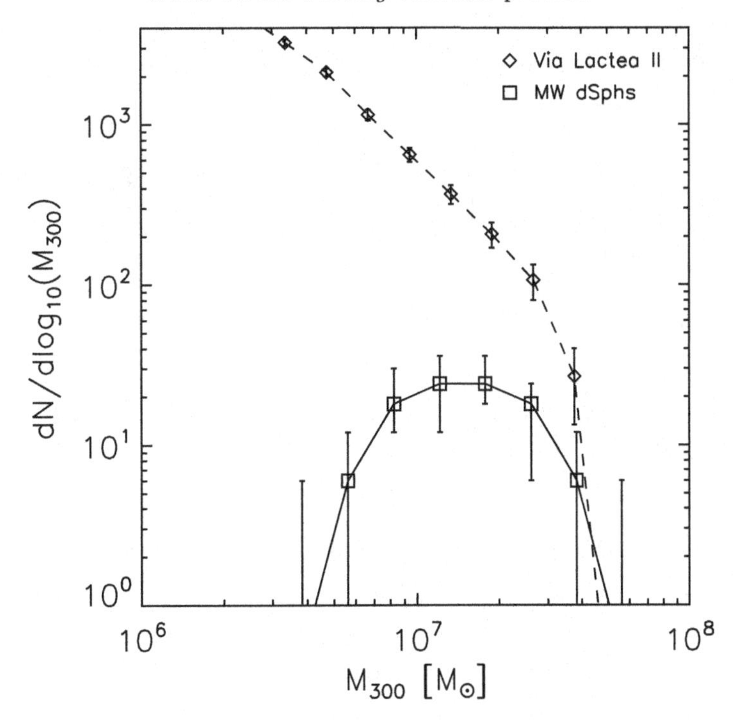

FIG. 3.6. Mass function for $M_{300} = M(< 300\,\text{pc})$ for MW dSph satellites and dark sub-halos in the Via Lactea II simulation within a radius of 400 kpc. The short-dashed curve is the sub-halo mass function from the simulation. The solid curve is the median of the observed satellite mass function. The error bars on the observed mass function represent the upper and lower limits on the number of configurations that occur 98% of the time (from Wolf *et al.*, in preparation). Note that the mismatch is about $\sim$1 order of magnitude at $M_{300} \simeq 10^7\,\text{M}_\odot$, and that it grows significantly toward lower masses.

a radius that is as close as possible to the median stellar radius for the galaxy population. By focusing on the mass within a particular (inner) radius, one is less reliant on cosmology-dependent assumptions about how the dark matter halo density profiles are extrapolated to V_{max}. This is particularly important if one is interested in simultaneously testing CDM models with variable σ_8 or if one is interested in more extreme variant models (e.g., dark matter from decays, Peter and Benson, 2010; or warm dark matter, Polisensky and Ricotti, 2010). At the time when Strigari *et al.* (2007b) made their suggestion, the smallest radius that was well resolved for sub-halos in numerical simulations was $r \simeq 600$ pc (Via Lactea I; Diemand *et al.*, 2007). This motivated a comparison using $M_{600} = M(r < 600\,\text{pc})$, which is reasonably well constrained for the 9 "classical" (pre-SDSS) dSph satellites of the Milky Way. They found that all but one of the classical Milky Way dSphs (Sextans) had an M_{600} mass consistent with (3–5) $\times 10^7\text{M}_\odot$. By comparing to results from the Via Lactea I simulation, Strigari *et al.* (2007b) concluded that the Milky Way dwarf masses were indicative of those expected if only the most massive halos (prior to accretion) were able to form stars.

An updated version of this comparison (from Wolf *et al.*, in preparation) is shown in Figure 3.6, where now we focus on integrated masses within 300 pc. This shift toward a smaller characteristic radius is motivated by an advance in simulation resolution, which now enables masses to be measured fairly accurately within 300 pc of sub-halo centers. It is fortuitous that 300 pc is also close to the median $r_{1/2}$ for Milky Way dSph galaxies with well-studied kinematic samples (Bullock *et al.*, 2010; Wolf *et al.*, 2010). The short dashed line in Figure 3.6 shows the M_{300} mass function from the Via Lactea II simulation (Diemand *et al.*, 2008), while the solid line shows M_{300} for the known Milky Way dSph

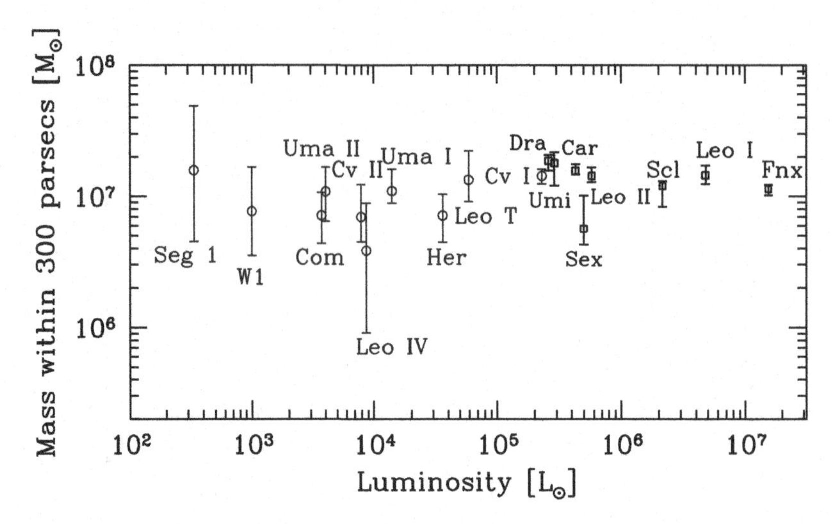

FIG. 3.7. Mass within 300 pc vs. V-band luminosity for classical (pre-SDSS) Milky Way dwarf satellites (black) and ultrafaint (post-SDSS) satellites (gray). The most important point to take away from this plot is that there is *no obvious trend between mass and luminosity.* The trend seems to demonstrate scatter, but the masses do not appear to vary systematically with luminosity.

galaxies. The masses are again indicative of a situation where only the most massive sub-halos host galaxies. Even at the point where the dSph mass function peaks ($M_{300} \simeq 10^7\,\mathrm{M_\odot}$) the simulation overpredicts the count by about a factor of 10. This order of magnitude mismatch between observed counts and predicted counts *at fixed mass* is a reasonably conservative statement of the MSP circa 2010.

The fact that the M_{300} mass function of Milky Way satellites peaks sharply (within a factor of $\sim$4) at a mass of $M_{300} \simeq 10^7\,\mathrm{M_\odot}$ is remarkable given that these galaxies span a factor of $\sim 10^5$ in luminosity. This point was highlighted by Strigari *et al.* (2008), who found that the M_{300} mass-luminosity relation for observed dwarfs is remarkably flat, with $M_{300} \propto L^{0.03\pm0.03}$. The associated plot from Strigari *et al.* (2008) is shown in Figure 3.7. The important point to take away from this plot is that there is *no detectable trend between galaxy mass and luminosity.* Galaxies with $L_V \simeq 4000\,\mathrm{L_\odot}$ (like Ursa Major II) demonstrate median M_{300} masses that are very close to those of galaxies at $L_V \simeq 10^7\,\mathrm{L_\odot}$ (like Fornax). Note that the Milky Way dwarfs do *not* all share *exactly* the same mass. For example, Sextans is clearly less massive than Carina, Leo II, and Draco. The mass of Hercules has been revised downward since the time that the plot in Figure 3.7 was published (Adén *et al.*, 2009) but this revision does not change the fact that there is a very weak relationship between mass and luminosity for the Milky Way dwarfs.

Segue 1 is arguably the most interesting case. With a luminosity of just $L_V \simeq 340\,\mathrm{L_\odot}$, the mass density of Segue 1 within its half-light radius ($r_{1/2} \simeq 38$ pc) is the highest of any Local Group dwarf: $\rho_{1/2} \simeq 1.6\ \mathrm{M_\odot\ pc^{-3}}$ (Martinez *et al.*, 2011). As can be seen by examining Figure 3.5 (red point, second from the left), a density this high at $\sim$40 pc can only be achieved within a sub-halo that is quite massive, with $V_{\max} > 10\ \mathrm{km\,s^{-1}}$ or with $M_{300} \gtrsim 5 \times 10^6\ \mathrm{M_\odot}$. Though the placement of Segue 1 on the Strigari plot requires significant extrapolation (from 40 pc to 300 pc), such an extrapolation is not unwarranted within the ΛCDM context because sub-halos that are this dense and this massive almost always extend to tidal radii larger than 300 pc (as can be inferred, for example, from Equation 3.5). With a luminosity some five orders of magnitude smaller than that of Fornax, Segue 1 is consistent with inhabiting a dark matter sub-halo with approximately the same potential well depth.

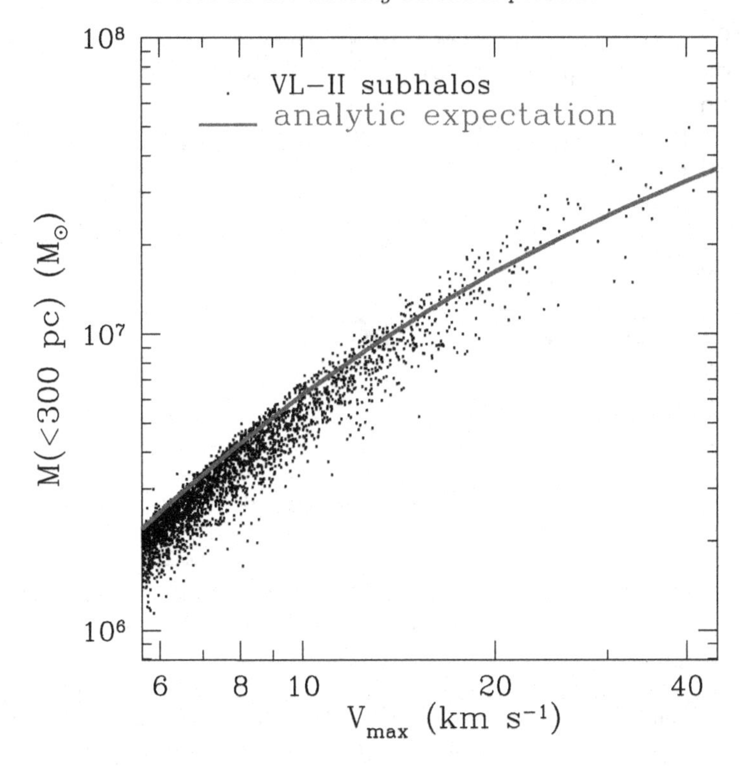

FIG. 3.8. Relationship between the mass within 300 pc and $V_{\max}$ for sub-halos in the Via Lactea II simulation (points from Diemand *et al.*, 2008) along with the analytic relationship expected for NFW halos that obey Equation 3.5 (solid line).

Given that the M_{300} mass variable is adopted for practical (not physical) reasons, it is worth examining its relationship to the more familiar measure $V_{\max}$. The relationship between the two variables is illustrated in Figure 3.8. The points are Via Lactea II sub-halos (kindly provided by the public release of Diemand *et al.*, 2008) and the solid line is the analytic estimate for NFW sub-halos that obey the $V_{\max}$–$r_{\max}$ relation given by Equation 3.5. For 60 $\mathrm{km\,s^{-1}} \gtrsim V_{\max} \gtrsim 15\ \mathrm{km\,s^{-1}}$, the mass within 300 pc correlates linearly with $V_{\max}$ as $M_{300} \simeq 10^7\,\mathrm{M_\odot}(\mathrm{V_{max}}/12.5\ \mathrm{km\,s^{-1}})$. For smaller halos, the relationship asymptotes to $M_{300} \propto V_{\max}^2$. The point to take away from Figure 3.8 is that the M_{300} variable is at least as sensitive to potential well depth as the more familiar variable $V_{\max}$.

3.3.4 *Making sense of the luminosity-mass relation*

How does the observed relationship between luminosity and mass compare to simple empirical expectations that have been gained from examining more massive halos? The symbols in Figure 3.9 show inferred $V_{\max}$ values for Milky Way dwarfs plotted as a function of luminosity. I have opted to use $V_{\max}$ as the scaling variable here (rather than M_{300}) in order to make direct contact with more traditional scaling relations. The error bars on each point are typically larger (in a relative sense) than they are in Figure 3.7 because the extrapolation to $V_{\max}$ is more uncertain than the extrapolation to M_{300} in most cases. Nevertheless, the same global trend holds: there is a very weak correlation between luminosity and $V_{\max}$ for the Milky Way dSph population. Note that I have extended the luminosity axis slightly in order to include $V_{\max}$ estimates for the LMC and SMC (from van den Bergh, 2000, who discusses the signifiant but unquantified uncertainties on mass estimates for the LMC and SMC). The lower red dashed line is

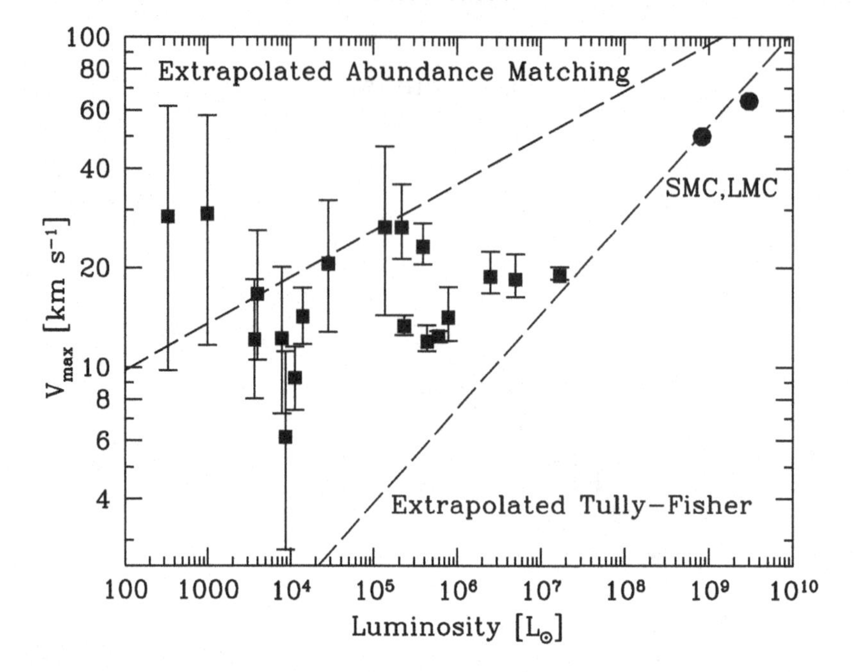

FIG. 3.9. The V_{max} vs. V-band luminosity relation for the Milky Way satellite population, as inferred from assuming that dSph galaxies sit within NFW dark matter halos that obey the same scaling relations as do sub-halos in ΛCDM N-body simulations. The lower dashed line is the Tully Fisher relation from (Courteau *et al.*, 2007) extrapolated to low luminosities and the upper dashed line is the relation one obtains from extrapolating the abundance matching power-law from Busha *et al.* (2010).

the extrapolated Tully Fisher relation[10] for brighter spiral galaxies from Courteau *et al.* (2007) and the upper blue dashed line is the relation one obtains from extrapolating the abundance matching power-law discussed in reference to Figure 3.1 (specifically the $V_{max} - L$ relation advocated by Busha *et al.*, 2010). The fact that the naive extrapolation of the abundance matching power-law provides a reasonable match at $L \simeq 10^4\,L_\odot$ is encouraging. Nevertheless, there are some surprising points of disagreement.

First, the data clearly demonstrate a flatter trend with luminosity than either of these scaling relations. The disagreement with the Tully Fisher line is less of a concern, as there is no reason to suspect that spheroidal galaxies should obey the same scaling relations as spiral galaxies. Abundance matching, on the other hand, is a global measure of the $V_{max} - L$ mapping that is *required* to produce the correct abundances of galaxies. Of particular interest in Figure 3.9 is the mismatch between abundance matching and dwarf properties for the *most luminous dwarfs*. The most luminous dSph galaxies at $L \sim 10^7\,L_\odot$ prefer $V_{max} \simeq 20$ km s^{-1}, while the abundance matching expectation at this luminosity is much higher $V_{max} \simeq 50$ km s^{-1}. Note that a completely independent analysis by Strigari *et al.* (2010) finds similarly low V_{max} values for the luminous dSph galaxies. Of course, one possible explanation is that the sub-halos that host these objects have lost a significant amount of dark matter mass, but several studies have looked at this effect in more detail and found that mass loss cannot easily account for this difference (Bullock *et al.*, 2010; Busha *et al.*, 2010). Indeed, accounting for the fact that dSph satellite galaxies are sub-halos simply lowers the normalization of the dashed line in Figure 3.9 by about 50%, without changing the slope to any appreciable agree. The reason for this is that significant

[10] The dSph satellites of the Milky Way also deviate from the Baryonic Tully Fisher relation (McGaugh and Wolf, 2010) as would be expected for systems that contain no detectable gas (except for Leo T; Grcevich and Putman, 2009).

mass loss tends to have occurred only in systems that were accreted at early times. At earlier times, larger $V_{\rm max}$ values are required to produce the same stellar luminosity (e.g., Behroozi *et al.*, 2010; Moster *et al.*, 2010), and this compensates for the fact that $V_{\rm max}$ tends to decrease with time once a sub-halo is accreted.

One possible explanation for the lack of trend between $V_{\rm max}$ (or M_{300}) and luminosity is that we are seeing evidence for a real scale in galaxy formation at $V_{\rm max} \simeq 15$ km s^{-1} or $M_{300} \simeq 10^7\,{\rm M}_\odot$, which is remarkably close to the scale imposed by photoionization suppression and the H I cooling limit (Li *et al.*, 2009; Macciò *et al.*, 2009; Okamoto and Frenk, 2009; Stringer *et al.*, 2010). This characteristic $V_{\rm max}/M_{300}$ scale is well above the scale that would be indicative of a halo that required H_2 to cool: these are unlikely to be pre-reionization fossils of first light star formation (Madau *et al.*, 2008; Ricotti, 2010).

A second explanation, which seems less likely, is that we are seeing a scale in dark matter clustering, which just happens to be very close to the mass scale where natural astrophysical suppression should kick in.

A third possibility, addressed in the next section, is that the lack of an observed trend between mass and luminosity is the product of selection bias: most ultrafaint galaxies do inhabit halos with $M_{300} \lesssim 10^7{\rm M}_\odot$, but they are too diffuse to have been discovered. If so, this implies that searches for the lowest mass fossil galaxies left over from reionization may be hindered by surface brightness limits (Bovill and Ricotti, 2009).

Finally, it is possible that the apparent flatness in the $V_{\rm max}/M_{300}$ – luminosity relationship for Milky Way dwarfs is simply an accident of small statistics or an artifact of misinterpretation of the data. Fortunately, as we now discuss, there is good reason to believe that the overall count of dwarf galaxies will grow significantly over the next decade. These discoveries should enable larger statistical samples of dwarfs with kinematically derived masses and a more stringent investigation of the trends that appear to be present in the data at the current time.

3.4 Empirical evidence for missing satellites

The Sloan Digital Sky Survey (SDSS; Adelman-McCarthy *et al.*, 2007) has revolutionized our understanding of the Milky Way's environment. In particular, searches in SDSS data have more than doubled the number of known Milky Way satellite galaxies over the last several years (Willman *et al.*, 2005; Grillmair, 2006, 2009; Zucker *et al.*, 2006; Belokurov *et al.*, 2007, 2009; Majewski *et al.*, 2007; Martin *et al.*, 2009), revealing a population of galaxies that were otherwise too faint to have been discovered (with $L \lesssim 10^5\,{\rm L}_\odot$). In addition to providing fainter detections, the homogeneous form of the SDSS has allowed for a much better understanding of the statistics of detection (Koposov *et al.*, 2008; Walsh *et al.*, 2009). The Milky Way satellite census is incomplete for at least two reasons: sky coverage and luminosity bias. A third source of incompleteness comes from the inability to detect and verify dwarf galaxies with very low surface brightness. We turn to this last point at the end of this section.

3.4.1 *Sky coverage*

Resolved star searches have covered approximately ~20% of the sky, so it is reasonable to expect that there is a factor of $c_{\rm sky} \simeq 5$ more satellites fainter than $L = 10^5\,{\rm L}_\odot$, bringing to the total count of Milky Way dwarf satellites to $N \simeq N_{\rm pre-SDSS} + c_{\rm sky}\, N_{\rm SDSS} \simeq 9 + 5 \times 12 \simeq 70$.

Note that the estimate $c_{\rm sky} \simeq 5$ assumes that there is no systematic angular bias in the satellite distribution. If the satellite distribution is anisotropic on the sky, the value of $c_{\rm sky}$ could be significantly different from 5. If, for example, the SDSS happens to be viewing a particularly underdense region of the halo, then the correction for covering the whole sky could be quite large $c_{\rm sky} \gg 5$. Tollerud *et al.* (2008) used sub-halos in Via Lactea I simulation (Diemand *et al.*, 2007) to estimate the variance in $c_{\rm sky}$ for SDSS-size

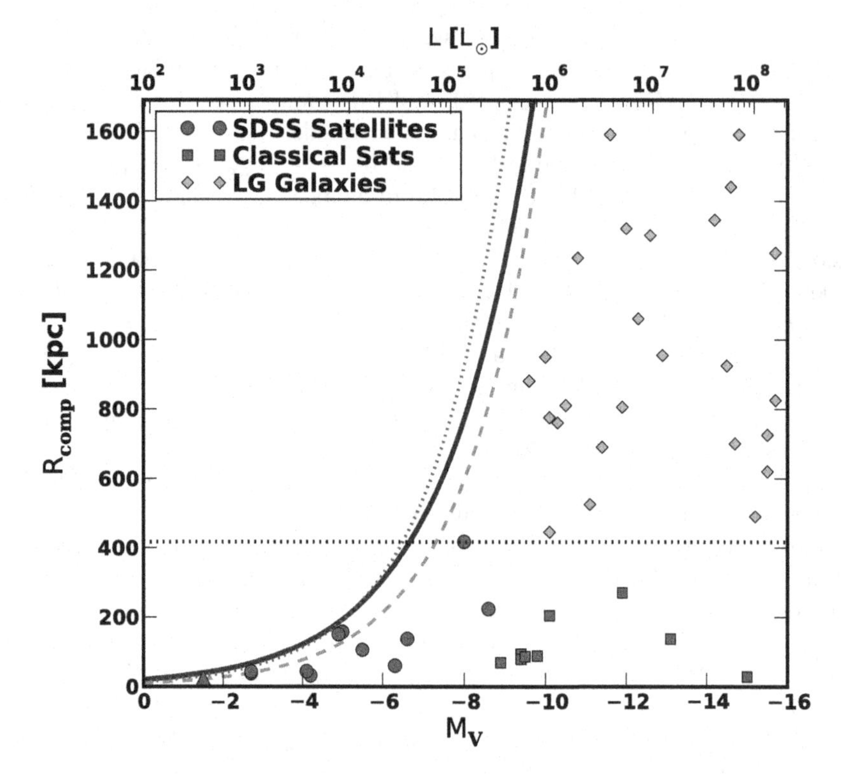

FIG. 3.10. The completeness radius for dwarf satellites from Tollerud *et al.* (2008). The three rising lines show the helio-centric distance, $R_{\rm comp}$, out to which dwarf satellites of a given absolute magnitude are complete within the SDSS DR5 survey. The solid is for the published detection limits in Koposov *et al.* (2008) and the other lines are described in Tollerud *et al.* (2008). The dotted black line at 417 kpc corresponds to generous characterization of the outer edge of the Milky Way halo. The data points are observed satellites of the Milky Way and the Local Group. The circles are the SDSS-detected satellites, the only galaxies for which the detection limits actually apply, although the detection limits nevertheless also delineate the detection zone for more distant Local Group galaxies (diamonds). Squares indicate classical Milky Way satellites. The faintest object (triangle) is Segue 1, which is outside the DR5 footprint.

pointings within the halo and found that $c_{\rm sky}$ can vary from just $\sim$3.5 up to $\sim$8.3 depending on the mock surveys orientation. In principle, then, one might expect as many as 110 satellites (or as few as 50) from the sky coverage correction alone.

3.4.2 *Luminosity bias*

The second source of number count incompleteness, luminosity bias, is more difficult to quantify because it depends sensitively on the (unknown) radial distribution of all satellites. Koposov *et al.* (2008) and Walsh *et al.* (2009) both found that detection thresholds are mostly governed by the distance to the object and the object luminosity, as long as the surface density of the dwarf was brighter than about $\mu = 30$ mag arcsec^{-2}.

Tollerud *et al.* (2008) used the Koposov *et al.* (2008) results to show that the SDSS is approximately complete down to a fixed apparent luminosity:

$$R_{\rm comp} \simeq 66\,{\rm kpc}\left(\frac{L}{1000\,{\rm L}_\odot}\right)^{1/2} \tag{3.11}$$

where $R_{\rm comp}$ is a spherical completeness radius beyond which a dSph of a particular luminosity will go undetected. The implied relationship between galaxy luminosity and corresponding heliocentric completeness radius is shown by the solid curve in Figure 3.10.

The horizontal dotted line in Figure 3.10 marks a generous estimate of $R_\mathrm{h} \simeq 417\,\mathrm{kpc}$ for the Milky Way halo edge. We see that only satellites brighter than $L \simeq 10^5\,\mathrm{L}_\odot$ are observable out to this radius. The fact that the faintest dwarf satellite galaxies known are more than 100 times fainter than this limit immediately suggests that there are many more faint satellite galaxies yet to be discovered.

Once armed with a completeness radius–luminosity relation like that given in Equation 3.11, one only needs to know the radial distribution of satellites $N(< R)$ in order to estimate the total count of faint galaxies out to some pre-defined edge of the Milky Way (we follow Tollerud *et al.*, 2008, and use $R_\mathrm{h} = 417\,\mathrm{kpc}$). For example, if we knew that there were N_obs Milky Way dwarfs within Δ L of some luminosity L_obs, then Equation 3.11 tells us that the census of these objects is complete to $R_\mathrm{comp}(L_\mathrm{obs})$. We may estimate the total number of galaxies between L_obs and $L_\mathrm{obs} + \Delta L$ using

$$N_\mathrm{tot}(L_\mathrm{obs}) \simeq c_\mathrm{sky}\, N_\mathrm{obs}\, \frac{N(< R_\mathrm{h})}{N(< R_\mathrm{comp}(L_\mathrm{obs}))}\,. \tag{3.12}$$

If we make the assumption that satellite galaxies are associated with sub-halos in a one-to-one fashion, then $N(< R_\mathrm{h})/N(< R)$ may be estimated from analyzing the radial distribution of ΛCDM sub-halos. Tollerud *et al.* (2008) showed that the implied *ratio* $N(< R_\mathrm{h})/N(< R)$ is almost independent of how the sub-halos are chosen. As an example, consider the correction implied for the $N_\mathrm{obs} = 2$ known Milky Way dwarfs that have $L_\mathrm{obs} \simeq 1000\,\mathrm{L}_\odot$. For this luminosity, we are complete to $R_\mathrm{comp} = 66\,\mathrm{kpc}$. The sub-halo distributions presented in Tollerud *et al.* (2008) obey $N(< 417\ \mathrm{kpc})/N(< 66\ \mathrm{kpc}) = 5$–10 for a wide range of sub-halo population choices. This suggests a total count of $L \sim 1000\,\mathrm{L}_\odot$ galaxies is $N_\mathrm{tot} \simeq 50$–100.

Figure 3.11 presents a more exacting correction of this kind from Tollerud *et al.* (2008). The lower (red) curve shows the observed cumulative luminosity function of Milky Way dSph galaxies. The middle (green) curve is corrected for SDSS sky coverage only and yields ~70 galaxies brighter than $L = 1000\ \mathrm{L}_\odot$. The upper curve has been corrected for luminosity completeness (using Via Lactea I sub-halos for $N(< R)$). The result is that we expect ~400 galaxies brighter than $1000\,\mathrm{L}_\odot$ within ~400 kpc of the Sun. The correction becomes more significant for lower luminosity systems because the completeness radius is correspondingly smaller. Note that this luminosity bias correction would be even more significant if one took into account the fact that the presence of a central disk will tend to deplete sub-halos (and associated galaxies) in central regions of the Galactic halo D'Onghia *et al.* (2010). This effect will tend to increase the fraction of galaxy sub-halos at large Galactocentric distances.

Referring back to Figure 3.3 and Equation 3.3, we see that $N \simeq 400$ satellites would be about the satellite count expected only for $V_\mathrm{max} > 10\,\mathrm{km\,s^{-1}}$ halos. It is encouraging that this is close to the V_max limit inferred directly from stellar kinematics, as illustrated in Figure 3.9. Of course, as discussed in Tollerud *et al.* (2008) and Walsh *et al.* (2009) , the overall luminosity bias correction is sensitive at the factor of ~2 level to the precise subset of sub-halos that are used to determine the radial distribution of galaxies. Generally, the best estimates suggest that there are hundreds of Milky Way satellite galaxies lurking in the outer reaches of the Milky Way.

There is real hope that these missing satellites will be detected as surveys like LSST (Ivezic *et al.*, 2008), DES (The Dark Energy Survey Collaboration, 2005), PanSTARRS (Kaiser *et al.*, 2002), and SkyMapper (Keller *et al.*, 2007) cover more sky and provide deeper maps of the Galactic environment. Figure 3.12 from Tollerud *et al.* (2008) provides a rough determination of the completeness radius for several planned surveys. According to this estimate, LSST will be able to detect objects as faint as $L \sim 500\,\mathrm{L}_\odot$ out to the edge of the Milky Way halo.

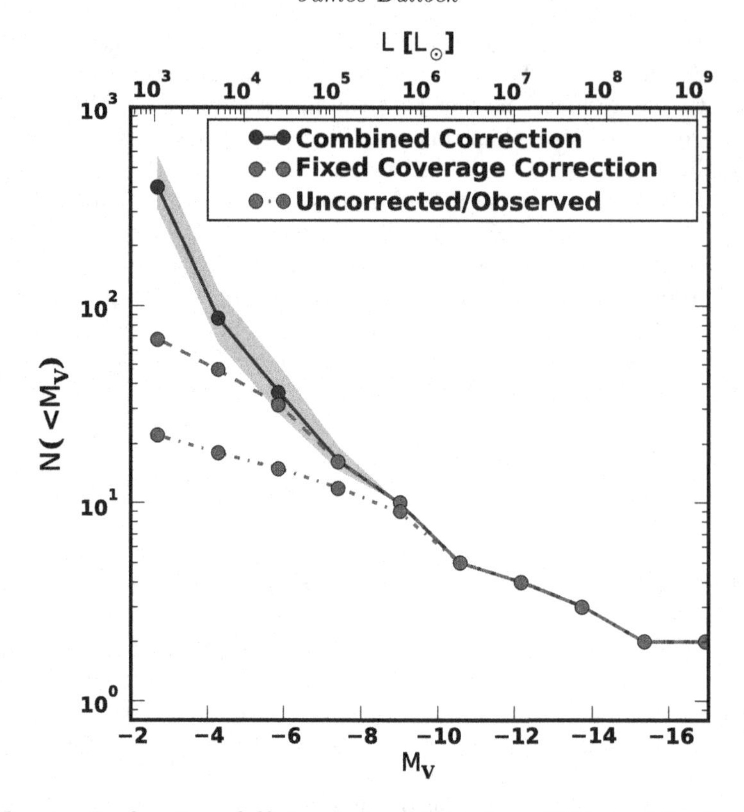

FIG. 3.11. Luminosity function of dSph galaxies within $R_{\rm h} = 417$ kpc of the Sun as observed (lower), corrected for only SDSS sky coverage (middle), and with luminosity completeness corrections from Tollerud *et al.* (2008) included (upper). Note that the brightest, classical (pre-SDSS) satellites are uncorrected, while new satellites have the correction applied. The shaded error region corresponds to the 98% spread over mock observation realizations within the Via Lactea I halo.

3.4.3 *Surface brightness limits and stealth galaxies*

Importantly, the luminosity-distance detection limits discussed previously only apply for systems with peak surface brightness obeying $\mu < 30$ mag arcsec^2 (Koposov *et al.*, 2008). Any satellite galaxy with a luminosity of $L \simeq 1000\ {\rm L}_\odot$ and a half-light radius $r_{1/2}$ larger than about 100 pc would have evaded detection with current star-count techniques regardless of its distance from the Sun. This phenomenon is illustrated in Figure 3.13 where I plot the 2D (projected) half-light radius $R_{\rm e}$ vs. L for Milky Way dSph galaxies. The solid line shows a constant peak (central) surface brightness of $\mu = 30$ mag arcsec^{-2}. The tendency for many of the fainter dwarfs to line up near the surface brightness detection limit is suggestive. There is nothing ruling out the presence of a larger population of more extended systems that remain undetected because of their low surface brightness. I refer to these undetected diffuse galaxies as *stealth galaxies*.

If a large number of stealth galaxies do exist, they are likely associated with low-mass dark matter sub-halos. If so, this will make it difficult to detect the lowest mass halos and will introduce a systematic bias that avoids low M_{300} masses in the Strigari plot (Figure 3.7).

One can understand this expectation by considering a galaxy with velocity dispersion σ_* and luminosity L embedded within a gravitationally dominant dark matter halo described by a circular velocity curve that increases with radius as an approximate power-law: $V_c(r) = V_{300}\,(r/300\,{\rm pc})^{\alpha}$. Equation 3.9 implies

$$r_{1/2} = 300\,{\rm pc}\,(\sqrt{3}\sigma_*/V_{300})^{1/\alpha}\,. \tag{3.13}$$

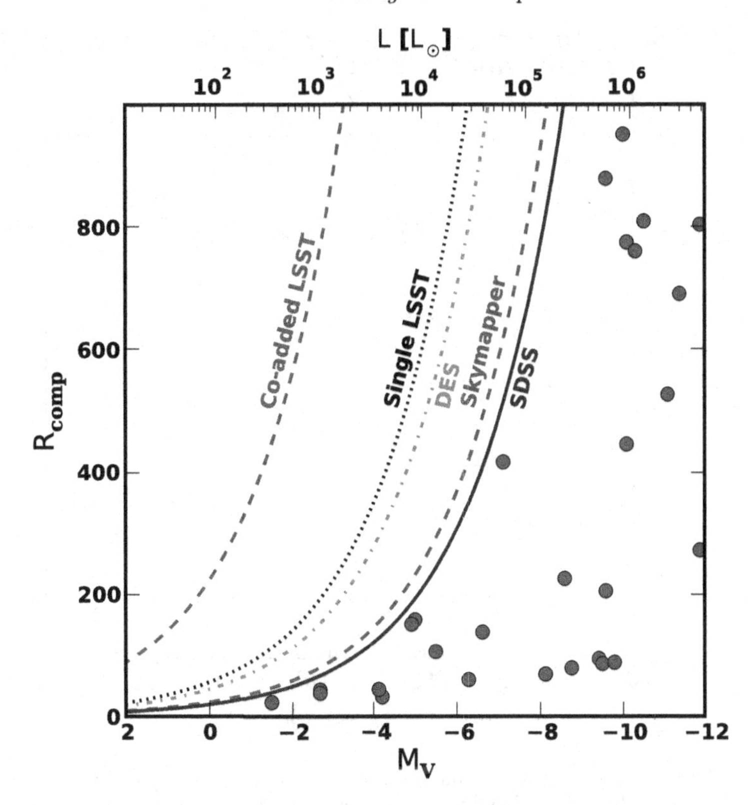

FIG. 3.12. Maximum radius for detection of dSphs as estimated by Tollerud *et al.* (2008) shown as a function of galaxy absolute magnitude for DR5 (assumed limiting r-band magnitude of 22.2) compared to a single exposure of LSST (24.5), co-added full LSST lifetime exposures (27.5), DES or one exposure from PanSTARRS (both 24), and the SkyMapper and associated Missing Satellites Survey (22.6). The data points are SDSS and classical satellites, as well as Local Group field galaxies.

For an NFW halo (Navarro *et al.*, 1997) with scale radius $r_{\max} \gg 600$ pc we have $\alpha = 1/2$ and

$$r_{1/2} \propto \left(\frac{\sigma_*}{V_{300}}\right)^2 \propto \frac{\sigma_*^2}{M_{300}}. \tag{3.14}$$

Clearly, $r_{1/2}$ increases as we decrease M_{300} at fixed σ_*. One implication is that if a galaxy has a stellar surface density $\Sigma_* \propto L/r_{1/2}^2$ that is just large enough to be detected, another galaxy with identical L and σ_* will be undetectable if it happens to reside within a slightly less massive halo.

Figure 3.14 from Bullock *et al.* (2010) provides a more explicit demonstration of how surface brightness bias can affect the mass-luminosity relation of dwarf galaxies. The points show M_{300} versus L for Milky Way dSph galaxies, with masses from Strigari *et al.* (2008) and luminosities updated as in Wolf *et al.* (2010). The region below the shaded band is undetectable for dwarf galaxies with $\sigma_* > 4$ km s^{-1}. The region below the dashed line is undetectable for dwarf galaxies with $\sigma_* > 5$ km s^{-1}.

This result suggests that surface brightness selection bias may play a role explaining the lack of observed correlation between luminosity and mass for Milky Way satellites. It also implies that searches for the lowest mass "fossil" galaxies left over from reionization may be hindered by surface brightness limits. This latter point was made earlier by Bovill and Ricotti (2009). According to estimates in Bullock *et al.* (2010), potentially half of several hundred satellite galaxies that could be observable by surveys like LSST are stealth. A complete census of these objects will require deeper sky surveys, 30m-class follow-up

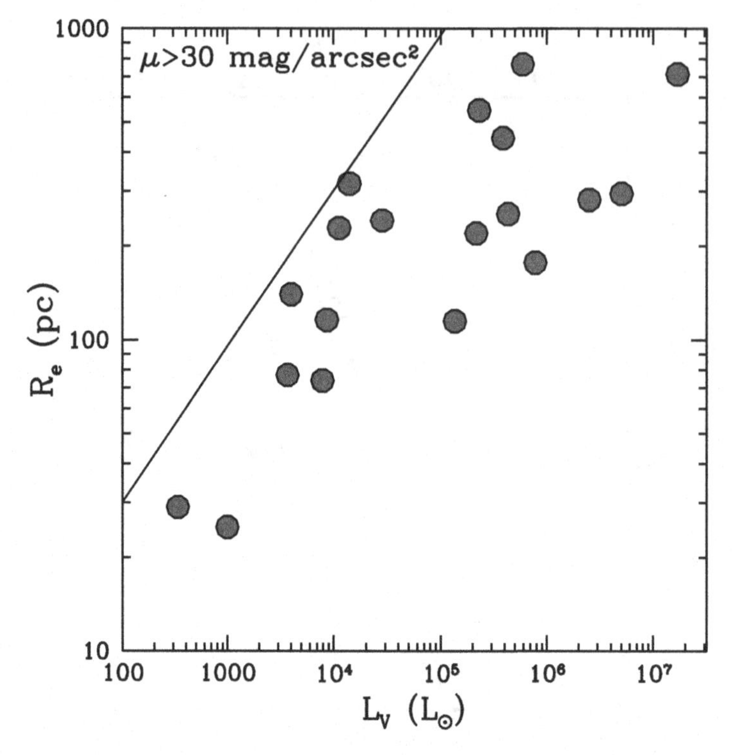

FIG. 3.13. Half-light radius R_e vs. V-band luminosity for Milky Way dSph galaxies. Galaxies above the solid line, with surface brightness fainter than $\mu = 30$ mag arcsec^{-2}, are currently undetectable.

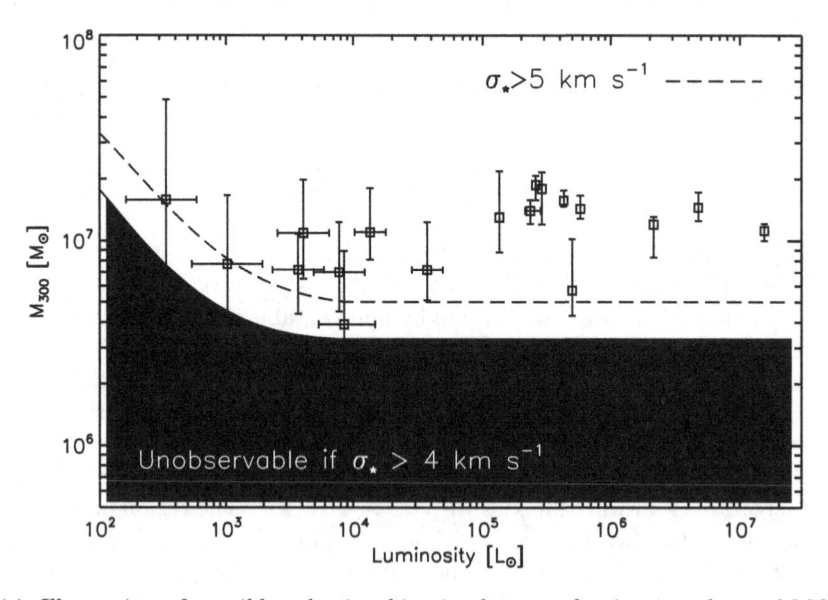

FIG. 3.14. Illustration of possible selection bias in the mass-luminosity plane of Milky Way dwarfs. The data points with errors reproduce the Strigari *et al.* (2008) M_{300} masses presented in Figure 3.7. Galaxies within the shaded region (below dashed line) will remain undetected ($\mu > 30$ mag arcsec^{-2}) if they have velocity dispersions $\sigma_* > 4\,\mathrm{km\,s^{-1}}$ ($\sigma_* > 5$ km s^{-1}).

telescopes, and more refined methods to identify extended, self-bound groupings of stars in the halo.

3.5 Summary

Advances in simulation technology have solidified the decade-old expectation that substructure should be abundant in and around CDM halos, with counts that rise steadily to the smallest masses. Properties of observed satellites in and around the Local Group provide an important means to test this prediction.

The substructure issue has gained relevance over the years because it touches a range of important issues that span many subfields, including the micro-physical nature dark matter, the role of feedback in galaxy formation, and star formation in the early universe. It is in fact difficult to state directly what it would take to solve the MSP without first associating it with a specific subfield. From a galaxy formation standpoint, one important question is to identify the primary feedback sources that suppress galaxy formation in small halos. An associated goal is to identify any obvious mass scale where the truncation in the efficiency of galaxy formation occurs.

It is particularly encouraging for ΛCDM theory that both direct kinematic constraints on the masses of Milky Way satellite galaxies and completeness correction estimates point to about the same mass and velocity scale, $V_{\rm max} \simeq 15\ {\rm km\,s^{-1}}$ or $M_{300} \simeq 10^7\,{\rm M_\odot}$. The luminosity bias correction discussed in association with Figure 3.11 suggests that the Milky Way halo hosts about 400 satellite galaxies with luminosities similar to the faintest dwarf galaxies known. As seen in either Figure 3.3 or Figure 3.2, a total count of 400 satellites is approximately what is expected at a minimum $V_{\rm max}$ threshold of $15\ {\rm km\,s^{-1}}$ – the same $V_{\rm max}$ scale evidenced in the kinematic measurements shown in Figure 3.9. These numbers did not have to agree. Interestingly, they both point to a mass scale that is close to the limit where cooling via atomic hydrogen is suppressed and where photoionization heating should prevent the accretion of fresh gas (see Dekel, 2005).

More puzzling is the overall lack of observed correlation between Milky Way satellite galaxy luminosities and their M_{300} masses or $V_{\rm max}$ values (see Figures 3.7 and 3.9). Most of the models that have been constructed to confront the mass-luminosity over-predict $V_{\rm max}$ values for the brightest dwarfs and under-predict them for the faintest dwarfs (Busha *et al.*, 2010; Li *et al.*, 2009; Macciò *et al.*, 2009; Okamoto and Frenk, 2009; Bullock *et al.*, 2010). These differences may reflect small-number statistics at the bright end, and a potential observational bias associated with surface brightness at the faint end.

Section 3.4.3 discussed the idea that there may be a population of low-luminosity satellite galaxies orbiting within the halo of the Milky Way that are too diffuse to have been detected with current star-count surveys, despite the fact that they have luminosities similar to those of known ultrafaint dSphs. These stealth galaxies should preferentially inhabit the smallest dark matter sub-halos that can host stars ($V_{\rm max} \lesssim 15\ {\rm km\,s^{-1}}$). One implication is that selection bias may play a role in explaining the lack of observed correlation between L and M_{300} for Milky Way satellite galaxies. This effect also implies that searches for the lowest mass "fossil" galaxies left over from reionization may be hindered by surface brightness limits. A large number (more than 100) stealth galaxies may be orbiting within the Milky Way halo. These systems have stellar distributions too diffuse to have been easily discovered so far, but new surveys and new techniques are in the works that may very well reveal hundreds of these missing satellite galaxies within the next decade.

Acknowledgments

I am most grateful to David Martínez-Delgado for organizing the lively and enriching twentieth Winter School and to the other lecturers and students who made the experience enjoyable. I would like to thank Peter Behroozi, Michael Kuhlen, and Joe Wolf for allowing me to include their unpublished figures here. I would also like to acknowledge my recent collaborators on MSP work for influencing my thoughts on the subject: Marla Geha, Manoj Kaplinghat, Greg Martínez, Quinn Minor, Josh Simon, Kyle Stewart, Louis Strigari, Erik Tollerud, Beth Willman, and Joe Wolf. I thank Erik Tollerud and Joe Wolf for providing useful comments on the manuscript.

REFERENCES

Abbott, T., Aldering, G., Annis, J., and 68 coauthors. 2005. The Dark Energy Survey Collaboration. The Dark Energy Survey, arXiv:astro-ph/0510346.

Adelman-McCarthy, J. K., and 153 colleagues. 2007. The fifth data release of the Sloan Digital Sky Survey. *ApJS*, **172**(Oct.), 634–644.

Adén, D., Wilkinson, M. I., Read, J. I., Feltzing, S., Koch, A., Gilmore, G. F., Grebel, E. K., and Lundström, I. 2009. A new low mass for the Hercules dSph: the end of a common mass scale for the Dwarfs? *ApJ*, **706**, L150.

Baldry, I. K., Glazebrook, K., and Driver, S. P. 2008. On the galaxy stellar mass function, the mass-metallicity relation and the implied baryonic mass function. *MNRAS*, **388**(Aug.), 945–959.

Behroozi, P. S. 2010. Private communication.

Behroozi, P. S., Conroy, C., and Wechsler, R. H. 2010. A comprehensive analysis of uncertainties affecting the stellar mass-halo mass relation for $0 < z < 4$. *ApJ*, **717**(July), 379–403.

Bell, E. F., and 17 colleagues. 2008. The accretion origin of the Milky Way's stellar halo. *ApJ*, **680**(June), 295–311.

Belokurov, V., and 9 colleagues. 2009. The discovery of Segue 2: a prototype of the population of satellites of satellites. *MNRAS*, **397**(Aug.), 1748–1755.

Belokurov, V., and 33 colleagues. 2007. Cats and dogs, hair and a hero: A quintet of new Milky Way companions. *ApJ*, **654**(Jan.), 897–906.

Bertschinger, E. 2006. Effects of cold dark matter decoupling and pair annihilation on cosmological perturbations. *Phys. Rev. D*, **74**(Sept.), 063509.

Blumenthal, G. R., Faber, S. M., Primack, J. R., and Rees, M. J. 1984. Formation of galaxies and large-scale structure with cold dark matter. *Nature*, **311**(Oct.), 517–525.

Bovill, M. S. and Ricotti, M. 2009. Pre-reionization fossils, ultrafaint dwarfs, and the missing galactic satellite problem. *ApJ*, **693**(Mar.), 1859–1870.

Bryan, G. L. and Norman, M. L. 1998. Statistical properties of X-ray clusters: analytic and numerical comparisons. *ApJ*, **495**(Mar.), 80.

Busha, M. T., Alvarez, M. A., Wechsler, R. H., Abel, T., and Strigari, L. E. 2010. The impact of inhomogeneous reionization on the satellite galaxy population of the Milky Way. *ApJ*, **710**(Feb.), 408–420.

Bullock, J. S. and Johnston, K. V. 2005. Tracing galaxy formation with stellar halos. I. Methods. *ApJ*, **635**(Dec.), 931–949.

Bullock, J. S. and Johnston, K. V. 2007. Dynamical evolution of accreted dwarf galaxies. *Island Universes – Structure and Evolution of Disk Galaxies*, Dordrechti Springer, 227.

Bullock, J. S., Kravtsov, A. V., and Weinberg, D. H. 2000. Reionization and the abundance of galactic satellites. *ApJ*, **539**(Aug.), 517–521.

Bullock, J. S., Kravtsov, A. V., and Weinberg, D. H. 2001. Hierarchical galaxy formation and substructure in the Galaxy's stellar halo. *ApJ*, **548**(Feb.), 33–46.

Bullock, J. S., Stewart, K. R., Kaplinghat, M., Tollerud, E. J., and Wolf, J. 2010. Stealth galaxies in the halo of the Milky Way. *ApJ*, **717**(July), 1043–1053.

Collins, M. L. M., and 10 colleagues. 2010. A Keck/DEIMOS spectroscopic survey of the faint M31 satellites AndIX, AndXI, AndXII and AndXIII†. *MNRAS*, **407**(Oct.), 2411–2433.

Conroy, C. and Wechsler, R. H. 2009. Connecting galaxies, halos, and star formation rates across cosmic time. *ApJ*, **696**(May), 620–635.

Cooper, A. P., and 11 colleagues. 2010. Galactic stellar haloes in the CDM model. *MNRAS*, **406**(Aug.), 744–766.

Courteau, S., Dutton, A. A., van den Bosch, F. C., MacArthur, L. A., Dekel, A., McIntosh, D. H., and Dale, D. A. 2007. Scaling relations of spiral galaxies. *ApJ*, **671**(Dec.), 203–225.

Davis, M., Efstathiou, G., Frenk, C. S., and White, S. D. M. 1985. The evolution of large-scale structure in a universe dominated by cold dark matter. *ApJ*, **292**(May), 371–394.

Dekel, A. 2005. Characteristic scales in galaxy formation. *Multiwavelength Mapping of Galaxy Formation and Evolution*,Renzini, A. and Bender, R., eds. , 269.

Dekel, A. and Silk, J. 1986. The origin of dwarf galaxies, cold dark matter, and biased galaxy formation. *ApJ*, **303**(Apr.), 39–55.

Diemand, J., Kuhlen, M., and Madau, P. 2007. Dark matter substructure and gamma-ray annihilation in the Milky Way halo. *ApJ*, **657**(Mar.), 262–270.

Diemand, J., Kuhlen, M., Madau, P., Zemp, M., Moore, B., Potter, D., and Stadel, J. 2008. Clumps and streams in the local dark matter distribution. *Nature*, **454**(Aug.), 735–738.

D'Onghia, E., Springel, V., Hernquist, L., and Keres, D. 2010. Substructure depletion in the Milky Way halo by the disk. *ApJ*, **709**(Feb.), 1138–1147.

Efstathiou, G. 1992. Suppressing the formation of dwarf galaxies via photoionization. *MNRAS*, **256**(May), 43P–47P.

Ferguson, A. M. N., Irwin, M. J., Ibata, R. A., Lewis, G. F., and Tanvir, N. R. 2002. Evidence for stellar substructure in the halo and outer disk of M31. *AJ*, **124**(Sept.), 1452–1463.

Geha, M., Willman, B., Simon, J. D., Strigari, L. E., Kirby, E. N., Law, D. R., and Strader, J. 2009. The least-luminous galaxy: spectroscopy of the Milky Way satellite Segue 1. *ApJ*, **692**(Feb.), 1464–1475.

Grcevich, J. and Putman, M. E. 2009. H I in Local Group dwarf galaxies and stripping by the galactic halo. *ApJ*, **696**(May), 385–395.

Grillmair, C. J. 2006. Detection of a 60 deg; long dwarf galaxy debris stream. *ApJ*, **645**(July), L37–L40.

Grillmair, C. J. 2009. Four new stellar debris streams in the Galactic halo. *ApJ*, **693**(Mar.), 1118–1127.

Gnedin, O. Y., Brown, W. R., Geller, M. J., and Kenyon, S. J. 2010. The mass profile of the Galaxy to 80 kpc. *ApJ*, **720**(Sept.), L108–L112.

Guhathakurta, P., and 9 colleagues. 2006. Dynamics and stellar content of the Giant Southern Stream in M31. I. Keck spectroscopy of red giant stars. *AJ*, **131**(May), 2497–2513.

Hayashi, E., Navarro, J. F., Taylor, J. E., Stadel, J., and Quinn, T. 2003. The structural evolution of substructure. *ApJ*, **584**(Feb.), 541–558.

Ibata, R., Martin, N. F., Irwin, M., Chapman, S., Ferguson, A. M. N., Lewis, G. F., and McConnachie, A. W. 2007. The haunted halos of Andromeda and Triangulum: a panorama of galaxy formation in action. *ApJ*, **671**(Dec.), 1591–1623.

Ivezić, Ž., and 42 colleagues 2000. Candidate RR Lyrae stars found in Sloan Digital Sky Survey Commissioning Data. *AJ*, **120**(Aug.), 963–977.

Ivezić, Ž., and 108 colleagues. 2008. LSST: from science drivers to reference design and anticipated data products. **(May)**, arXiv:0805.2366.

Kaiser, N., and 25 colleagues. 2002. Pan-STARRS: a Large Synoptic Survey Telescope array. *Proc. SPIE*, **4836**(Dec.), 154–164.

Kalirai, J. S., and 9 colleagues 2010. The SPLASH Survey: internal kinematics, chemical abundances, and masses of the Andromeda I, II, III, VII, X, and XIV dwarf spheroidal galaxies. *ApJ*, **711**(Mar.), 671–692.

Kauffmann, G., White, S. D. M., and Guiderdoni, B. 1993. The formation and evolution of galaxies within merging dark matter haloes. *MNRAS*, **264**(Sept.), 201.

Kazantzidis, S., Łokas, E. L., Callegari, S., Mayer, L., and Moustakas, L. A. 2011. On the efficiency of the tidal stirring mechanism for the origin of dwarf spheroidals: dependence on the orbital and structural parameters of the progenitor disky dwarfs. *ApJ*, **726**(Jan.), 98.

Kazantzidis, S., Mayer, L., Mastropietro, C., Diemand, J., Stadel, J., and Moore, B. 2004. Density profiles of cold dark matter substructure: implications for the missing-satellites problem. *ApJ*, **608**(June), 663–679.

Keller, S. C., and 12 colleagues. 2007. The SkyMapper Telescope and the Southern Sky Survey. *PASA*, **24**(May), 1–12.

Kirby, E. N., Simon, J. D., Geha, M., Guhathakurta, P., and Frebel, A. 2008. Uncovering extremely metal-poor stars in the Milky Way's ultrafaint dwarf spheroidal satellite galaxies. *ApJ*, **685**(Sept.), L43–L46.

Klypin, A., Gottlöber, S., Kravtsov, A. V., and Khokhlov, A. M. 1999b. Galaxies in N-body simulations: overcoming the overmerging problem. *ApJ*, **516**(May), 530–551.

Klypin, A., Kravtsov, A. V., Valenzuela, O., and Prada, F. 1999a. Where are the missing Galactic satellites? *ApJ*, **522**(Sept.), 82–92.

Komatsu, E., and 20 colleagues 2011. Seven-year Wilkinson Microwave Anisotropy Probe (WMAP) observations: cosmological interpretation. *ApJS*, **192**(Feb.), 18.

Koposov, S., and 10 colleagues 2008. The luminosity function of the Milky Way satellites. *ApJ*, **686**(Oct.), 279–291.

Kravtsov, A. 2010. Dark matter substructure and dwarf galactic satellites. *Advances in Astronomy*, **2010**, 1–21.

Kravtsov, A. V., Berlind, A. A., Wechsler, R. H., Klypin, A. A., Gottlöber, S., Allgood, B., and Primack, J. R. 2004a. The dark side of the halo occupation distribution. *ApJ*, **609**(July), 35–49.

Kravtsov, A. V., Gnedin, O. Y., and Klypin, A. A. 2004b. The tumultuous lives of galactic dwarfs and the missing satellites problem. *ApJ*, **609**(July), 482–497.

Kuhlen, 2010. Private communication.

Kuhlen, M., Weiner, N., Diemand, J., Madau, P., Moore, B., Potter, D., Stadel, J., and Zemp, M. 2010. Dark matter direct detection with non-Maxwellian velocity structure. *J. Cosmology Astropart. Phys.*, **2**(Feb.), 30.

Li, Y.-S., Helmi, A., De Lucia, G., and Stoehr, F. 2009. On the common mass scale of the Milky Way satellites. *MNRAS*, **397**(Sept.), L87–L91.

Loeb, A. and Zaldarriaga, M. 2005. Small-scale power spectrum of cold dark matter. *Phys. Rev. D*, **71**(May), 103520.

Macciò, A. V., Kang, X., and Moore, B. 2009. Central mass and luminosity of Milky Way satellites in the Λ Cold Dark Matter model. *ApJ*, **692**(Feb.), L109–L112.

Madau, P., Kuhlen, M., Diemand, J., Moore, B., Zemp, M., Potter, D., and Stadel, J. 2008. Fossil remnants of reionization in the halo of the Milky Way. *ApJ*, **689**(Dec.), L41–L44.

Majewski, S. R., and 11 colleagues. 2007. Discovery of Andromeda XIV: a dwarf spheroidal dynamical rogue in the Local Group?. *ApJ*, **670**(Nov.), L9–L12.

Majewski, S. R., Skrutskie, M. F., Weinberg, M. D., and Ostheimer, J. C. 2003. A Two Micron All Sky Survey view of the Sagittarius Dwarf Galaxy. I. Morphology of the Sagittarius core and tidal arms. *ApJ*, **599**(Dec.), 1082–1115.

Martin, N. F., and 12 colleagues 2009. PAndAS' CUBS: discovery of two new dwarf galaxies in the surroundings of the Andromeda and Triangulum galaxies. *ApJ*, **705**(Nov.), 758–765.

Martin, N. F., Ibata, R. A., Chapman, S. C., Irwin, M., and Lewis, G. F. 2007. A Keck/DEIMOS spectroscopic survey of faint Galactic satellites: searching for the least massive dwarf galaxies. *MNRAS*, **380**(Sept.), 281–300.

Martínez, G. D., Bullock, J. S., Kaplinghat, M., Strigari, L. E., and Trotta, R. 2009. Indirect dark matter detection from dwarf satellites: joint expectations from astrophysics and supersymmetry. *J. Cosmology Astropart. Phys.*, **6**(June), 14.

Martínez, G. D., Minor, Q. E., Bullock, J., Kaplinghat, M., Simon, J. D., and Geha, M. 2011. A complete spectroscopic survey of the Milky Way satellite Segue 1: dark matter content, stellar membership, and binary properties from a bayesian analysis. *ApJ*, **738**(Sept.), 55.

Mateo, M. L. 1998. Dwarf galaxies of the Local Group. *ARA&A*, **36**, 435–506.

McConnachie, A. W., and 28 colleagues. 2009. The remnants of galaxy formation from a panoramic survey of the region around M31. *Nature*, **461**(Sept.), 66–69.

McGaugh, S. S. and Wolf, J. 2010. Local Group dwarf spheroidals: correlated deviations from the baryonic Tully-Fisher relation. *ApJ*, **722**(Oct.), 248–261.

Moore, B., Ghigna, S., Governato, F., Lake, G., Quinn, T., Stadel, J., and Tozzi, P. 1999. Dark matter substructure within Galactic halos. *ApJ*, **524**(Oct.), L19–L22.

Moster, B. P., Somerville, R. S., Maulbetsch, C., van den Bosch, F. C., Macciò, A. V., Naab, T., and Oser, L. 2010. Constraints on the relationship between stellar mass and halo mass at low and high redshift. *ApJ*, **710**(Feb.), 903–923.

Navarro, J. F., Frenk, C. S., and White, S. D. M. 1997. A universal density profile from hierarchical clustering. *ApJ*, **490**(Dec.), 493.

Newberg, H. J., and 18 colleagues. 2002. The ghost of sagittarius and lumps in the halo of the Milky Way. *ApJ*, **569**(Apr.), 245–274.

Okamoto, T. and Frenk, C. S. 2009. The origin of failed sub-haloes and the common mass scale of the Milky Way satellite galaxies. *MNRAS*, **399**(Oct.), L174–L178.

Peñarrubia, J., Navarro, J. F., and McConnachie, A. W. 2008. The tidal evolution of Local Group dwarf spheroidals. *ApJ*, **673**(Jan.), 226–240.

Peter, A. H. G. and Benson, A. J. 2010. Dark-matter decays and Milky Way satellite galaxies. *Phys. Rev. D*, **82**(Dec.), 123521.

Polisensky, E. and Ricotti, M. 2010. Constraints on the dark matter particle mass from the number of Milky Way satellites. *Bulletin of the American Astronomical Society*, **42**(Jan.), #408.02.

Press, W. H. and Schechter, P. 1974. Formation of galaxies and clusters of galaxies by self-similar gravitational condensation. *ApJ*, **187**(Feb.), 425–438.

Profumo, S., Sigurdson, K., and Kamionkowski, M. 2006. What mass are the smallest protohalos?. *Physical Review Letters*, **97**(July), 031301.

Reid, B. A., and 29 colleagues 2010. Cosmological constraints from the clustering of the Sloan Digital Sky Survey DR7 luminous red galaxies. *MNRAS*, **404**(May), 60–85.

Ricotti, M. 2010. The first galaxies and the likely discovery of their fossils in the Local Group. *Advances in Astronomy*, **2010**, 1–21.

Ricotti, M., Gnedin, N. Y., and Shull, J. M. 2001. Feedback from galaxy formation: production and photodissociation of primordial H_2. *ApJ*, **560**(Oct.), 580–591.

Simon, J. D., and 11 colleagues 2011. A complete Spectroscopic survey of the Milky Way satellite Segue 1: The darkest galaxy. *ApJ*, **733**(May), 46.

Simon, J. D. and Geha, M. 2007. The kinematics of the ultrafaint Milky Way satellites: solving the missing satellite problem. *ApJ*, **670**(Nov.), 313–331.

Springel, V., and 8 colleagues. 2008. The Aquarius Project: the sub-haloes of galactic haloes. *MNRAS*, **391**(Dec.), 1685–1711.

Stadel, J., Potter, D., Moore, B., Diemand, J., Madau, P., Zemp, M., Kuhlen, M., and Quilis, V. 2009. Quantifying the heart of darkness with GHALO – a multibillion particle simulation of a galactic halo. *MNRAS*, **398**(Sept.), L21–L25.

Stoehr, F., White, S. D. M., Tormen, G., and Springel, V. 2002. The satellite population of the Milky Way in a ΛCDM universe. *MNRAS*, **335**(Oct.), L84–L88.

Strigari, L. E., Bullock, J. S., and Kaplinghat, M. 2007a. Determining the nature of dark matter with astrometry. *ApJ*, **657**(Mar.), L1–L4.

Strigari, L. E., Bullock, J. S., Kaplinghat, M., Diemand, J., Kuhlen, M., and Madau, P. 2007b. Redefining the missing satellites problem. *ApJ*, **669**(Nov.), 676–683.

Strigari, L. E., Bullock, J. S., Kaplinghat, M., Simon, J. D., Geha, M., Willman, B., and Walker, M. G. 2008. A common mass scale for satellite galaxies of the Milky Way. *Nature*, **454**(Aug.), 1096–1097.

Strigari, L. E., Frenk, C. S., and White, S. D. M. 2010. Kinematics of Milky Way satellites in a Lambda cold dark matter universe. *MNRAS*, **408**(Nov.), 2364–2372.

Stringer, M., Cole, S., and Frenk, C. S. 2010. Physical constraints on the central mass and baryon content of satellite galaxies. *MNRAS*, **404**(May), 1129–1136.

Tegmark, M., Silk, J., Rees, M. J., Blanchard, A., Abel, T., and Palla, F. 1997. How small were the first cosmological objects? *ApJ*, **474**(Jan.), 1.

The Dark Energy Survey Collaboration 2005. The Dark Energy Survey. (Oct.), arXiv:astro-ph/0510346.

Tollerud, E. J., Bullock, J. S., Graves, G. J., and Wolf, J. 2011. From galaxy clusters to ultrafaint dwarf spheroidals: a fundamental curve connecting dispersion-supported galaxies to their dark matter halos. *ApJ*, **726**(Jan.), 108.

Tollerud, E. J., Bullock, J. S., Strigari, L. E., and Willman, B. 2008. Hundreds of Milky Way satellites? Luminosity bias in the satellite luminosity function. *ApJ*, **688**(Nov.), 277–289.

Tinker, J., Kravtsov, A. V., Klypin, A., Abazajian, K., Warren, M., Yepes, G., Gottlöber, S., and Holz, D. E. 2008. Toward a halo mass function for precision cosmology: the limits of universality. *ApJ*, **688**(Dec.), 709–728.

van den Bergh, S. 2000. Updated information on the Local Group. *PASP*, **112**(Apr.), 529–536.

van der Marel, R. P., Magorrian, J., Carlberg, R. G., Yee, H. K. C., and Ellingson, E. 2000. The velocity and mass distribution of clusters of galaxies from the CNOC1 Cluster Redshift Survey. *AJ*, **119**(May), 2038–2052.

Viel, M., Becker, G. D., Bolton, J. S., Haehnelt, M. G., Rauch, M., and Sargent, W. L. W. 2008. How cold is cold dark matter? Small-scales constraints from the flux power spectrum of the high-redshift Lyman-α forest. *Physical Review Letters*, **100**(Feb.), 041304.

Walker, M. G., Mateo, M., and Olszewski, E. W. 2009a. Stellar velocities in the Carina, Fornax, Sculptor, and Sextans dSph galaxies: data from the Magellan/MMFS Survey. *AJ*, **137**(Feb.), 3100–3108.

Walker, M. G., Mateo, M., Olszewski, E. W., Peñarrubia, J., Wyn Evans, N., and Gilmore, G. 2009b. A universal mass profile for dwarf spheroidal galaxies?. *ApJ*, **704**(Oct.), 1274–1287.

Walsh, S. M., Willman, B., and Jerjen, H. 2009. The invisibles: a detection algorithm to trace the faintest Milky Way satellites. *AJ*, **137**(Jan.), 450–469.

Watkins, L. L., and 10 colleagues. 2009. Substructure revealed by RRLyraes in SDSS Stripe 82. *MNRAS*, **398**(10/2009.), 1757–1770.

White, S. D. M. and Rees, M. J. 1978. Core condensation in heavy halos – a two-stage theory for galaxy formation and clustering. *MNRAS*, **183**(May), 341–358.

Willman, B. 2010. In pursuit of the least luminous galaxies. *Advances in Astronomy*, **2010**, 1–11.

Willman, B., and 14 colleagues. 2005. A new Milky Way dwarf galaxy in Ursa Major. *ApJ*, **626**(June), L85–L88.

Wolf, J., *et al.* in preparation.

Wolf, J., Martinez, G. D., Bullock, J. S., Kaplinghat, M., Geha, M., Muñoz, R. R., Simon, J. D., and Avedo, F. F. 2010. Accurate masses for dispersion-supported galaxies. *MNRAS*, **406**(Aug.), 1220–1237.

Xue, X. X., and 16 colleagues 2008. The Milky Way's circular velocity curve to 60 kpc and an estimate of the dark matter halo mass from the kinematics of 2400 SDSS Blue Horizontal-Branch stars. *ApJ*, **684**(Sept.), 1143–1158.

Zentner, A. R. and Bullock, J. S. 2003. Halo substructure and the power spectrum. *ApJ*, **598**(Nov.), 49–72.

Zucker, D. B., and 19 colleagues. 2004. A new giant stellar structure in the outer halo of M31. *ApJ*, **612**(Sept.), L117–L120.

Zucker, D. B., and 32 colleagues. 2006. A new Milky Way dwarf satellite in Canes Venatici. *ApJ*, **643**(June), L103–L106.

4. The Milky Way satellite galaxies as critical tests of contemporary cosmological theory

P. KROUPA

4.1 Introduction

Our understanding of the cosmological world relies on two fundamental assumptions: (1) The validity of General Relativity, and (2) conservation of matter since the Big Bang. Both assumptions yield the standard cosmological model according to which dark matter structures form first and then accrete baryonic matter that fuels star formation in the emerging galaxies. One important way to test assumption one is to compare the phase-space properties of the nearest galaxies with the expectations of the standard cosmological model.

Although the possibility of the existence of dark matter (DM) was first evoked more than 85 years ago (Einstein, 1921; Oort, 1932; Zwicky, 1933) and has been under heavy theoretical and experimental scrutiny (Bertone *et al.*, 2005) since the discovery of flat galactic rotation curves by Rubin and Ford (1970) and their verification and full establishment by Bosma (1981), the DM particle candidates still elude both direct and indirect detection (Lingenfelter *et al.*, 2009; Latronico and for the Fermi LAT Collaboration, 2009). Indeed, it appears that also the cryogenic dark matter search (CDMS) experiment fails to find significant evidence for the existence of cold dark matter (CDMS II Collaboration *et al.*, 2010). Favored today is dark matter made of non-relativistic ("cold") particles (cold DM, CDM) as it allows the correct degree of large-scale structure formation. Less-massive particles can perhaps account for the observed structures as long as the particles are not too light, leading to Warm DM (WDM) models, while light, relativistic ("hot") particles (Hot DM, HDM) are excluded because structures on galactic scales cannot form sufficiently rapidly. In the following, the term "DM" is used when referring to generic properties or consequences of the CDM and WDM hypothesis.

According to the DM hypothesis, galaxies have been assembling via accretion and numerous mergers of smaller DM halos. Therefore, galaxies like the MW should be swarmed by hundreds and thousands of such halos (Moore *et al.*, 1999; Diemand *et al.*, 2008), whereby the number of sub-halos is slightly smaller in WDM than in CDM models. Furthermore, the triaxial nature of the flow at formation would make it impossible to destroy halo substructure by violent relaxation (Boily *et al.*, 2004). These should be distributed approximately isotropically about their host and have a mass function such that the number of sub-halos in the mass interval $M_{\rm vir}, M_{\rm vir} + dM_{\rm vir}$ is approximately $dN \propto M_{\rm vir}^{-1.9}\, dM_{\rm vir}$ (Gao *et al.*, 2004).

In contrast to this expectation only a few dozen shining satellites have been found around both disk galaxies with bulges, the MW, and Andromeda (M31), whereas the next largest disk galaxy in the Local Group, M33, has no known satellites and also no bulge. Observations of the internal kinematics (velocity dispersion) of the satellites indeed suggest they are the most DM-dominated galaxies known (e.g., figure 15 in Simon and Geha 2007). That is, the motions of their stars seem to be defined by an unseen mass component by moving much faster than can be accounted for by their luminous matter. The known satellites have therefore been taken to be the luminous "tip of the iceberg" of the vast number of dark sub-halos orbiting major galaxies such as the MW. One of the many problems this interpretation faces is that the solution of the Jeans equations led to DM density profiles within the dwarf-spheroidal (dSph) satellites that are typically consistent with having large cores with radii comparable to the luminous radii of the galaxies (Gilmore *et al.*, 2007) rather than the cusped profiles expected from numerical models of the formation of DM structures (Navarro *et al.*, 1997; but see Del Popolo and Kroupa, 2009).

Observations of spiral and low-surface brightness galaxy rotation curves, interestingly, also point toward DM halos having a large, kpc-scale, constant density core (de Blok and Bosma, 2002; Gentile *et al.*, 2004). The latest ΛCDM simulations of the AQUARIUS project by Navarro *et al.* (2010) point toward Einasto profiles asymptotically reaching a zero-slope at the very center: however, the difference with cusped profiles is on too small a scale (20% difference in kinematics at 5% of the scale radius) to solve the problems of galaxy rotation curves.

Moreover, DM simulations are not even close to producing large enough galactic disks nor bulgeless disk galaxies due to the predicted low angular momentum within the simulations (Combes, 2004, 2010). Real disk galaxies are large, and ultrathin and bulgeless disk galaxies are common in nature. Heating of the ambient gas through feedback energy from the central super-massive blackhole and from supernovae has been invoked to slow accretion and thus allow larger disks to grow (e.g., Weil *et al.*, 1998; Piontek and Steinmetz, 2009), but this mechanism does not work in bulgeless disk galaxies. Another solution to this long-standing problem includes continuous gas infall controlled by efficient cooling (Dekel *et al.*, 2009), a strong hint that baryons come to dominate the formation of disk galaxies. However, while explaining the formation of large and massive thick galactic disks, the progenitors of present-day early-type galaxies, this scenario has yet to demonstrate its ability to create thin lower-mass galactic disks. In general, it has not been possible to theoretically arrive at the observed large fraction of thin disk galaxies and bulgeless galaxies within the DM framework.

Much theoretical effort has been invested in trying to understand why the number of luminous satellites is so much smaller than the number of DM-halos predicted by the currently favored concordance ΛCDM hypothesis. It has been suggested that stellar feedback and heating processes limit baryonic growth, that reionization stops low-mass DM halos from accreting sufficient gas to form stars, and that tidal forces from the host halo limit growth of the DM sub-halos and lead to truncation of DM sub-halos (Dekel and Silk, 1986; Dekel and Woo, 2003; Kirby *et al.*, 2009; Koposov *et al.*, 2009; Macciò *et al.*, 2009, 2010; Okamoto and Frenk, 2009; Shaya *et al.*, 2009; Busha *et al.*, 2010). Moreover, the recent discovery of new (ultrafaint) dwarf spheroidals around the MW has also been presented as a great improvement regarding this missing satellite problem. Such lines of reasoning have generally led to claims that within the ΛCDM cosmology no serious small-scale issues are apparent (e.g., Tollerud *et al.*, 2008; Primack, 2009).

An additional but mostly ignored challenge for understanding the origin and nature of the observed satellite galaxies comes from their being significantly anisotropically distributed about the MW, and possibly about Andromeda also. The existence of closely aligned streams of satellites was realized soon after the discovery of the satellites (Lynden-Bell, 1976; Kunkel, 1979; Bell, 1983) and has been discussed also in more recent times (Majewski, 1994; Palma *et al.*, 2002; Koch and Grebel, 2006). Notably, Lynden-Bell and Lynden-Bell (1995) pointed out the probable existence of more than one stream, each of which contains globular clusters and dSph satellites, with most of the streams being quite aligned. The disagreement of the MW system with the expectations from the cosmological DM hypothesis, however, was emphasized only recently by Kroupa *et al.* (2005). They pointed out that the observed satellite system of the MW was incompatible at the 99.5% significance level with the theoretical distribution expected if the satellites were DM sub-halos tracing an isotropic DM host halo. Until then, the prediction within the DM hypothesis was that the distribution of sub-halos ought to be nearly spherical and tracing the shape of the host DM halo. For example, Aubert *et al.* (2004) show an MW-type DM halo to have an infall asymmetry of about 15%. The sub-halos enter the host halo along filaments, these having a width comparable to the diameter of the growing host halo.

It had been speculated that the highly anisotropic spatial satellite distribution maps a highly prolate DM halo of the MW which would need to have its principal axis oriented nearly perpendicularly to the MW disk (Hartwick, 2000). The recent measurement of the shape of the DM halo of the MW within 60 kpc by Law *et al.* (2009), based on the Sagittarius stream, suggests the DM halo to be prolate, but with major and minor axes lying within the plane of the MW disk. But, the 11 "classical" (brightest) MW satellites form a pronounced Disk-of-Satellites (DoS) which lies nearly perpendicular to the MW disk. During the past 15 years, 13 additional "new" and mostly ultrafaint satellite galaxies have been found largely with the Sloan Digital Sky Survey (SDSS).[1] The majority of the new dwarfs also follow this distribution (Metz *et al.*, 2009). Subsequent mathematically and computationally involved statistical tests of data distributions on spheres confirmed the highly significant disagreement between prolate, isotropic, and oblate DM parent halos and the observed distribution if the dSph satellites are interpreted to be sub-halos orbiting within the parent halo (Metz *et al.*, 2007, 2008, 2009). The DM halo of the MW and the satellites therefore do not trace a similar three-dimensional structure, unless the major axis of the MW halo changes its orientation by about 90 degrees beyond 60 kpc. The observed distribution of the ultrafaint satellites could be biased by the small fraction of the sky scanned in the SDSS-DR6 photometric catalog: a more definitive conclusion will be reached when the sky is more fully covered. However, the already covered volume does in fact contain regions that are empty of satellites, whereas other regions contain new satellites, which is the basis of the result reported by Metz *et al.* (2009) that the new MW satellites follow the MW DoS. Indeed, it appears implausible that the not yet detected satellites can re-instate a near-isotropy because these putative (hyperfaint) satellites would then have to have a different spatial distribution than the 11 brighter ("classical") satellites and the new ultrafaint ones discussed here. If true, this would suggest two populations of satellites; but this has not been predicted within the framework of the DM hypothesis. Deep follow-up observations of fainter ultra-low-luminosity satellite candidates discovered recently (Walsh *et al.*, 2009) show these to be spurious detections, suggesting that there are no more lower luminosity satellites ($M_V < -2$, $D < 150\,\mathrm{kpc}$) in the SDSS footprint, i.e., an area covering half of the Northern hemisphere (Jerjen *et al.*, in preparation), in contradiction to the ΛCDM prediction that there should be about 50 additional satellites in a quarter celestial sphere (Koposov *et al.*, 2009; Cooper *et al.*, 2010). Also, as demonstrated in Section 4.5, the new satellites do in fact define a DoS by themselves, and this DoS is virtually the same as the DoS defined by the 11 classical satellites.

It thus seems very difficult to reconcile the presently observed planar distribution of satellites with the expected spheroidal distribution of DM sub-halos around the MW. This problem is reinforced by a striking degree of phase-space correlation between the inner and outer satellites: the normal vector of the DoS is mostly defined by the distribution of the outermost satellites, while the average satellite orbital angular momentum is defined mostly by the inner satellites, because proper motions are available only for these. It is striking that both the normal and orbital angular momentum vectors are co-aligned in such a way that the DoS is to be envisioned as a system of satellites that are orbiting mostly in one sense about the MW, i.e., a rotating flattened structure (Palma *et al.*, 2002; Metz *et al.*, 2008). So perhaps this questions the current paradigm of structure formation according to which the dSph satellites are dark matter sub-halos.

This currently popular paradigm relies on the existence of (up to now) undetected DM particles, but there exist alternative theories of gravity developed to explain observations without resorting to the hypothesis of the existence of dark matter. These would, however, imply a violation of fundamental assumption 1 stated earlier, at least in the strict

[1] For convenience, the 11 brightest satellite galaxies are here referred to as being the "classical" satellites. these include the LMC and the SMC. The other satellites are fainter than the faintest "classical" satellites (UMi and Draco), and these are called the "new" or the "ultrafaint" satellites or dwarfs.

sense of Einstein's General Theory of Relativity (GR). Finding a definitive test that distinguishes between these two different solutions to the problem of galactic dynamics and cosmological structure formation is hard. Both DM and modified gravity are designed to solve similar problems, so the test must rely on subtle differences between the models and the observational data. The greatest differences between the two competing approaches are expected in the weak gravitational regime where the subtleties of non-Newtonian dynamics are most pronounced, which is why the constituents of outer surroundings of galaxies allow the most stringent tests.

In essence, the aim is thus to test gravitation in the weak field limit: Does the currently widely accepted/used model (GR plus DM and Λ as a vacuum energy contribution) account for the observed phase-space population of visible matter in the outer fringes of galaxies? Given that a substantial number of independent research groups working in the traditional ΛCDM and WDM approaches have by now made firm statements on dwarf satellite galaxies, the DM plus GR hypothesis can be tested sensitively on these scales.

Here the most recent state-of-the art models that have been calculated by a number of independent research groups within the DM framework aimed at explaining the properties of the faint satellite galaxies are examined critically. In doing so three new problems for DM cosmology are uncovered: the absence of a mass-luminosity relation for dwarf spheroidal satellites of the Milky Way (Section 4.2), the wrong mass-distribution of luminous-satellite halo-masses (Section 4.3), and the observed relation between the bulge mass and the number of satellites of Local Group galaxies (Section 4.4). The question of whether the DoS exists, and if in fact the latest MW satellite discoveries follow the DoS, or whether the existence of the DoS is challenged by them is addressed in Section 4.5. In Section 4.6 a possible solution to these problems is proposed, also solving the DoS problem, namely, that most dwarf galaxies have a tidal origin. Conclusions regarding the consequences of this with respect to DM and modified gravity are drawn in Section 4.7.

4.2 The satellite mass–luminosity relation

Our understanding of the physical world relies on some fundamental physical principles. Among them, energy conservation implies it to be increasingly more difficult to unbind subcomponents from a system with increasing binding energy.

Within the DM hypothesis, the principle of energy conservation therefore governs how DM potentials fill up with matter. There are two broadly different physical models exploring the consequences of this, namely, models of DM halos based on internal energy sources (mostly stellar feedback), and models based on external energy input (mostly ionization radiation). In the following, first the observational mass luminosity data for the known satellite galaxies are discussed, and these are then compared to the theoretical results that are calculated assuming that dark matter exists such that Newtonian dynamics is valid.

4.2.1 *The observational data*

Based on high-quality measurements of individual-stellar line-of-sight velocities in the satellite galaxies and assuming Newtonian dynamics to be valid, Strigari *et al.* (2008) (hereinafter S08) calculate dynamical masses, $M_{0.3\,\mathrm{kpc}}$, within the inner 0.3 kpc of 18 MW dSph satellite galaxies over a wide range of luminosities ($10^3 \lesssim L \lesssim 10^7\ \mathrm{L}_\odot$). The LMC and SMC are excluded, as is Sagittarius because it is currently experiencing significant tidal disturbance. They significantly improve the previous work by using larger stellar data sets and more than doubling the number of dwarf galaxies and applying more detailed mass modeling. Their results confirm the earlier suggestion by Mateo *et al.* (1993); Mateo (1998); Gilmore *et al.* (2007), and Peñarrubia *et al.* (2008) that the satellites share a common DM mass scale of about 10^7 M$_\odot$, "and conclusively establish" this common mass scale.

The finding of S08 can be quantified by writing

$$\log_{10} M_{0.3\,\mathrm{kpc}} = \log_{10} M_0 + \kappa \log_{10} L, \tag{4.1}$$

and by evaluating the slope, κ, and the scaling, M_0. S08 report $\kappa = 0.03 \pm 0.03$ and $M_0 \approx 10^7$ M$_\odot$ (i.e., $M_{0.3\,\mathrm{kpc}}$ is essentially constant because κ is close to 0). Using the Dexter Java application of Demleitner *et al.* (2001), a nonlinear, nonsymmetric error weighted least-squares fit to the S08 measurements reproduces the common mass and slope found by S08, as can be seen from the parameters listed in Table 4.1. By excluding the least-luminous dSph data point one obtains the same result (Table 4.1).

It follows from Equation 4.1 that

$$M_{0.3\,\mathrm{kpc}}^{1/\kappa} = M_0^{1/\kappa}\, L. \tag{4.2}$$

This central mass of the DM halo can be tied through high-resolution CDM simulations to the total halo virial mass before its fall into the host halo (S08, see also Section 4.3),

$$M_{\mathrm{vir}} = M_{0.3\,\mathrm{kpc}}^{1/0.35} \times 10^{-11} \mathrm{M}_\odot, \tag{4.3}$$

yielding $M_{\mathrm{vir}} = 10^9$ M$_\odot$ for $M_{0.3\,\mathrm{kpc}} = 10^7$ M$_\odot$ (the common-mass scale). Thus, substituting $M_{0.3\,\mathrm{kpc}}$ in Equation 4.3 using Equation 4.2,

$$M_{\mathrm{vir}}^{0.35/\kappa} = M_0^{1/\kappa} \times 10^{-(11\times 0.35)/\kappa}\, L. \tag{4.4}$$

Note that this value of the halo mass near 10^9 M$_\odot$ for the satellites in the S08 sample is confirmed by a new analysis assuming, like S08, Newtonian Jeans equations, in which Wolf *et al.* (2010) show that the mass can be derived from a velocity dispersion profile within the de-projected 3D half light profile with minimal assumptions on the velocity anisotropy. They thus obtain a very robust mass estimator.

The observed five-sigma lower value for $0.35/\kappa \equiv \eta$ is thus 2.06 (with $\kappa = 0.02 + 5 \times 0.03$ from Table 4.1).

4.2.2 *Model type A: Internal energy sources*

Dekel and Silk (1986) and Dekel and Woo (2003) studied models according to which star formation in DM halos below a total halo mass of $M_{\mathrm{vir}} \approx 10^{12}$ M$_\odot$ is governed by thermal properties of the in flowing gas that is primarily regulated by supernova feedback. These models show that the mass-to-light ratio of sub-halos follows $M_{\mathrm{vir}}/L \propto L^{-2/5}$ (Equation 24 in Dekel and Woo 2003; see also Equation 33 in Dekel and Silk 1986). This approximately fits the observed trend for dSph satellite galaxies (Mateo, 1998).

These models thus imply

$$M_{\mathrm{vir}}^{\eta_{\mathrm{th}}} = \zeta\, L, \tag{4.5}$$

where L is the total luminosity, M_{vir} is the virial DM halo mass, $\eta_{\mathrm{th}} = 5/3$, and ζ is a proportionality factor. In essence, this relation contains the physical reality that more-massive halos have a larger binding energy such that it becomes more difficult to remove matter from them in comparison to less massive halos.

Comparing with Equation 4.4 it follows that the observed five-sigma lower value for $0.35/\kappa \equiv \eta = 2.06$ is in clear violation of Equation 4.5 where $\eta_{\mathrm{th}} = 5/3 = 1.67$.

4.2.3 *Model type B1, B2: external energy source*

Busha *et al.* (2010) follow a different line of argument for explaining the dSph satellite population by employing the DM halo distribution from the *Via Lactea* simulation. Here the notion is that reionization would have affected DM halos variably, because of an

TABLE 4.1. Fitted parameters for Equation 4.1. Our fit to $\kappa = 0.35/\eta$: data 1–4 are observational values, data A–F are models (see Section 4.2). Notes: 1: our fit to S08 (who give central 300 pc masses, 18 satellites, their figure 1); 2: our fit to S08 without Seg.1 (faintest satellite, i.e., 17 satellites, their figure 1); 3: our fit to S08 without Seg.1 and without Hercules (i.e., 16 satellites, their figure 1); 4: our fit to the obs.data plotted by Okamoto and Frenk (2009) (who give central 600 pc masses, only 8 satellites, their figure 1); A: Dekel and Silk (1986); Dekel and Woo (2003), stellar feedback (Equation 4.5); B1: our fit to Busha *et al.* (2010), their SPS model; B2: our fit to Busha *et al.* (2010), in-homogeneous reionization model; C: our fit to Macciò *et al.* (2010), semi-analytical modeling (SAM), fit is for $L_V > 3 \times 10^5$ $L_{V,\odot}$; D: our fit to Okamoto and Frenk (2009) (Aq-D-HR). E1: our fit to the 1 keV WDM model of Macciò and Fontanot (2010); E2: our fit to the 5 keV WDM model of Macciò and Fontanot (2010); F: our fit to the Aquarius sub-halo-infall models of Cooper *et al.* (2010). The entries with an asterisk are for the central 600 pc radius region.

The slope of the mass–luminosity relation of dSph satellite galaxies

data	κ	radius/pc	$M_0/10^7$ $M_\odot$
Empirical:			
1	$+0.02 \pm 0.03$	300	1.02 ± 0.39
2	$+0.02 \pm 0.03$	300	1.01 ± 0.40
3	$+0.01 \pm 0.03$	300	1.09 ± 0.44
*4	-0.03 ± 0.05	600	6.9 ± 4.9
DM Models:			
A: feedback	0.21	300	—
B1: reionization, SPS	0.15 ± 0.02	300	0.24 ± 0.06
B2: reionization	0.17 ± 0.01	300	0.18 ± 0.02
C: SAM	0.42 ± 0.02	300	2.0 ± 0.9
*D: Aq-D-HR	0.17 ± 0.02	600	0.41 ± 0.14
E1: 1keV(WDM)	0.23 ± 0.04	300	0.069 ± 0.045
E2: 5keV(WDM)	0.12 ± 0.02	300	0.43 ± 0.081
F: Aq-infall	0.13 ± 0.01	300	0.32 ± 0.022

inhomogeneous matter distribution. A given DM halo must grow above a critical mass before reionization in order to form stars or accrete baryons. Thus the inhomogeneous reionization model (Busha *et al.*, 2010, their figure 6) implies, upon extraction of the theoretical data and using the same fitting method as above, theoretical κ-values of 0.15–0.17. These disagree, however, with the observational value of 0.02 with a significance of more than 4 sigma (i.e., 99.99%; Table 4.1).

Busha *et al.* (2010) suggest that adding scatter into the theoretical description as to how DM halos are filled with luminous baryons would reduce the discrepancy, but it is difficult to see how this can be done without violating the actual scatter in the observed $M_{0.3\,\mathrm{kpc}} - L$ relation.

4.2.4 *Model type C: semi-analytical modeling (SAM)*

How the multitude of DM halos are filled with baryons given the above processes combined is investigated by Macciò *et al.* (2010) with semi-analytical modeling (SAM) of DM sub-halos based on N-body merger tree calculations and high-resolution recomputations. Their theoretical mass–luminosity data are plotted in their figure 7, and a fit to the redshift $z = 0$ data for $L_V > 3 \times 10^5$ $L_{V,\odot}$ satellites is listed in Table 4.1. The theoretical SAM data set shows a steep behavior, $\kappa = 0.42$. Given the observational data, this model is ruled out with a confidence of more than 10 sigma.

4.2.5 *Model type D: high-resolution baryonic physics simulations (Aq-D-HR)*

The satellite population formed in a high-resolution Nbody ΛCDM resimulation with baryonic physics of one of the MW-type "Aquarius" halos is studied by Okamoto and Frenk (2009). The treatment of baryonic processes include time-evolving photo-ionization, metallicity dependent gas cooling, photo-heating, supernova (SN) feedback, and chemical enrichment through SN Ia, SN II, and AGB stars. reionization is included and the galactic winds driven by stellar feedback are assumed to have velocities proportional to the local velocity dispersion of the dark matter halo. In these models 100% of SNII energy is deposited as thermal energy. Galactic winds are thus produced even for the least-massive dwarf galaxies. Winds are observed in strong star bursts induced through interactions, rather than in self-regulated dwarf galaxies that may pose a problem for this ansatz (Ott *et al.*, 2005). The details of the simulations are provided by Okamoto *et al.* (2010). The resultant sub-halo population with stars is suggested to reproduce the S08 common-mass scale.

Following the same procedure as for the previous models, this claim is tested obtaining κ for the Okamoto and Frenk (2009) models from their figure 1 (upper panel, red asterisks), comparing it to the observational data also plotted in their figure 1 (note that Okamoto and Frenk 2009 plot the masses within 600 pc rather than 300 pc as used above). For the observational data, it follows from their plot, which only includes central-600 pc masses for the eight most luminous satellites, $\kappa_{\rm obs,OF} = -0.03 \pm 0.05$. This is nicely consistent with the full S08 sample (18 satellites) discussed earlier. For their model data $\kappa = 0.17 \pm 0.02$, i.e., the model can be discarded with a confidence of three sigma or 99.7%.

4.2.6 *Model type E1, E2: warm dark matter*

Macciò and Fontanot (2010) present theoretical distributions of satellite galaxies around a MW type host halo for different cosmological models, namely, ΛCDM and warm-dark matter (WDM) with three possible DM-particle masses of $m_w = 1$, 2, and 5 keV. They perform numerical structure formation simulations and apply semi-analytic modeling to associate the DM sub-halos with luminous satellites. The luminosity function and mass luminosity data of observed satellites is suggested to be reproduced by the WDM models implying a possible lower limit to the WDM particle of $m_w \approx 1\,$keV.

The model and observational mass luminosity data are compared in their figure 5 for $m_w = 1$ and 5 keV. The slopes of these model data are listed in Table 4.1. From Table 4.1 it follows that the WDM model with $m_w \approx 1$ keV is ruled out with very high confidence (four sigma or 99.99%), and in any case has too few satellites fainter than $M_V \approx -8$ (their figure 4). WDM models with $m_w \approx 5$ keV are excluded at least with a three sigma or 99.7% confidence, and, as is evident from their figure 4, the models contain significantly too few satellites brighter than $M_v = -11$.

4.2.7 *Model type F: infalling and disrupting dark matter satellite galaxies*

Cooper *et al.* (2010) study CDM model satellites in individual numerical models of dark matter halos computed within the Aquarius project. Semi-analytical modeling is employed to fill the sub-halos with visible matter, and the orbits of the infalling satellites are followed. General agreement with the observed satellites is claimed.

Much as the other models, in this numerical CDM model of substructure and satellite formation in an MW-type host halo, the MW sub-halos fall in stochastically and therefore do not agree with the observed phase-space correlated satellites, i.e., with the existence of the DoS (Section 4.5). Furthermore, the presented model mass-luminosity data (their figure 5) lead to a too steep slope (Table 4.1) compared to the observations. Indeed, the dark matter based model is excluded with a confidence of at least 99.7%. Additionally, figure 5 of Cooper *et al.* (2010) shows a very significant increase in the number of model

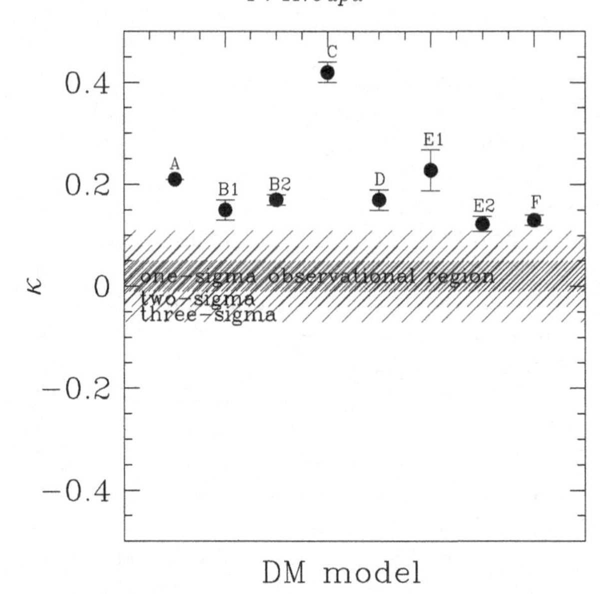

FIG. 4.1. The slope of the mass–luminosity relation, κ (Equation 4.1), for the models listed in Table 4.1. The observational constraints with confidence intervals are depicted as hatched regions. Satellites with a larger dark matter mass are on average more luminous such that the mass–luminosity relation has $\kappa > 0$. However, the observational constraints lie in the region $\kappa \approx 0$ (see Table 4.1). The hypothesis that the data are consistent with any one of the models can be discarded with very high (at least three sigma, or more than 99.7%) confidence.

satellites with a similar brightness as the faintest known satellite (Segue 1). This is in contradiction with the failure to find any additional satellites of this luminosity in the most recent data mining of the existing northern SDSS survey data, as emphasized in Section 4.1.

4.2.8 *Model A vs B vs C vs D vs E vs F vs observed slope*

In Figure 4.1 the latest theoretical ansatzes A–F to solve the cosmological substructure problem are compared with the latest observational limit on the slope κ, as a visualization of the already reached conclusion that theoretical expectation of the observed κ fails. This is a result of energy conservation: because binding energy is an obstacle for baryons trying to leave a DM halo, a natural theoretical slope, $\kappa > 0$, appears in ΛCDM, such that less-massive halos end up being less luminous, on average. Because there is a well-defined mapping at any redshift between the virial mass and the mass within some radius (see Section 4.3), it follows that the dSph satellites ought to show a clear mass luminosity relation *if* the DM hypothesis were a correct description of reality.

A point of contention for dark matter models of dSph satellite galaxies is that the DM halos supposedly grow at various rates and are also truncated variously due to tidal influence. For example, Macciò *et al.* (2009) reduce manually, by a factor of two, the characteristic density of *all* sub-halos with concentration less than 10 at the moment of infall, meaning that *all* massive sub-halos have been significantly stripped, which is unlikely. The highly complex interplay between dark matter accretion and orbit-induced accretion truncation leads to the power-law mass function of DM halos, and at the same time would imply that the outcome in which all luminous DM sub-halos end up having the same mass were incompatible with the ΛCDM-theoretical expectations (Section 4.3).

While the theoretical results always lead to a trend of luminosity with halo mass as a result of energy conservation, the observed satellites do not show this trend according to Mateo (1998), Peñarrubia *et al.* (2008), and Strigari *et al.* (2008). From Figure 4.1

we note that 7 ΛCDM models of the satellites deviate 4σ or more from the data, while only one (the WDM model with $m_w = 5\,\text{keV}$) deviates more than 3σ from the data. The likelihood[2] that any one of them describes the data is thus less than 0.3%. Interestingly, Dekel and Woo (2003) had already identified an issue with fitting the data below a stellar mass of $3 \times 10^7\ \mathrm{M}_\odot$ (their figure 5) and volunteered an explanation for this in their section 7 which, in essence, acknowledges the failure of the CDM prediction.

As a caveat, the observed absence of a mass-luminosity relation partially depends on the data for the ultrafaint dwarfs: indeed, for the classical (most luminous) dSphs, Serra *et al.* (2010) argue that there may be a trend, $\kappa > 0$, essentially due to their suggested increase for the mass of the Fornax dSph satellite. It is, on the other hand, plausible that the ultrafaint dwarfs do not possess any dark halo (see Section 4.6), and that the derivation of their enclosed mass is due to observational artifacts, but in that case, they should not be used as a possible improvement for the missing satellite problem, thus in any case putting the ΛCDM model in serious trouble.

Indeed, Adén *et al.* (2009) suggest that for the Hercules dSph satellite interloper stars need to be removed from the observational sample, leading to a revision of the mass within 300 pc to the value $M_{0.3\,\mathrm{kpc}} = 1.9^{+1.1}_{-1.6} \times 10^6\ \mathrm{M}_\odot$ (instead of the value $M_{0.3\,\mathrm{kpc}} = 7.2^{+0.51}_{-0.21} \times 10^6\ \mathrm{M}_\odot$ derived by S08). This new mass measurement, however, now lies more than $4\,\sigma$ away from all ΛCDM models considered earlier (Table 4.1). Hercules would thus not be understandable in terms of a DM-dominated model, and Adén *et al.* (2009) state that DM-free models cannot be excluded (note also Figure 4.6), or that Hercules may be experiencing tidal disturbance in the outer parts. Tidal disturbance, however, would have to be very significant for its inner structure to be affected, because for conformity with the theoretical DM models its $M_{0.3\,\mathrm{kpc}}$ mass would have to have been much higher and similar to the value derived by S08. In fact, given the current Galactocentric distance of Hercules of 138 kpc and the result that the inner region of a satellite is only affected by tides after significant tidal destruction of its outer parts (Kazantzidis *et al.*, 2004), this scenario is physically implausible. There are therefore three possibilities: (i) Hercules is a DM-dominated satellite. This, however, then implies that no logically consistent solution within the CDM framework is possible because its mass luminosity datum would lie well away from the theoretical expectation. (ii) Hercules has no DM. This implies that it cannot be used in the mass luminosity data analysis explained earlier and would also imply the existence of an additional type of DM-free satellites that, however, share virtually all observable physical characteristics with the putatively DM filled satellites. (iii) Hercules has been significantly affected by tides. This case is physically implausible, but it also implies that Hercules cannot be used in the mass-luminosity analysis earlier (just as Sagittarius is excluded because of the significant tidal effects it is experiencing). Omitting Hercules from the data thus leads to a revised observational slope $\kappa = 0.01 \pm 0.03$ such that none of the conclusions reached here on the performance of the DM models are affected.

4.3 The mass function of CDM halo masses

One of the predictions of the ΛCDM hypothesis is the self-similarity of DM halos down to (at least) the mass range of dwarf galaxies, i.e., that massive halos contain sub-halos of lesser mass, but the same structure in a statistical sense (Moore *et al.*, 1999; for a major review, see Del Popolo and Yesilyurt, 2007). The mass function of these sub-halos is, up to a critical mass M_{crit}, very well approximated by

$$\xi_{\mathrm{sub}}(M_{\mathrm{vir}}) = \frac{dN}{dM_{\mathrm{vir}}} \propto M_{\mathrm{vir}}^{-1.9} \tag{4.6}$$

[2] The likelihood $= 1-$confidence/100 gives an indication of how likely the data can be accounted for given a model.

where dN is the number of sub-halos in the mass interval $M_{\rm vir}, M_{\rm vir} + dM_{\rm vir}$ (Gao *et al.*, 2004). $M_{\rm crit}$ is given by $M_{\rm vir} \approx 0.01 M_{\rm h}$ where $M_{\rm h}$ is the virial mass of the hosting CDM halo. The virial mass, $M_{\rm vir}$, is defined by

$$M_{\rm vir} = \frac{4\pi}{3} \Delta_{\rm vir} \rho_0 r_{\rm vir}^3 \tag{4.7}$$

where ρ_0 is the critical density of the universe and $\Delta_{\rm vir}$ is a factor such that $\Delta_{\rm vir}\rho_0$ is the critical density at which matter collapses into a virialized halo, despite the overall expansion of the universe. The virial radius $r_{\rm vir}$ is thereby determined by the density profile of the collapsed CDM halo. For $M_{\rm vir} > 0.01\, M_{\rm h}$, the mass function steepens (Gao *et al.*, 2004), so that it is effectively cut off at a mass $M_{\rm max}$, $M_{\rm crit} < M_{\rm max} < M_h$. Note that $M_{\rm max}$ is the mass at which the mass function is cut off (see Equation 4.8). It is reasonable to identify it with the mass of the most massive halo, which must be larger than $M_{\rm crit}$, where the mass function begins to deviate from Equation 4.6. $M_{\rm max}$, being the mass of the most massive sub-halo, must be smaller than than M_h.

Thus, a halo with $M_{\rm vir} \approx 10^{12}$ $M_\odot$, like the one that is thought to be the host of the MW, should have a population of sub-halos spanning several orders of magnitude in mass. It is well known that, in consequence, a steep sub-halo mass function like Equation 4.6 predicts many more low-mass sub-halos than the number of observed faint MW satellites (Klypin *et al.*, 1999; Moore *et al.*, 1999), a finding commonly referred to as the "missing satellites problem." Efforts to solve this problem rely on physical processes that could either clear CDM halos of all baryons or inhibit their gathering in them in the first place, and that would affect low-mass halos preferentially (e.g., Moore *et al.*, 2006; Li *et al.*, 2010, Section 4.2). More specifically, Li *et al.* (2010) find that the mass function of luminous halos, $\xi_{\rm lum}(M_{\rm vir})$, would essentially be flat for 10^7 $M_\odot \leq M_{\rm vir} < 10^9$ $M_\odot$. All sub-halos with $M_{\rm vir} \geq 10^9$ $M_\odot$ would keep baryons and therefore $\xi_{\rm lum}(M_{\rm vir}) = \xi_{\rm sub}(M_{\rm vir})$ in this mass range. Thus, the mass function of *luminous sub-halos* can be written as

$$\xi_{\rm lum}(M_{\rm vir}) = k k_i M_{\rm vir}^{-\alpha_i} \tag{4.8}$$

with

$$\begin{array}{lll} \alpha_1 = 0, & k_1 = 1, & 10^7 \leq \frac{M_{\rm vir}}{M_\odot} < 10^9, \\ \alpha_2 = 1.9, & k_2 = k_1\,(10^9)^{\alpha_2 - \alpha_1}, & 10^9 \leq \frac{M_{\rm vir}}{M_\odot} \leq M_{\rm max} \end{array}$$

where the factors k_i ensure that $\xi_{\rm vir}(M_{\rm vir})$ is continuous where the power changes and k is a normalization constant chosen such that

$$\int_{10^7}^{M_{\rm max}} \xi_{\rm vir}(M_{\rm vir})\, dM_{\rm vir} = 1. \tag{4.9}$$

From a mathematical point of view, Equation 4.8 is the probability distribution of luminous sub-halos. Note that the luminous sub-halo mass function proposed in Moore *et al.* (2006) is very similar to the one in Li *et al.* (2010). In the high-mass part, it has the same slope as the mass function for all sub-halos and flattens in the low-mass part (cf. figure 3 in Moore *et al.*, 2006)). The lower mass limit for luminous halos is, however, suggested to be $M_{\rm vir} \approx 10^8$ $M_\odot$ in Moore *et al.* (2006). Note also that the mass function of *all sub-halos* has $\alpha_1 \approx \alpha_2 \approx 1.9$ (Gao *et al.*, 2004).

4.3.1 *NFW halos*

It is well established by now that the density profiles of galaxy-sized CDM halos are similar to a universal law proposed by Navarro *et al.* (1997). It is given as

$$\rho_{\rm NFW}(r) = \frac{\delta_c \rho_0}{r/r_{\rm s}\,(1 + r/r_{\rm s})^2}, \tag{4.10}$$

where r is the distance from the center of the halo and ρ_0 is the critical density of the universe, while the characteristic radius r_s and δ_c are mass-dependent parameters.

By integrating $\rho_{\rm NFW}(r)$ over a volume, the total mass of CDM within this volume is obtained. Thus,

$$M(r) = \int_0^r \rho(r') 4\pi r'^2 \, dr' \tag{4.11}$$

is the mass of CDM contained within a sphere with radius r around the center of the CDM halo, and $M(r) = M_{\rm vir}$ for $r = r_{\rm vir}$. Performing the integration on the right-hand side of Equation 4.11 and introducing the concentration parameter $c = r_{\rm vir}/r_s$ leads to

$$M(r) = \frac{4\pi\rho_0\delta_c r_{\rm vir}^3}{c^3}\left[\frac{r_{\rm vir}}{r_{\rm vir} + c\,r} + \ln\left(1 + \frac{c\,r}{r_{\rm vir}}\right) - 1\right]. \tag{4.12}$$

Note that δ_c can be expressed in terms of c,

$$\delta_c = \frac{\Delta_{\rm vir}}{3}\frac{c^3}{\ln(1+c) - c/(1+c)}, \tag{4.13}$$

as can be verified by setting $r = r_{\rm vir}$ in Equation 4.12 and substituting $M(r_{\rm vir}) = M_{\rm vir}$ by Equation 4.7.

If the halo is luminous, it is evident that $M(r)$ is smaller than the *total* mass included within r, M_r. However, assuming that the MW satellites are in virial equilibrium and that their dynamics is Newtonian, the mass-to-light ratios calculated for them are generally high and imply that they are DM dominated and thus, $M(r) = M_r$ would be a good approximation. It is therefore adopted for the present discussion. Note in particular that $M(r = 0.3\,{\rm kpc}) = {\rm M}_{0.3\,{\rm kpc}}$ in this approximation.

In principle, the parameters ρ_0 (Navarro *et al.*, 1997), c (Bullock *et al.*, 2001), and $\Delta_{\rm vir}$ (Macciò *et al.*, 2003) depend on the redshift z, but for the purpose of this chapter only $z = 0$ needs to be considered, as this is valid for the local universe. Thus,

$$\rho_0 = \frac{3H_0^2}{8\pi G} \tag{4.14}$$

with the Hubble constant $H_0 = 0.71\,(100\,{\rm km\,s^{-1}\,Mpc^{-1}})$ (Spergel *et al.*, 2007), $\Delta_{\rm vir} \simeq 98$ for ΛCDM-cosmology (Macciò *et al.*, 2003), and

$$\log_{10}(\overline{c}) = 2.31 - 0.109\log_{10}\left(\frac{M_{\rm vir}}{{\rm M}_\odot}\right) \tag{4.15}$$

where $\overline{c}$ is the expectation value of c as a function of $M_{\rm vir}$. Thus, $\overline{c}$ decreases slowly with $M_{\rm vir}$, while the scatter of the actual c is rather large, being

$$\sigma_{\log_{10} c} = 0.174 \tag{4.16}$$

(Macciò *et al.*, 2007). Note that the only caveat here is that we used the NFW profile to integrate the mass, and we expect the now-preferred Einasto profile (Navarro *et al.*, 2010) to make only a small difference in the central parts.

4.3.2 *Probing the ΛCDM hypothesis with $M_{0.3\,{\rm kpc}}$*

S08 use stellar motions in 18 MW satellites for calculating their mass within the central 300 pc, $M_{0.3\,{\rm kpc}}$. They assume the satellites to be in virial equilibrium and that Newtonian dynamics can be applied to them. The sample from S08 can be enlarged to 20 satellites by including the Large Magellanic Cloud (LMC) and the Small Magellanic Cloud (SMC), since van der Marel *et al.* (2002) estimated the mass of the LMC within the innermost 8.9 kpc, $M_{\rm LMC}$, using the same assumptions as S08. This implies $M_{\rm LMC} = (8.7 \pm 4.3) \times 10^9\,{\rm M}_\odot$, of which the major part would have to be DM. Equations (4.7), (4.12), (4.13), and (4.15) allow us to create tabulated expectation values of $M(r)$ for NFW halos with different $M_{\rm vir}$ and it can thereby be seen that for a typical NFW halo with $M(r = 8.9\,{\rm kpc}) = 8.7 \times 10^9\ {\rm M}_\odot$, $M(r = 0.3\,{\rm kpc}) = 2.13 \times 10^7\ {\rm M}_\odot = M_{0.3\,{\rm kpc}}$, and $M_{\rm vir} = 1.2 \times 10^{11}\ {\rm M}_\odot$. Noting that the SMC has about one tenth of the mass of the LMC (Kallivayalil *et al.*, 2006), the virial mass of its halo can be estimated as $M_{\rm vir} = 1.2 \times 10^{10}\ {\rm M}_\odot$, corresponding to $M_{0.3\,{\rm kpc}} = 1.51 \times 10^7\ {\rm M}_\odot$.

In order to test the shape of the MW satellite distribution function against the shape of the distribution of the $M_{0.3\,{\rm kpc}}$ values of the MW satellites, artificial samples of 10^6 $M_{0.3\,{\rm kpc}}$ masses are generated in concordance with the ΛCDM hypothesis, using Monte-Carlo simulations. As noted in Section 4.3.1, $M_{0.3\,{\rm kpc}}$ is well approximated by $M(r = 0.3\,{\rm kpc})$ in a CDM-dominated galaxy, $M(r = 0.3\,{\rm kpc})$ can be calculated if $M_{\rm vir}$ and c are given, and the expectation value for c is a function of $M_{\rm vir}$. The first step is therefore to choose a value for $M_{\rm vir}$ using uniform random deviates and the probability distribution of luminous halos given in Equation 4.8 (see, e.g., chapter 7.2 in Press *et al.*, 1992, for details). The next step is to attribute a value for $\log_{10}(c)$ to the chosen $M_{\rm vir}$. This is done by multiplying Equation 4.16 with a Gaussian random deviate and adding the result to the value for $\log_{10}(\bar{c})$, which is calculated from Equation 4.15. After transforming $\log_{10}(c)$ to c, $M_{0.3\,{\rm kpc}} = M(r = 0.3\,{\rm kpc})$ of the given halo can be calculated from Equation 4.12, using Equation 4.7, and Equation 4.13. These steps are repeated, until a sample of 10^6 $M_{0.3\,{\rm kpc}}$ values is generated.

If two samples are given, the maximum distance between their cumulative distribution functions, D, can be calculated. Performing the KS-test, this quantity allows us to estimate how likely it is that they are drawn from the same distribution function. Our null hypothesis is that the observed satellite galaxies are drawn from the theoretically calculated mass function of luminous halos; the parent distribution is thus assumed to be the ΛCDM sub-halo mass function of luminous sub-halo $M(0.3\,{\rm kpc})$ values. Setting $M_{\rm max}$ in Equation 4.8 to $10^{11}\ {\rm M}_\odot$, which is approximately the mass estimated for the CDM halo of the LMC and with $M_{\rm min} = 10^7\ {\rm M}_\odot$ leads to $D = 0.333$. According to the KS-test, given the parent distribution the probability for an even larger distance is 0.018. This means that the null hypothesis can be excluded with 98.2% confidence. Both cumulative distributions are shown in Figure 4.2.

Omitting the LMC and SMC from the observational sample but keeping $M_{\rm min} = 10^7\ {\rm M}_\odot$ and $M_{\rm max} = 10^{11}\ {\rm M}_\odot$ in the theoretical sample yields $D = 0.294$ leading to exclusion of the null hypothesis with a confidence of 92.9%. Additionally, setting $M_{\rm max} = 4 \times 10^{10}\ {\rm M}_\odot$, which is the $M_{\rm vir}$ that corresponds to the most massive $M_{0.3\,{\rm kpc}}$ in the S08 sample (i.e., the most massive remaining sub-halo), yields $D = 0.301$, leading to exclusion of the null hypothesis with a confidence of 93.9%. The latter two tests comprise a homogeneous mass sample of observed satellites as compiled by S08.

The fact that the mass function is expected to steepen at $M_{\rm crit} = 0.01\,M_{\rm h}$ even increases the discrepancy between the ΛCDM hypothesis and the observations. Returning the LMC and SMC back into the observational sample and cutting off $\xi_{\rm sub}(M_{\rm vir})$ at $M_{\rm max} = 10^{10}\ {\rm M}_\odot$ (with $M_{\rm min} = 10^7\ {\rm M}_\odot$), which would be close to $M_{\rm crit}$ for the CDM halo of the MW (see Section 4.3), and one order of magnitude below the estimated mass of the CDM halo of the LMC, implies $D = 0.359$ and an exclusion with 99.2% confidence.

On the other hand, setting $M_{\rm max} = 10^{12}\ {\rm M}_\odot$ (with $M_{\rm min} = 10^7\ {\rm M}_\odot$) leads to $D = 0.329$ and an exclusion with 98.0% confidence. Any reasonable uncertainty to the actual value of $M_{\rm max}$ can therefore be excluded as an explanation for the discrepancy between the observed sample of $M_{0.3\,{\rm kpc}}$ and a sample generated based on the ΛCDM hypothesis. As a consequence, the same is true for the uncertainty to the actual mass of the halo of the MW, $M_{\rm h}$, as $M_{\rm max}$ is linked to $M_{\rm h}$ (see Section 4.3).

Thus $M_{\rm max}$ is kept at $10^{11}\ {\rm M}_\odot$ in the following. Setting the lower limit of $\xi_{\rm lum}(M_{\rm vir})$ from $10^7\ {\rm M}_\odot$ to $10^8\ {\rm M}_\odot$ then leads to $D = 0.319$ and an exclusion of the null hypothesis with a confidence of 97.8% confidence. $10^8\ {\rm M}_\odot$ is the $M_{\rm vir}$ suggested by the lowest $M_{0.3{\rm kpc}}$ in the sample from S08. Note that the likelihood decreases with decreasing $M_{\rm max}$. This is

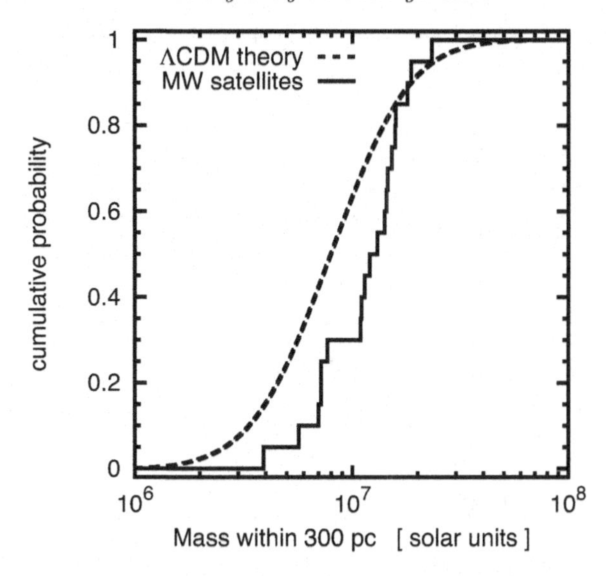

FIG. 4.2. The "overpredicting luminous satellite problem." The cumulative distribution function for the mass within the central 300 pc, $M_{0.3\,\mathrm{kpc}}$, of the MW satellites (solid line) and the cumulative distribution function for $M_{0.3\,\mathrm{kpc}}$ of a sample of 10^6 CDM halos picked from the distribution Equation 4.8 (dashed line). This is the parent distribution with the hypothesis being that the MW satellite $M_{0.3\,\mathrm{kpc}}$ masses are drawn from the parent distribution. The maximum distance between the two curves is 0.333 so that the null hypothesis that the data stem from the ΛCDM model distribution can be excluded with 98.2% confidence.

due to the overabundance of $M_{0.3\,\mathrm{kpc}} \approx 10^7\ \mathrm{M}_\odot$ halos becoming more prominent in the observational sample.

S08 suggest that $\xi_{\mathrm{lum}}(M_{\mathrm{vir}})$ might even be cut off below a mass of $\approx 10^9\ \mathrm{M}_\odot$, either because halos below that mass do not contain baryons or do not form at all.[3] Indeed, modifying $\xi_{\mathrm{lum}}(M_{\mathrm{vir}})$ given by Equation 4.8 accordingly results in an agreement between the theoretical distribution and the data ($D = 0.188$ with exclusion confidence of only 56%). A $\xi_{\mathrm{lum}}(M_{\mathrm{vir}})$ with a lower mass limit of $10^9\ \mathrm{M}_\odot$ is, however, in disagreement with the ΛCDM hypothesis, since the limiting mass below which all CDM halos are dark ought to be two orders of magnitude lower according to Li *et al.* (2010). Such a lower mass limit would be in line with WDM. But models based on WDM are discarded with high confidence in Section 4.2.6.

In *summary*, the mass distribution of the putative DM halos of observed satellites can be understood in terms of the ΛCDM hypothesis at worse than 5% confidence. Assuming the dSph satellites are in virial equilibrium and Newtonian dynamics to be valid, the observationally deduced DM halo masses of the MW satellites show a significant overabundance of $M_{0.3\,\mathrm{kpc}} \approx 10^7\ \mathrm{M}_\odot$ halos and a lack of less massive values compared to the theoretically calculated distribution for luminous sub-halos, despite much effort to solve the "common-mass-scale" problem (Section 4.2).

As a final note, the recently newly derived reduced mass of Hercules (see end of Section 4.2.8) does not affect the calculated likelihoods nor the conclusions reached here.

4.4 The bulge mass versus satellite number relation

According to the DM hypothesis, the fact that more massive DM host halos have a larger number of luminous satellites is a simple consequence of substructure formation because the number of sub-halos above a low-mass threshold increases with host halo

[3] A scenario in which halos with $M_{\mathrm{vir}} < 10^9\ \mathrm{M}_\odot$ do not form at all implies that DM would be warm. WDM models are tested in Section 4.2.6.

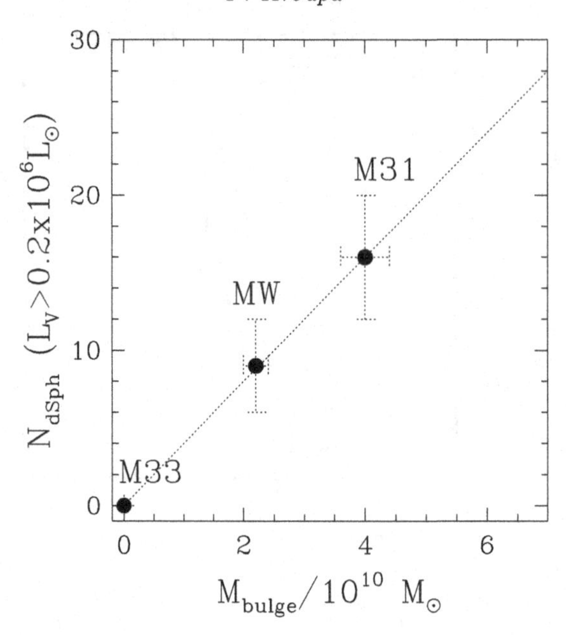

FIG. 4.3. The number of dSph and dE satellite galaxies more luminous than 0.2×10^6 L$_\odot$ is plotted versus the bulge mass of the host galaxy (MW: Zhao 1996, M31: Kent 1989, M33: Gebhardt *et al.* 2001). The dotted line is the fit $N_{\mathrm{dSph}} = 4 \times M_{\mathrm{bulge}}/(10^{10}\ \mathrm{M}_\odot)$ (Equation 4.17). Only satellites within a distance of 270 kpc of the MW and M31 are used.

mass. However, there is no predicted direct relation between the halo mass and the bulge mass, since pairs of galaxies with and without a bulge (such as M31, Rubin and Ford 1970, and M101, Bosma *et al.*, 1981, respectively) but with the same rotation velocity can be found. Bulge masses, on the other hand, relate to the accretion and merging history of the host galaxy. Noteworthy is that none of the approaches A–F that have been explicitly constructed to account for the dSph satellite population of MW-type hosts (Section 4.2) predict a correlation between the host galaxy bulge mass and the number of luminous satellites.

Karachentsev *et al.* (2005) note the existence of a correlation between the bulge luminosity and the number of associated satellite galaxies such that galaxies without a bulge have no known dSph companions, such as M101. Also, Karachentsev *et al.* (2005) point out that the number of known dSph satellites increases with the tidal environment.

The existence of such a correlation can be tested in the Local Group, where we have very good estimates of the number of satellites within the nominal virial radii of the respective hosts and of the stellar bulge masses of the three major galaxies (MW, M31, and M33).

We considered only the satellites brighter than $L_V = 0.2 \times 10^6$ L$_\odot$ ($M_V < -8.44$), given that the census of fainter satellites is incomplete for the MW (notably in the southern hemisphere), and also for M31, and M33 given their distances. We display the result in Figure 4.3: it appears that, in the Local Group at least, there may be a linear correlation between the bulge mass and the number of early-type satellites. A least-squares linear fit to the data indeed yields

$$N_{\mathrm{dSph}} = (4.03 \pm 0.04) \times M_{\mathrm{bulge}}/(10^{10}\mathrm{M}_\odot). \tag{4.17}$$

While this result appears challenging for DM (see Section 4.2.8), it could indicate a natural relation between the formation of the bulge of the host galaxy and the formation of early-type satellite galaxies around it. This, together with the absence of a mass-luminosity relation for the satellites (Section 4.2), could severely affect our present understanding of how satellite dwarf galaxies form and evolve. However, this is a linear fit to only three points, so it will be mandatory to check this correlation by surveying disk galaxies in the Local Volume with different bulge masses for a deep and exhaustive counting of faint early-type satellite galaxies. In terms of the present day stellar mass fraction, the dSph satellites of the MW add up to at most a few times 10^7 $M_\odot$, so that they amount to about 0.15% of the mass of the MW bulge.

4.5 The disk of satellites (DoS): the new dwarfs

The DoS is a well-established feature of the Milky Way satellite system (Metz *et al.*, 2009), and a similar structure has been reported for the Andromeda system (Koch and Grebel, 2006) for which, however, the distance uncertainties are larger and the satellite population is richer. In the case of the very well studied MW, the DoS is very pronounced for the classical (11 brightest) satellites including the LMC and SMC (Kroupa *et al.*, 2005). But how are the new satellites, the ultrafaint ones, distributed?

Recently, Watkins *et al.* (2009) and Belokurov *et al.* (2010) reported the discovery of two new MW satellite galaxies, Pisces I and II, respectively, enlarging the total MW satellite system to 24 satellites. Pisces I and II were found in the southern part of the SDSS survey area, making them the two first non-classical satellite galaxies found in the southern Galactic hemisphere. Furthermore, distances to a number of the already known satellite galaxies have been updated in recent years, most notably the new distance estimate for Boo II by Walsh *et al.* (2008), which changes the distance from 60 to 42 kpc.

Metz *et al.* (2007, 2009) have previously employed a sophisticated fitting routine to find the DoS. Do the new satellites fall into the DoS, substantiating the disk's existence in the southern Galactic hemisphere? Here, an intuitive disk-fitting and disk-testing method that are easy to follow are introduced. They lead to perfect agreement with the results obtained before.

4.5.1 *DoS parameters*

Here a simple and straightforward method is described to calculate the DoS parameters $l_{\rm MW}$, $b_{\rm MW}$, $D_{\rm p}$, and Δ, which are, respectively, the directions of the plane-normal in Galactic longitude and latitude, the smallest distance of the plane to the Galactic center and the thickness of the DoS.

The positions of the satellites are adopted from Metz *et al.* (2007, 2009), the positions for Pisces I and II are taken from the corresponding papers. The positions of satellites on the sky and their radial distances are transformed into a Galactocentric, Cartesian coordinate system assuming the distance of the Sun to the center of the MW to be 8.5 kpc. The z-coordinate points into the direction of the Galactic north pole.

The 3D coordinates are projected into two dimensions, plotting z against a projection onto a plane defined by the Galactic longitude $l_{\rm MW}$. This resembles a view of the MW satellite system as seen from infinity. The view of the satellite system is rotated in steps of 1°. For each step, a line is fitted to the projected satellite distribution. The line is determined using the least-squares method, demanding the satellite-distances, as measured perpendicularly to the fitted line, to become minimal. This line constitutes a plane seen edge-on in the current projection. The two free parameters of the fit are the closest distance from the MW center, $D_{\rm P}$, and the inclination $b_{\rm MW}$ of the normal vector to the line from the z-axis. The plane-normal-vector's longitude is $l_{\rm MW}$, given by the projection. The fits are performed for each angle $l_{\rm MW}$ between 0° and 360°. After a half rotation, the view is identical to the one 180° before, mirrored along the z-axis.

For each angle $l_{\rm MW}$ the root mean square (RMS) thickness Δ of the satellite distribution around the fitted line is determined. The normal vector to the best-fitting disk (the DoS) to the full three-dimensional distribution of the MW satellites is then given by those $l_{\rm MW}$ and $b_{\rm MW}$ that give the smallest RMS thickness $\Delta_{\rm min}$.

To account for the distance uncertainties of the satellites, the major source of error, the procedure is repeated 100 times. Each time, the radial position of each satellite is randomly chosen from a normal distribution centered on the satellite's radial distance. It has a standard deviation given by the distance uncertainties to the satellite galaxies. Once a realization with varied radial distances is set up, the coordinate-transformation into the Galactic coordinate system is performed. The parameters of the best fits are determined for each realization. Their final values are determined by averaging the results of all realizations; the standard deviations of their values are adopted as the uncertainties of our fits. For the two new satellites Pisces I and II no distance uncertainties are given. We estimate them to be 15% of the distance, in agreement with the usual distance uncertainties to the faint satellite galaxies.

Fitting all 24 currently known MW satellite galaxies, the minimum disk thickness is found to be $\Delta_{\rm min} = 28.8 \pm 1.3$ kpc. This is more than eight sigma away from the maximum thickness of $\Delta_{\rm max} = 56.5 \pm 1.8$ kpc obtained at a 90° projection of the same data. *Thus, the DoS is highly significant.* The position of the minimum thickness gives the best-fitting disk. The normal vector defining the DoS points to $l_{\rm MW} = 154^\circ.6 \pm 1^\circ.9$ and has an inclination of $b_{\rm MW} = -3^\circ.2 \pm 0^\circ.9$, i.e., is almost polar. As the position of the Sun is at $l_{\rm MW} = 0^\circ$, we see the DoS almost face on, tilted by merely 25°. $D_{\rm P}$, the closest distance of the plane fit from the MW center, is 6.4 ± 0.8 kpc.

4.5.2 *A novel disk test*

Another test to determine whether the satellite galaxies are distributed in a disk can be performed by comparing the number of satellites nearby the plane to the number farther away: Let $N_{\rm in}$ be the number of all satellites that have a perpendicular distance of less than 1.5 times the minimal disk thickness $\Delta_{\rm min}$ from the line-fit. $N_{\rm out}$ correspondingly counts all satellites farther away. $N_{\rm in}$ and $N_{\rm out}$ are determined for each rotation angle, measuring distances from the line (i.e., plane viewed edge-on in the given projection) that fits the distribution in the given projection best. This is illustrated in Figure 4.4. It shows an edge-on view of the best-fitting plane, along with a view rotated by 90°. Both views see the disk of the MW edge-on.

Figure 4.5 shows the ratio of galaxies found within the DoS to those outside (solid black line): $\mathcal{R} = N_{\rm in}/N_{\rm out}$. The situation is shown for the unvaried distances. If the MW satellites are distributed in a disk, $\mathcal{R}$ approaches a maximum when looking edge-on, while it will rapidly decrease once the projection is rotated out of the disk plane. It is a good test to discriminate a disk-like distribution from a spheroidal one. The latter would not lead to much variation in the ratio.

It can be seen that $\mathcal{R}$ approaches a maximum close to the best-fit $l_{\rm MW}$. At the maximum, only two of the 24 satellite galaxies are found outside of the demanded distance from the disk. The maximum $\mathcal{R}$ is thus 11.0, situated only a few degrees away from the $l_{\rm MW}$ that gives the smallest thickness. This has to be compared to the broad minimum of $\mathcal{R} \approx 1$. The disk-signature is obvious, proving the existence of a DoS that incorporates the new satellites found in the SDSS.

4.5.3 *Classical versus new satellites: is there a DoS in both cases?*

In addition to the situation for all 24 known MW satellites, the analysis is also carried out for two distinct subsamples: the 11 classical, most luminous satellite galaxies and the 13 new satellites discovered mostly in the SDSS. Each of them uses a minimal thickness, given by the subsample distribution, in determining $\mathcal{R}$. If all satellite galaxies follow the same distribution, given by the DoS, a separate fitting should lead to similar parameters. If, on the other hand, the new (mostly ultrafaint) satellites follow a different distribution,

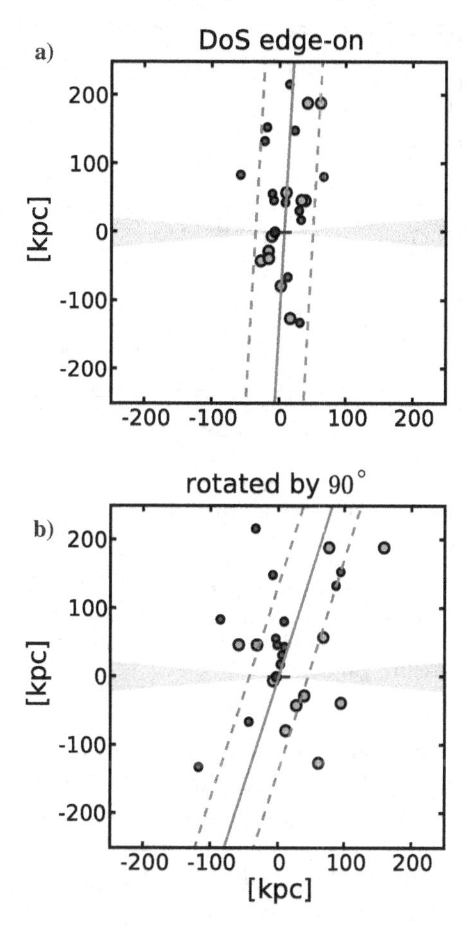

FIG. 4.4. The MW DoS: the 3-D distribution of the MW satellite galaxies. The 11 classical satellites are shown as large gray circles, the 13 new satellites are represented by the smaller dark gray dots, whereby Pisces I and II are the two southern dots. The black near-central line in the center shows the MW-disk orientation, the Sun is located at the position of the black dot. The obscuration-region of $\pm 5°$ from the MW disk is shown by the gray areas. The gray solid line shows the best fit plane (edge-on) to the satellite distribution at the given projection, the dashed gray lines define the region $\pm 1.5 \times \Delta_{\min}$, $\Delta_{\min}$ being the RMS-thickness of the thinnest DoS ($\Delta_{\min} = 28.8$ kpc in both panels). (a) shows an edge-on view of the DoS. Only three of the 24 satellites are outside of the dashed lines, giving $N_{\rm in} = 21$, $N_{\rm out} = 3$ and thus a ratio of $\mathcal{R} = 7.0$. (b) shows a view rotated by 90°, the DoS is seen face-on. Now, only 13 satellites are close to the best fit, 11 are outside, resulting in $\mathcal{R} = 1.2$. Note the absence of satellites *in the SDSS survey volume* (upper left and right regions of the upper panel, see also Figure 1 in Metz *et al.*, 2009 for the SDSS survey regions).

then this would challenge the existence of a DoS. *It is worth emphasizing that while the brightest satellites in a ΛCDM model of an MW-type halo may exceptionally form a weak disk-like structure (Libeskind et al., 2009), none of the existing standard-cosmological models predicts the whole satellite population to be disk-like.*

Furthermore, comparing the results for the classical 11 satellites with the ones obtained by the more sophisticated fitting technique used by Metz *et al.* (2007) is a good test to check whether the technique applied here gives reliable results.

The graphs for both subsamples are included in Figure 4.5, the results for classical satellites are represented by the dashed yellow, the new (SDSS) satellite galaxies by the dashed green line. Both are in good agreement not only with the combined sample, but also with each other. They peak at their best-fit $l_{\rm MW}$, with each of them having an $N_{\rm out}$ of only one galaxy at the peak.

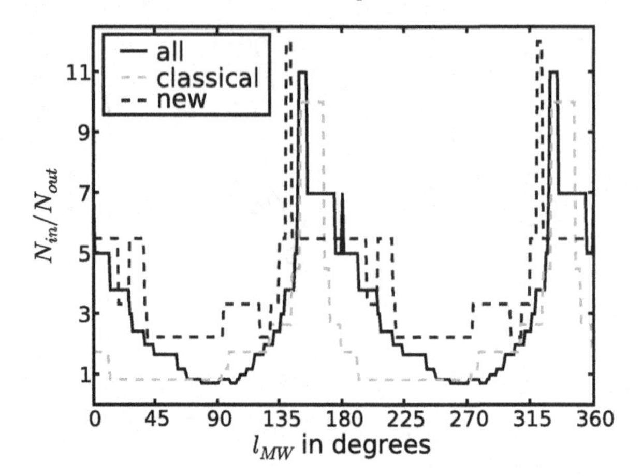

FIG. 4.5. Testing for the existence of a satellite disk. The behavior of $\mathcal{R}$ for each view of the MW, given by the Galactic longitude of the normal vector for each plane fit. $\mathcal{R}$ is the ratio of satellites within $1.5 \times \Delta_{\min}$ ($\Delta_{\min} = 28.8$ kpc), $N_{\rm in}$, to those farther away from the best-fitting line, $N_{\rm out}$, calculated for all 24 known satellites, as well as for the fits to the 11 classical and the 13 new satellites separately. The disk-like distribution can be clearly seen as a strong peak close to $l_{\rm MW} = 150°$. Note that the positions of the peaks are close to each other for both subsamples separately. This shows that the fainter satellite galaxies found recently independently define the same DoS as the classical satellite galaxies.

Applying the technique presented in Section 4.5.1 to calculate the DoS parameters, the new satellites have a best-fitting disk with a normal vector pointing to $l_{\rm MW} = 149°.1 \pm 5°.2$, only five degrees away from the direction that was obtained by considering all known MW satellites. The inclination is $b_{\rm MW} = 8°.0 \pm 3°.1$, again an almost perpendicular orientation of the DoS, and only 10 degrees away from the previously determined value. The inclination, in contrast to the longitude, is less well defined as the sample has only two galaxies in the southern Galactic hemisphere. The derived RMS thickness is $\Delta_{\min} = 28.7 \pm 2.4$ kpc, identical to the one given by all satellite galaxies. The minimum distance from the MW center is $D_{\rm P} = 15.9 \pm 3.35$ kpc, almost 10 kpc more than in the case of all satellites. Similar to $b_{\rm MW}$ this value is more influenced by the small number of satellites in the southern hemisphere.

The fitting to the 11 classical satellites leads to results that are in very good agreement, too. The best-fitting position for the 11 classical satellites is $l_{\rm MW} = 157°.2 \pm 1°.2$ and $b_{\rm MW} = -12°.7 \pm 0°.5$, the thickness is found to be $\Delta = 18.5 \pm 0.6$ kpc, and the closest distance to the MW center is $D_{\rm P} = 8.2 \pm 0.6$ kpc. This is in excellent agreement with the results from Metz *et al.* (2007). In that paper, the authors reported $l_{\rm MW} = 157°.3$, $b_{\rm MW} = -12°.7$, $\Delta_{\min} = 18.5$ kpc and $D_{\rm P} = 8.3$ kpc. This illustrates that, even though we use a more simple disk-finding technique, our results are extremely accurate.

The agreement of the fit parameters for the two subsamples *separately* is impressive. Two populations of MW satellite galaxies (bright vs faint) define the same DoS. This shows that the new, faint satellites fall close to the known, classical, DoS ($\equiv$DoS$_{\rm cl}$). Even without considering the classical satellite galaxies, the new satellites define a disk, DoS$_{\rm new}$, that has essentially the same parameters. This is a dramatic confirmation for the existence of a common DoS$\approx$DoS$_{\rm new}$ $\approx$DoS$_{\rm cl}$.

4.5.4 *The origin of the DoS*

A pronounced DoS is therefore a physical feature of the MW system. But what is its origin? Is the existence of the classical satellite DoS$_{\rm cl}$ and the existence of the new satellite DoS$_{\rm new}$, such that DoS$_{\rm new}$ $\approx$ DoS$_{\rm cl}$, consistent with standard cosmological theory?

Li and Helmi (2008) and D'Onghia and Lake (2008) proposed a solution to this *satellite phase-space correlation problem*: they suggest the correlation to be due to the infall of groups of DM-dominated dwarf galaxies. However, this proposition is challenged by the fact that all known nearby groups of dwarf galaxies are spatially far too extended to account for the thinness of the DoS (Metz *et al.*, 2009). It may be speculated that the groups that fell in correspond to compact dwarf groups that do not exist any longer because they already merged. But this is compromised by the observation that their putative merged counterparts in the field do not exist (Metz *et al.*, 2009). Indeed, Klimentowski *et al.* (2010) model an MW-type system and deduce, "Based on our result we conclude that such a disk is probably not an effect of a group infall unless it happened very recently" (their section 4.2.2). Furthermore, this notion would imply dwarf galaxy groups to be full of dSph galaxies whereas the pristine (before group infall) MW halo would have formed none, in violation of the observed morphology density relation (e.g., Okazaki and Taniguchi, 2000).

On the other hand, according to a recent suggestion by Libeskind *et al.* (2009), the MW constitutes an improbable but possible constellation of CDM-dominated satellites about an MW-type disk galaxy that have independent infall and accretion histories. They analyze an N-body sample of 30946 MW-mass DM host halos with mass in the range 2×10^{11} $M_{\odot}$ to 2×10^{12} $M_{\odot}$ for the properties of their sub-structure distribution. They first select from this sample only those halos that host a galaxy of similar luminosity as the MW. From this remaining sample of 3,201 hosts they select those that contain 11 or more luminous satellites leaving 436 (1.4%) host halos that compare favorably with the MW luminosity and number of luminous satellites. From this sample they find that a fraction of 30% host halos have luminous satellites distributed anisotropically similar to the observed MW case. Thus, only 0.4% of all existing MW-mass CDM halos would host an MW-type galaxy with the right satellite spatial distribution. This probability of 4×10^{-3} that the DM model accounts for the observed MW-type satellite system would be smaller still if proper motion measurements of additional satellites affirm the orbital angular momentum correlation highlighted by Metz *et al.* (2008) or if the satellites that might be discovered on the southern hemisphere by the *Stromlo Missing Satellite Survey*[4] lie within the DoS. We have seen that the Pisces I and II ultrafaint satellites, newly found in the southern stripes of the SDSS, do indeed conform and in fact enhance the DoS.

It needs to be emphasized that the DM-based models have so far not been able to account for the observed fact that the DoS lies nearly perpendicular to the MW disk; the groups reporting the DM-based models need to *postulate* that this occurs naturally. The combined probability that a DM-based model accounts for the observed MW-type system, which has the properties that the satellites have correlated angular momenta and form a DoS highly inclined to the baryonic disk of the host galaxy, cannot be assessed, but is, by logical implication, smaller than 4×10^{-3}. What's more, the normal to the DoS is defined mostly by the outermost satellites, whereas the direction of the average orbital angular momentum vector is defined by the innermost satellites for which proper motions have been measured. Both the normal and the average orbital angular momentum vector are nearly co-aligned, implying a strong degree of phase-space correlation between the satellites such that the DoS is rotating (Metz *et al.*, 2008). But perhaps the MW is but a very special system, an outlier within the DM-based framework?

Apart from accounting neither for the near-to-perpendicular orientation of the DoS relative to the MW disk nor for the co-alignment of the (outer satellite) normal and (inner satellite) angular momentum vectors, this possibility is challenged by the observed fact that the nearest similar DM halo hosts a similar disk galaxy, Andromeda, with a similar satellite system as the MW. M31, while having a richer and more complex satellite population and a larger bulge mass (Figure 4.3), may also have a DoS (Koch and Grebel, 2006, see also figure 4 in Metz *et al.*, 2009) suggesting that such satellite distributions may not be uncommon among MW-type DM halos, in contrast to the standard CDM-based result as highlighted by Libeskind *et al.* (2009).

[4] http://www.mso.anu.edu.au/~jerjen/SMS_Survey.html.

The Libeskind *et al.* (2009) analysis also shows that about 10% of MW-type DM halos would host an MW-luminous galaxy; the 90% others would presumably host galaxies with different luminosities suggesting a very large variation between DM halo and luminous galaxy properties. This, however, appears to be in conflict with the properties of observed disk galaxies. Using a principal component analysis on hundreds of disk galaxies, Disney *et al.* (2008) demonstrate that observed disk galaxies are very simple systems defined by one underlying parameter rather than about six if the galaxies were immersed in DM halos. Citing additional similar results, van den Bergh (2008) and Gavazzi (2009) reach the same conclusion, as well as Gentile *et al.* (2009) and Milgrom (2009). The small amount of variation of disk galaxies thus appears to be incompatible with the very large variation inherent in the DM model, as quantified by the Libeskind *et al.* (2009) analysis: 90% of MW-mass DM halos would have disk galaxies that differ substantially in luminosity from the MW if the DM-ansatz were correct, and yet our closest neighbor, Andromeda, is very similar to the MW.

4.6 A possible physical interpretation of the observations: a tidal origin for dwarf galaxies?

It has been shown previously that the CDM hypothesis has a likelihood of less than

- 0.3% to account for the satellite mass luminosity data (Section 4.2);
- 5% to account for the mass function of visible sub-halos (Section 4.3);
- 0.4% to account for the DoS (Section 4.5).

The CDM hypothesis can thus account for the MW satellite population with a combined likelihood of less than 6×10^{-7}:

It has also been shown that the WDM hypothesis has a likelihood of less than

- 0.3% to account for the satellite mass luminosity data (Section 4.2.6);
- 0.4% to account for the DoS (Section 4.5, since the same argument on the DoS applies in WDM as in CDM cosmology).

The WDM hypothesis can therefore account for the MW satellite population with a combined likelihood of less than 1.2×10^{-5}.

It therefore appears that the DM hypothesis needs to be rejected at the 99.999% confidence level. The solution to the origin of the MW satellite galaxies, and by implication of the satellites of other galaxies, and to the observed non-Newtonian motions of matter in the very weak field limit thus needs to be sought with non-Newtonian theories of gravitation.

Indeed, a different (non-DM based) scenario for the origin of dSph satellite galaxies had been suggested already at a time before the concept of DM was taken seriously, namely, that they may be ancient tidal-dwarf galaxies (TDGs) (Lynden-Bell, 1976; Kunkel, 1979; Bell, 1983). This proposition can naturally account for their correlated phase-space distribution in the form of a rotating disk-like distribution (Section 4.5).

4.6.1 *The bulge mass versus satellite number relation*

If the MW dSph satellite galaxies are ancient TDGs that formed during a possible interaction of the very young MW and another very young galaxy at high redshift, then it would be plausible to expect a correlation between the bulge mass of the host galaxy and the number of such satellites (see Section 4.6.1), because bulges may form as a result of gas-dissipational galaxy collisions.

If both the bulge and the satellites have their origin in the same early merger(s), the link between bulge mass and number of satellites (Section 4.4) becomes natural. Taking 10^7–10^8 $M_\odot$ to be the mass of an average young TDG and making the rough guess that only half the presentday bulge masses of the MW and M31 formed during the early collision (the rest growing due to secular dynamical processes), then the data presented in Section 4.4 suggest that perhaps a fraction $\gamma \approx 0.8$ to 8% of the mass of a collisionally formed bulge may be distributed in TDGs formed during the collision. This is a very rough estimate of γ, and it is to be expected that γ depends on the encounter conditions and the gas fraction of the involved galaxies. Indeed, Wetzstein *et al.* (2007) demonstrate in a series of numerical models that the number of TDGs increases with the gas fraction of the host. This is relevant for galaxy encounters at a high redshift where merging is expected to have been frequent. To refine this picture, it will be necessary to survey disk galaxies in the Local Volume with different bulge masses for the presence of early-type satellite galaxies to test whether this correlation holds true, and to perform numerical experiments of colliding late-type gas-rich galaxies at high redshift to theoretically constrain γ as a farther handle on the possible number of TDGs that may be contributing to the overall faint-satellite galaxy population.

4.6.2 *The origin of dwarf elliptical galaxies and implications*

The natural way to explain the phase-space correlation (disk of satellites) as well as the new problems reported here (Section 4.2–4.5) suggests the dSph satellite galaxies of the MW to be a population of TDGs that probably formed during a gas-rich encounter between the early MW and another galaxy. But if they all formed at the same time, how can the different chemical properties and star-formation histories of the different dwarf galaxies be explained within this scenario? It is known that the satellite galaxies all have ancient populations of an indistinguishable age (Grebel, 2008), perhaps being created when the TDGs were born. TDGs may also be formed with globular clusters as long as the star-formation rate surpasses a few $M_\odot$ yr^{-1} for 10 Myr (Weidner *et al.*, 2004). The chemo-dynamical modeling by Recchi *et al.* (2007) has shown that once a TDG (without DM) forms it is not natural for it to blow out the gas rapidly. Rather, the rotationally supported small gas-rich disks of young TDGs begin to evolve through self-regulated star formation (Pflamm-Altenburg and Hensler, 2011) until their gas is either consumed or removed through ram-pressure stripping. Consequently, their internal evolution through star formation can be slow and individual, such that TDGs that formed during one collision event can have different chemical properties many Gyr after their formation. Removal of the interstellar medium from the TDG through ram-pressure takes about half an orbital time, which is typically one to a few Gyr after formation. This timescale is consistent with the observed cessation of star-formation in most dSph satellites (Grebel, 1999). TDGs that do not lose or consume their gas because they remain at large distance from the host remain as dIrr galaxies (Hunter *et al.*, 2000). Within Newtonian dynamics, dynamical modeling over many orbits around the MW DM halo has demonstrated that even low-mass satellites do not easily disrupt unless they are on virtually radial orbits (Kroupa, 1997; Metz and Kroupa, 2007). In non-Newtonian dynamical theories, disruption will be even more difficult, because gravity effectively strengthens in the very weak field limit. Once formed, TDGs cannot fall back onto their hosts since dynamical friction is insignificant for them. In summary, the physics of TDG formation and evolution is sufficiently well understood to make it clear that *once formed at a sufficient distance from the host, they cannot simply vanish.*

The MW dSph satellites might therefore be understood as ancient TDGs that formed even within a DM universe. But this interpretation leads to the problem that the very extensive modeling within the ΛCDM hypothesis strictly implies that MW-luminous galaxies must be accompanied by hundreds of shining albeit faint satellites (Macciò *et al.*, 2010; Busha *et al.*, 2010; Koposov *et al.*, 2009). For example, Tollerud *et al.* (2008) conjecture that “there should be between 300 and 600 satellites within 400 kpc of the Sun that are brighter than the faintest known dwarf galaxies and that there may be as many as 1,000, depending on assumptions.” The most recent very deep search for additional ultrafaint satellites in the northern hemisphere suggests the bottom of the barrel has already been reached with the presently catalogued objects. No additional new fainter satellites are emerging (Jerjen *et al.*, in preparation, Section 4.1). Therefore, the validity of the DM theoretical framework must again be questioned, because the dSph satellites cannot be two types of object at the same time, namely, DM-dominated substructures and ancient TDGs.

If the dSph satellites are ancient TDGs stemming from an early gas-rich encounter involving the proto-MW and probably contributing a collision product to the MW bulge (see Sections 4.4 and 4.6.1), then this would mean that the MW would have a severe substructure problem as there would not be any satellites with DM halos less massive than about 10^{10} $M_\odot$ with stars, in clear violation of CDM predictions as provided by, for example, Diemand *et al.* (2008); Koposov *et al.* (2009); Busha *et al.* (2010); Macciò *et al.* (2010). Perhaps a few dSph satellites are ancient TDGs, such as the classical or nine brightest satellites, and the rest are the DM-dominated sub-halos. This possibility is unlikely, because the new satellites are affirmed to be associated with the DoS and because they share many of the physical properties of the brighter satellites (e.g., S08).

Indeed, this disagreement between the DM hypothesis and observation extends, by logical implication, to the overall faint-galaxy population. The production of TDGs would have been ubiquitous in the early universe such that all dwarf elliptical (dE) galaxies may be evolved TDGs. As shown by Okazaki and Taniguchi (2000), the production of TDGs calculated in a standard cosmology naturally matches the observed number of dE galaxies in various environments. The result of Okazaki and Taniguchi (2000) is rather striking, since they find that within the standard cosmological framework only one to two long-lived (i.e., bright) TDGs need to be produced on average per gas-dissipational encounter to cater for the population of dE galaxies and the density-morphology relation in the field and in galaxy groups and clusters.[5]

Viewing dE galaxies as old TDGs would be consistent with their deviating from the mass-radius, $M(r)$, relation of pressure-supported (early-type) stellar systems. dE galaxies follow a $r \propto M^{1/3}$ sequence reminiscent of tidal-field-dominated formation. *All* other pressure-supported galactic systems (elliptical galaxies, bulges, and ultra-compact dwarf galaxies) with stellar mass $M > 10^6$ $M_\odot$ follow instead the relation $r \propto M^{0.60\pm0.01}$ (figure 2 in Dabringhausen *et al.*, 2008, see also figure 7 in Forbes *et al.*, 2008, and figure 11 in Graham and Worley, 2008), which may result from opacity-limited monolithic collapse (Murray, 2009). Viewing dE galaxies as TDGs would also be consistent with the observation that they have essentially stellar mass-to-light ratios similar to globular clusters (Bender *et al.*, 1992; Geha *et al.*, 2003; Dabringhausen *et al.*, 2008; Forbes *et al.*, 2008). If dE (baryonic mass $> 10^8$ $M_\odot$) and dSph (baryonic mass $< 10^8$ $M_\odot$) are old TDGs, why do they appear as different objects? That the dE and dSph galaxies differ in terms of their baryonic-matter density may be a result of the fact that below 10^8 $M_\odot$, spheroidal objects on the $r \propto M^{1/3}$ relation cannot hold their warm gas and consequently they must expand (Pflamm-Altenburg and Kroupa, 2009), becoming more susceptible to tides from their host a la Kroupa (1997).

Dwarf elliptical galaxies are pressure-supported stellar systems, while young TDGs are rotationally supported (Bournaud *et al.*, 2008). With a mass of less than typically 10^9 $M_\odot$ the rotational velocity becomes smaller than the velocity dispersion of the stellar

[5] Note that Bournaud (2010) misquotes Okazaki and Taniguchi (2000) as having deduced that about 10 long-lived TDGs need to be made per encounter, while Okazaki and Taniguchi (2000) write: "Adopting the galaxy interaction scenario proposed by Silk and Norman (1981), we find that if only a few dwarf galaxies are formed in each galaxy collision, we are able to explain the observed morphology-density relations for both dwarf and giant galaxies in the field, groups of galaxies, and clusters of galaxies." They also state, "The formation rate of TDGs is estimated to be ~1–2 in each galaxy interaction," and proceed to compare this number with the actually observed number of TDGs born in galaxy encounters.

population (of the order of 20 $\mathrm{km\,s^{-1}}$). The fact that a sizable fraction of dE galaxies show rotation, some even with spiral structure (Jerjen *et al.*, 2000; Barazza *et al.*, 2002; Geha *et al.*, 2003; Ferrarese *et al.*, 2006; Beasley *et al.*, 2009; Chilingarian, 2009), is thus also consistent with their possible origin as TDGs. For an excellent review of dE galaxies, see Lisker (2009).

The formation of TDGs underlies well-tested fundamental principles (angular momentum and energy conservation) such that their emergence is an intrinsic process in any gravitational framework. Given the results obtained by Okazaki and Taniguchi (2000) who demonstrate that, within standard hierarchical structure formation, very few long-lived TDGs need to form per gas-dissipational encounter to account for all dE galaxies; the results by Kroupa (1997) who demonstrates that tidal effects do not easily destroy even low-mass TDGs; and by Recchi *et al.* (2007) who find that young TDGs do not blow themselves apart through a starburst, it becomes difficult to avoid the conclusion that dE galaxies ought to be identified as old TDGs rather than dark matter halos. The consequence of this would be that either DM halos with masses $< 10^{10}\ \mathrm{M}_\odot$ do not form stars, in clear violation of the vast amount of theoretical results available today within the DM framework, or that such halos simply do not exist, in clear contrast to the requirement of their existence within the framework of the DM hypothesis.

The following assertions can now be stated:

(i) If dSph galaxies are DM substructures, then the DM hypothesis is wrong.

The reason is that the occurrence of DoSs about MW-type disk galaxies would be negligible, while two similar systems (MW and M31 with DoSs) are present in the Local Group. To invalidate this assertion one would need to show that dSph galaxies do not contain DM. This is straightforward in non-Newtonian dynamics. The reliance on the existence of DoSs can be removed without significantly affecting this deduction, because an MW-type halo has a 1.4% likelihood of hosting an MW-type galaxy with 11 or more luminous satellites (Section 4.5.4), whereas in the Local Group both MW-type halos have MW-type galaxies with similar numbers of satellites.

(ii) If dSph galaxies are old TDGs, then the DM hypothesis is wrong.

The reason is that in this case the substructure crisis would be disastrous. To invalidate this assertion, one would have to show that TDG formation can be neglected. This appears to be difficult in view of the available theoretical and observational results on TDG formation, evolution, and survival.

(iii) If dE galaxies are DM substructures, then the DM hypothesis is wrong.

The reason is that the formation of TDGs within the standard DM cosmological framework would lead to a population of dE galaxies which originate as TDGs and are comparable in number to the known dE galaxies. Such dE galaxies are not consistent with the current interpretation of the observations (as all known dEs are understood to be DM dominated, e.g., Moore *et al.*, 1999). To invalidate this assertion one would need to show that dE galaxies do not contain DM. This appears to be suggested in view of their generally low M/L ratios (Bender *et al.*, 1992; Geha *et al.*, 2003).

(iv) If dE galaxies are old TDGs, then the DM hypothesis is wrong.

The reason is that in this case the substructure crisis would be disastrous. To invalidate this assertion, one would have to show that TDG formation can be neglected. This appears to be difficult in view of the available theoretical and observational results on TDG formation, evolution, and survival.

Given that all four assertions, if true, imply the DM hypothesis to be incorrect as a representation of the real world, the need to introduce non-Newtonian dynamics emerges. Note that these arguments are largely independent of those presented at the beginning of Section 4.6 that lead to the same conclusion.

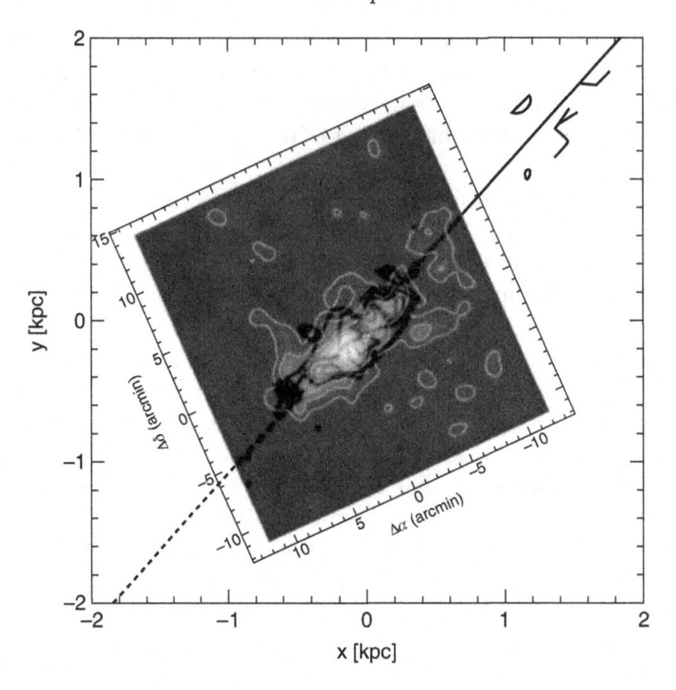

FIG. 4.6. Model RS1-5 from (Kroupa, 1997) (on the kpc grid) is plotted over the surface brightness contours of Hercules by (Coleman *et al.*, 2007) (celestial coordinate grid). The dashed and dotted curve are, respectively, the past and future orbit of RS1-5.

4.6.3 *Implications for gravitational dynamics*

If the dSph satellites are ancient TDGs, then understanding their internal kinematics remains a challenge because TDGs do not contain significant amounts of DM (Barnes and Hernquist, 1992; Bournaud *et al.*, 2007; Gentile *et al.*, 2007; Milgrom, 2007; Wetzstein *et al.*, 2007). However, the inferred large M/L ratios of dSph satellites (and especially of the ultrafaints) may not be physical values but may be a misinterpretation of the stellar phase-space distribution within the satellite. If this were to be the case then the absence of a "mass"–luminosity relation (Section 4.2) for dSph satellites may be naturally understood.

The following gedanken-experiment illustrates what could be the case: an unbound population of massless stars on similar orbits, each slightly phase-shifted and inclined relative to the other orbits, will reconfigure at apogalacticon, and an observer would clearly see a stellar phase-space density enhancement and would also observe a velocity dispersion. The M/L ratio calculated from the observed velocity dispersion would not be a true physical M/L ratio. Indeed, fully self-consistent Newtonian N-body models have already demonstrated that unphysically large M/L ratios arise if TDGs are allowed to orbit a host galaxy sufficiently long such that the remaining stellar population within the ancient TDG adopts a highly non-isotropic phase-space distribution function (Kroupa, 1997; Klessen and Kroupa, 1998; Metz and Kroupa, 2007). Such models suggest that it may be wrong to use an observed velocity dispersion to calculate a mass for the dSph satellites. Thus, tidal shaping of TDGs over a Hubble time can produce remnant objects that have internal highly anisotropic stellar phase-space distributions that would be falsely interpreted by an observer to have a high M/L ratio, as explicitly shown by Kroupa (1997).

It is indeed remarkable how model RS1-5 of that paper and shown here as a snapshot (Figure 4.6) is an essentially perfect match to the dSph satellite Hercules (figure 2 in Coleman *et al.*, 2007) discovered 10 years later by Belokurov *et al.* (2007). The half-light radius is 180 pc in the model and 168 pc in reality, RS1-5 has a velocity dispersion of about 2.8 $\mathrm{km\,s^{-1}}$ (table 2 in Kroupa, 1997), whereas Hercules has a measured velocity dispersion of $3.72 \pm 0.91\ \mathrm{km\,s^{-1}}$ (Adén *et al.*, 2009a), and the inferred mass-to-light ratio that one would deduce from velocity dispersion measurements under the assumption of equilibrium is about 200 in both cases. RS1-5 and Hercules have luminosities agreeing within one order of magnitude (the model being the brighter one), but RS1-5 has no DM.

The TDG models for dSph satellites presented by Kroupa (1997) and Klessen and Kroupa (1998), which have been computed within the standard cosmological framework, thus lead to a population of ancient TDGs that are in reasonable agreement with the observed luminosities, dimensions, and M/L ratios of dSph satellites (Metz and Kroupa, 2007). These model-dSph satellites require no fine-tuning of parameters but only assume the formation of purely baryonic $10^7\ \mathrm{M_\odot}$ TDGs about 10 Gyr ago. This model does not imply any relation between luminosity and (wrongly inferred) "dynamical mass," in agreement with the lack of such a relation (Section 4.2). Also, this model would naturally explain why the mass function of luminous DM sub-halos cannot account for the observations (Section 4.3).

But the model fails in one important aspect: the model galaxies have a larger spread in velocity dispersions (a few to about 30 $\mathrm{km\,s^{-1}}$) than the near-constant observed value in the range of 7 to 11 $\mathrm{km\,s^{-1}}$ for the classical dSph satellites. Thus, although a pronounced DoS with correlated orbital angular momenta is a strong indication that the dSph satellites are not DM-dominated substructures and an attempt to model them as ancient TDGs within the standard DM framework is successful in explaining their luminous structures, the detailed kinematical description fails.

However, Newtonian dynamics, which is the fundamental basis for the DM framework, cannot be applicable, a conclusion reached earlier. The next step in exploring this alternative model for the origin of dwarf galaxies will thus be to seek solutions within non-Newtonian frameworks in the weak-field limit, which is where the dSph-satellite dynamics is relevant.

Alternatives to Newtonian dynamics comprise the increasingly popular modified--Newtonian-dynamics (MOND) approach resting on a modification of the Newtonian acceleration in the weak field limit (Milgrom, 1983; Bekenstein and Milgrom, 1984; Sanders and McGaugh, 2002; Bekenstein, 2004; Famaey and Binney, 2005; Famaey *et al.*, 2007; McGaugh, 2008; Nipoti *et al.*, 2008; Sanders, 2008a,b; Tiret and Combes, 2008; Bruneton *et al.*, 2009), and a modified-gravity (MOG) adding a Yukawa-like force in the weak-field limit (Moffat and Toth, 2009a,b, and references therein).

Some serious theoretical approaches trying to embed MOND within a Lorentz-covariant framework (Bekenstein, 2004; Sanders, 2005; Bruneton and Esposito-Farèse, 2007; Zlosnik *et al.*, 2007; Zhao, 2008; Blanchet and Le Tiec, 2009; Skordis, 2009; Esposito-Farèse, 2011) are currently under deep scrutiny. Most of these theories do not (yet) explain why the acceleration threshold, a_0, which is the single parameter of MOND (adjusted by fitting to one single system), is of the order of $c\sqrt{\Lambda}$ (where Λ is the cosmological constant and c the speed of light). They also require a transition function, $\mu(x)$ (e.g., Equation 4.20) from the Newtonian to the modified regime, which is not (yet) specified by the theories.

A possible explanation for this coincidence $a_0 \approx c\sqrt{\Lambda}$ and a theoretically based transition function are suggested by Milgrom (1999). In a Minkowski (flat) space-time, an accelerated observer sees the vacuum as a thermal bath with a temperature proportional to the observer's acceleration (Unruh, 1976). This means that the inertial force in Newton's second law can be defined to be proportional to the Unruh temperature. On the other hand, an accelerated observer in a de Sitter universe (curved with a positive cosmological constant Λ) sees a non-linear combination of the Unruh (1976) vacuum radiation and of the Gibbons and Hawking (1977) radiation due to the cosmological horizon in the presence of a positive Λ. Milgrom (1999) then defines inertia as a force driving such an observer back to equilibrium as regards the vacuum radiation (i.e., experiencing only the Gibbons-Hawking radiation seen by a non-accelerated observer). Observers experiencing a very small acceleration would thus see an Unruh radiation with a small temperature close to the Gibbons-Hawking one, meaning that the inertial resistance defined by the difference between the two radiation temperatures would be smaller than in Newtonian dynamics, and thus the corresponding acceleration would be larger. This precisely gives the MOND formula of Milgrom (1983) with a well-defined transition-function $\mu(x)$, and $a_o = c\,(\Lambda/3)^{1/2}$. Unfortunately, no covariant version (if at all possible) of this approach has been developed yet.

The theoretical basis for the MOG approach relies on chosen values for integration constants in solving the equations of the theory. This approach seems to work well from an observational point of view, and it's fundamental basis needs farther research, as is the case also for MOND. It is perhaps noteworthy that a formulation of MOG in terms of Scalar, Vector, and Tensor fields (Moffat, 2006) may be hinting at a possible convergence with the Bekenstein (2004) Tensor-Vector-Scalar theory of gravity.

Both the MOND and MOG approaches have been applied to the satellite galaxy problem with appreciable success (Milgrom, 1995; Brada and Milgrom, 2000; Angus, 2008; Moffat and Toth, 2008). It has already been conclusively demonstrated that galaxy rotation curves, are very naturally and also exceedingly well described in MOND purely by the baryon distribution without any parameter adjustments (Sanders and McGaugh, 2002; McGaugh, 2004, 2005a; Sanders and Noordermeer, 2007), and MOG is reported to also do well on this account (Brownstein and Moffat, 2006). In contrast, the DM approach can only poorly reproduce the vast variety of rotation curves, and cannot explain the amazing regularities found in them (McGaugh, 2004; McGaugh *et al.*, 2007; Gentile *et al.*, 2009; Milgrom, 2009). Noteworthy is that Bosma (1981) had already discovered that the ratio between the surface density of all matter and the HI surface density is constant for radii larger than about 5 kpc (his figure 7). With this he had demonstrated that the dark matter surface density distribution has the same distribution as that of baryonic matter. Indeed, that the rotation curves would be purely defined by the baryonic matter distribution in non-DM models would naturally explain the later finding based on a large sample of galaxies by Disney *et al.* (2008), Gentile *et al.* (2009), and Milgrom (2009) that disk galaxies appear to be governed by a single parameter.

4.6.4 *Purely baryonic but non-Newtonian dynamics?*

The TDG scenario thus explains the origin of dSph and dE galaxies, and, given the tests presented in Sections 4.2–4.5, Newtonian dynamics would have to be replaced by another theory.

If it were true that physical reality is non-Newtonian, then a simple test helps as a consistency check: high dynamical mass-to-light $[(M/L)_{\rm dyn}]$ ratios (derived assuming Newtonian dynamics) would not be due to DM but due to the dynamics being non-Newtonian in the weak-field limit and/or be due to objects being unbound non-equilibrium systems (Section 4.6.3). Thus, assuming MOND were valid (the currently available simplest alternative to Newtonian dynamics), all systems with non-stellar $(M/L)_{\rm dyn}$ values (as derived in Newtonian gravity) would have to have internal accelerations roughly below the MONDian value $a_o = 3.9\ {\rm pc/Myr^2}$. That is, all pressure-supported (spheroidal) stellar systems that appear to be dominated dynamically by DM would need to have an internal acceleration $a < a_o$. Figure 4.7 shows the acceleration a star inside a pressure-supported system experiences at the effective radius, r_e, of its host system with luminosity spanning 10^4 to $10^{12}\ {\rm L}_\odot$,

$$a(r_e) = G\,\frac{M}{r_{\rm e}^2}. \tag{4.18}$$

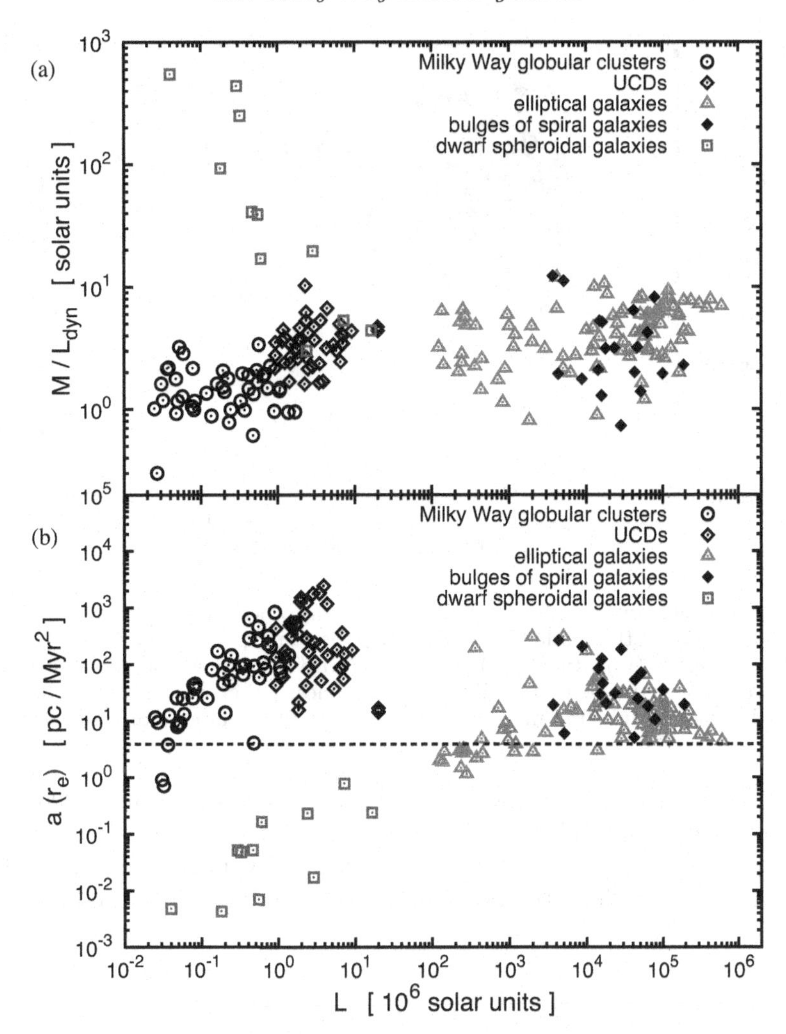

FIG. 4.7. (a): The dynamical $(M/L)_{\rm dyn}$ ratio (calculated assuming Newtonian dynamics to be valid) in dependence of the luminosity, L, for pressure-supported stellar systems following Dabringhausen *et al.* (2008) and Dabringhausen *et al.* (2009) (note that here dE and E galaxies are both plotted with the same elliptical galaxy symbol). (b): The Newtonian acceleration (Equation 4.18) of a star located at the effective radius within the host system in dependence of the host luminosity. The dashed line is a_0. In both panels, UCD = ultra compact dwarf galaxy.

Here $M = 0.5\,\Upsilon\,L$ is the stellar mass within r_e and L is the absolute V-band luminosity in solar units. We adopt for the stellar mass-to-light ratio in the V-band $\Upsilon = 3$ for collisionless systems (two-body relaxation time longer than a Hubble time), while $\Upsilon = 1.5$ for collisional systems (Kruijssen and Lamers, 2008; Kruijssen and Mieske, 2009), i.e., for systems that have evaporated a significant fraction of their faint low-mass stars through the process of energy equipartition. Values of $(M/L)_{\rm dyn}$ as large as 10 can be expected for purely baryonic systems if these retain their stellar remnants and hot gas. For example, the mass of an E galaxy may be comprised of only 30% stars with the rest made up of stellar remnants and gas that cannot cool to form new stars (Parriott and Bregman, 2008; Dabringhausen *et al.*, 2009). On comparison of the two panels it emerges that roughly only those systems with $a < a_o$ show non-baryonic $(M/L)_{\rm dyn}$ values. This is more clearly shown in Figure 4.8 where the MOND prediction for the range of dynamical mass-to-light ratios measured by a Newtonist living in a MOND Universe is plotted as a function

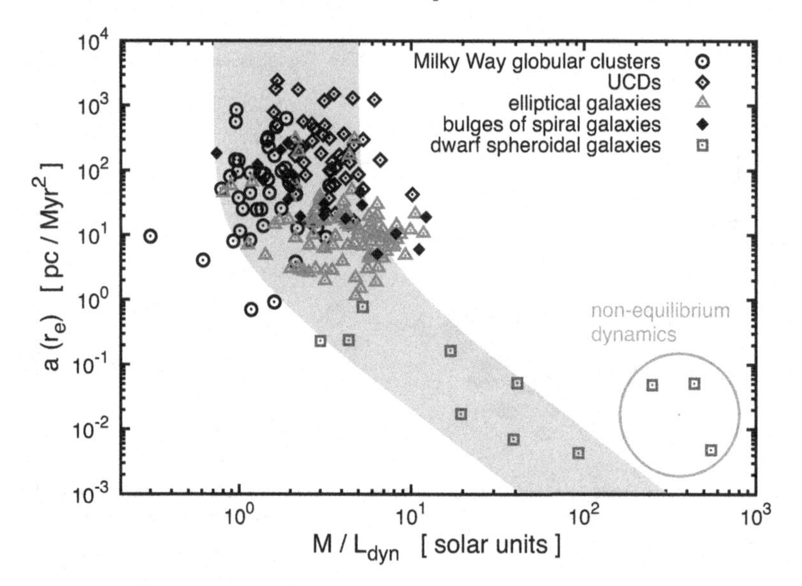

FIG. 4.8. Correlation between $a(r_e)$ and $(M/L)_{\rm dyn}$ for the objects shown in Figure 4.7. The lower and upper boundaries of the shaded region indicate the range predicted by MOND without any parameter adjustments (Equation 4.19). Encircled dwarf spheroidals outside this range (UMa, Dra, and UMi) may indicate non-equilibrium dynamics, either because the whole system is unbound, or because of unbound interloper stars among the member stars (see Section 4.6.4).

of acceleration. Adopting a conservative value (Kruijssen and Lamers, 2008; Kruijssen and Mieske, 2009) of the baryonic mass-to-light ratio $\Upsilon_{\rm bar}$ between 0.7 (for a globular cluster with an old metal-poor population depleted in low-mass stars) and 5 (for an old metal-rich population), the prediction of MOND inside the effective radius is (Famaey and Binney, 2005; Angus *et al.*, 2010)

$$(M/L)_{\rm dyn\,mond} = 0.5 \times \Upsilon_{\rm bar} \times \left(1 + \sqrt{1 + 4a_o/a}\right). \qquad (4.19)$$

Writing customarily $x = g/a_o$, where g is the actual full acceleration experienced by a ballistic particle (in MOND), Equation 4.19 follows from the following particular form of the transition MOND function (Milgrom, 1983),

$$\mu(x) = x/(1 + x) \qquad (4.20)$$

which is valid up to $x \approx 10$. The theoretical transition derived by Milgrom (1999) and mentioned in Section 4.6.3 would yield virtually the same result.

Again, the elliptical galaxies with large dynamical mass-to-light ratios could harbor a significant amount of gas (Parriott and Bregman, 2008), meaning that $\Upsilon_{\rm bar} = 5$ would be an underestimation in that case, whereas UCDs may have large stellar M/L values due to a bottom-heavy IMF (Mieske and Kroupa, 2008) or a top-heavy IMF (Dabringhausen *et al.*, 2009). The three classical dwarfs that lie outside the predicted MOND range for $(M/L)_{\rm dyn}$ on Figure 4.8 are UMa, Draco, and UMi. UMa may have an anisotropic velocity dispersion (Angus, 2008); Draco is known to be a long-standing problem for MOND, but the recent technique of interloper removal by Serra *et al.* (2010) could probably solve the problem, although this particular case remains open to debate; UMi is a typical example of a possibly out of equilibrium system, as it is elongated with substructure and evidence for tidal tails (D. Martinez-Delgado, private communication). Ultrafaint dwarf spheroidals are expected to be increasingly subject to such non-equilibrium dynamics, as shown to be the case even in Newtonian dynamics (Kroupa 1997, Section 4.6.3).

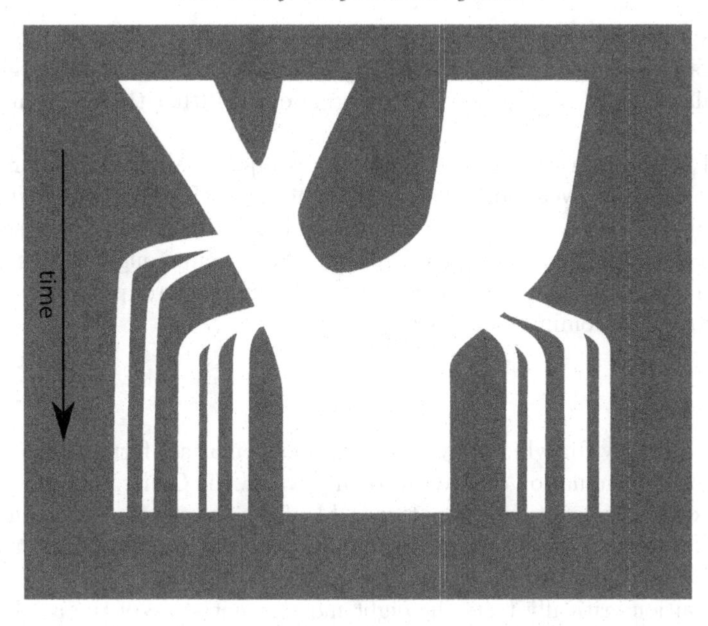

FIG. 4.9. A new cosmological structure formation framework: the mangrove merger tree. In a modified-Newtonian framework purely baryonic galaxies merge thereby spawning new dwarf galaxies giving rise to the morphology-density relation (adapted from Metz, 2008).

Thus, whereas in DM cosmology the association of highly non-stellar $(M/L)_{\rm dyn}$ values with $a < a_o$ would be coincidental as it is not built into the theory, it is natural in a MONDian universe for observers who believe they live in a Newtonian world. Noteworthy is that the same statement can be made for the Tully–Fisher scaling relation for rotationally supported galaxies (Tully and Fisher, 1977; McGaugh, 2005; Combes, 2009) as well as the newly found scaling relation of Gentile *et al.* (2009) and Milgrom (2009). Notably, this very recent realization that the ratio between DM mass and baryonic mass within the DM core radius is constant despite the very large variation of the DM–to–baryonic-matter ratio globally within galaxies cannot be understood within the DM hypothesis. A constant ratio within that radius implies that the distribution of baryonic matter is indistinguishable from that of the supposedly present DM, whereas outside that radius the effects of DM should become noticeable. In MOND models that behavior of gravity comes naturally.

The mass-deficit seen in very young TDGs constitutes entirely independent empirical evidence toward this same statement. Young tidal dwarf galaxies (TDG), which should be devoid of collisionless DM, appear to nevertheless exhibit a mass-discrepancy in Newtonian dynamics. This is a huge problem for the DM hypothesis, but it is perfectly explained by MOND (Gentile *et al.*, 2007; Milgrom, 2007).

4.6.5 *A new structure formation scenario*

It so emerges that the nearby feeblest and smallest galaxies hold perhaps the strongest clues as to the nature and composition of the Universe. Figure 4.9 schematically depicts a possible structure formation scenario in such a modified Newtonian framework: whereas purely baryonic galaxies would merge, these events would spawn new dwarf galaxies such that the density-morphology relation would be established (more dE galaxies in denser environments, Okazaki and Taniguchi, 2000). Cosmological structure formation computations will quantify this tree within a non-Newtonian gravity framework.

The MONDian modeling by Tiret and Combes (2008) and Combes and Tiret (2010) has already shown that TDGs are produced during gas-dissipational galaxy mergers, and that the interaction times between galaxies are much longer, whereas the number of mergers is smaller compared to the situation in a DM universe. So, the number of observed galaxy mergers would result from the long timescale of merging and thus more close galaxy-galaxy encounters per merging event rather than to a high number of mergers.

This would in addition naturally explain why compact galaxy groups are observed to not evolve statistically over more than a crossing time (Pompei and Iovino, 2010). In contrast, assuming DM-Newtonian dynamics to hold, the merging timescale is of the order of one crossing time such that compact galaxy groups ought to disappear over a crossing time. The lack of significant evolution of compact groups would appear to not be explainable if DM dominates galaxy dynamics.

4.7 Conclusions

We inhabit a reality for which physicist seek mathematical formulations. An extraordinarily successful formulation of gravitational physics, the General Theory of Relativity, requires the existence of exotic dark matter (DM) if the observed rotation curves of galaxies and other dynamical effects are to be explained in that theory, which has Newtonian dynamics as its weak-field limit.

This contribution critically tests the eight most recent state-of-the art DM models of the dSph MW satellite galaxy population computed by a number of independently working research groups designed explicitly to explain the observed MW satellite galaxies. Three new problems for DM cosmology arise, adding to the list including the problems to form large disks and bulgeless disks within DM halos (the angular momentum problem), the predicted and unobserved cuspiness of DM halos (the cusp problem), the lack of observed substructures around galaxies (the missing satellite problem), the rotation curves of young TDGs which have been observed to reach too high a velocity for their baryonic mass content (as they cannot contain DM, the rotation curves must be manifestations for non-Newtonian dynamics), and the planar distribution of Milky Way satellites (the disk of satellites problem, or phase-space correlation problem). The new problems are these:

- The mass-luminosity relation: In Section 4.2 it is shown that assuming the Mateo *et al.* (1993), Mateo (1998), Gilmore *et al.* (2007), Peñarrubia *et al.* (2008), and S08 results on the mass-luminosity relation slope, $\kappa \approx 0$, to hold implies a significant failure of DM-based models to account for the lack of a DM-mass-luminosity relation for dSph satellites. The slopes of *all eight dark matter based models* that were calculated in attempts to match the observed data are more than 3σ away from the data (Figure 4.1). That is, the DM-based models behave physically correct in that more massive model DM halos contain more baryons due to the higher binding energy. The analysis, however, demonstrates that physical reality does not follow this trend. The hypothesis that DM with Newtonian gravity describes the reality of the mass-luminosity relation can therefore be discarded with more than 99.7% confidence, and in fact, for seven of the eight models the rejection confidence is 99.99% or even larger. The very recently revised mass of one of the satellites, Hercules, does not alter this conclusion.
- The overpredicting luminous satellite problem: The fact that the shape of the predicted mass function of visible sub-halos is inconsistent with the observed mass function was demonstrated for the first time in Section 4.3. The hypothesis that the satellites are luminous CDM sub-halos can be excluded with a confidence of better than 93%. Despite all efforts to solve the "missing satellites problem," a significant "*overpredicting luminous sub-halo problem*" thus persists.
- The bulge-mass–satellite number relation: Additionally, Section 4.4 highlights a linear correlation between the bulge mass and the number of early-type satellites. This together with the correlated phase-space distribution of the MW and M31 satellites

suggests a physical link between the formation of the bulge of the host galaxy and the formation of early-type satellite galaxies around it.

- It is also demonstrated that the 13 new (ultrafaint) satellite galaxies (including Pisces I and II) form a disk-of-satellites (DoS_{new}) that is virtually identical to the well-established and very pronounced DoS_{cl} made by the 11 classical (bright, including the LMC and SMC) satellite galaxies ($DoS_{new} \approx DoS_{cl} \approx DoS$). The missing satellite searches are sufficiently well advanced and surveyed regions of the Galactic sky are sufficiently empty (e.g., Figure 4.4) to exclude the possibility that a more isotropic satellite distribution may be re-instated by even fainter not yet discovered satellites. Furthermore, as demonstrated elsewhere (Metz *et al.*, 2008), the average orbital angular momentum vector, which is defined by the inner classical satellites through measured proper motions, is co-aligned with the normal vector of the DoS suggesting a high phase-space correlation of the satellites (i.e., a rotating DoS).

Within the ΛCDM hypothesis, only 0.4% of MW-type dark matter halos are expected to host a population of luminous satellites similar in number and spatial arrangement to that of the actual MW, given the currently available proper motion measurements (Section 4.5.4). If the other satellites turn out to also have such correlated orbital angular momenta directions, or if satellites to be discovered with the *Stromlo Missing Satellite* Survey will be found to also lie within the DoS, then the likelihood that the DM model accounts for the observed systems becomes smaller than 4×10^{-3}. In contrast, in the Local Group the MW *and* Andromeda have similar dark matter halos and both have a similar number of satellites, arranged in a DoS (MW), and perhaps partially in a DoS (Andromeda).

Combining the two tests described for which confidence statements on CDM models can be made (the mass-luminosity and the mass function tests), the hypothesis that the luminous satellites are CDM sub-halos can be discarded with 99.999% confidence. That is, the likelihood that the observed satellites stem from CDM objects is 0.001%. This likelihood is only based on the analysis of the mass-luminosity data and the mass function, and does not contain an assessment of the likelihood of obtaining, within the ΛCDM hypothesis, a spatial satellite distribution as observed. This, by itself, is improbable (0.4%), as stated. A combined probability that CDM-based models do account for the MW and its satellite system is thus smaller than 10^{-6}. If we concentrate only on the mass-luminosity and the spatial-distribution tests (ignoring the mass function test), then the combined probability that the dark matter based models account for the MW satellites is still nevertheless only 0.3% times 0.4% $= 1.2 \times 10^{-5}$. The confidence of excluding WDM models is therefore near 100% as well.

The implication would therefore be that the Newtonian-dynamics plus dark matter ansatz must be abandoned in galaxies, and that most phenomena associated with DM must be explained with modified gravity.[6] Thus, when we study the properties of the Local Group, a strong test of gravitational physics in the extreme weak-field limit is arrived at such that General Relativity appears to be ruled out in this regime.

An alternative scenario capable of naturally solving the problems for the DM model presented in this chapter, as well as the phase-space correlation problem and the density-morphology relation of dwarf early-type galaxies is proposed in Section 4.6. In this scenario, dSph satellites would be ancient TDGs that formed during a very early merger involving the proto-MW. According to this scenario the distribution of MW satellites contains important information on the early merger events that shaped the Galaxy. Metz *et al.* (2009) identified a pronounced minor filamentary structure, positioned perpendicular to the super-galactic plane and aligned with the MW DoS. This may be the original

[6] Indeed, this result ought not to be surprising given that Newton derived the gravitational $1/r^2$ law over a very limited physical range (Solar system), while with the Local Group we can probe gravitational physics on a vastly larger scale and in a much weaker field regime.

structure from which the infalling galaxy came about 10 or more Gyr ago that interacted with the young MW to form the TDGs evident today as dSph satellites.

In Section 4.6 it is emphasized that the number of TDGs produced in standard DM cosmology is easily able to account for all dE galaxies (Okazaki and Taniguchi, 2000), leaving little room for shining DM halos with masses $< 10^{10}$ $M_{\odot}$.

Thus, a physically consistent picture emerges: standard DM cosmology implies most if not all dE galaxies to be evolved TDGs, implying a catastrophic substructure problem for DM models. At the same time, the tests discussed previously on the MW satellite galaxies independently imply failure of the DM hypothesis and that the satellites are most naturally explained as ancient TDGs. A convergence is thus discernible leading to abandonment of the DM hypothesis and therewith of Newtonian dynamics.

The implied alternative scenario (non-Newtonian universe and MW satellites and dE galaxies being TDGs) appears to be the most simple and natural solution to many problems of observational galactic and extra-galactic astrophysics. For example, the very recent realization by Gentile *et al.* (2009) and Milgrom (2009) that the ratio between DM mass and baryonic mass within the putative DM core radius is constant despite the very large variation of the DM to baryonic-matter ratio globally within galaxies cannot be understood within the DM hypothesis. But this is the expected behavior in MOND. The so-called DM core radius is merely that radius at which the acceleration falls below $a_0/2$, so that non-Newtonian dynamics (appearing, for a Newtonian observer, as the emergence of DM) becomes dominant outside this radius.

Remarkably, the breakdown of General Relativity (GR) on cosmic scales has been deduced with a 98% confidence level by Bean (2009) based on an analysis of the growth of large-scale structure constrained by the Integrated Sachs-Wolfe effect and its cross-correlation with the galaxy distribution and the new COSMOS weak lensing shear field. According to Bean (2009) this would mean that "dark energy might be a modification to GR, rather than Λ." This is completely independent evidence on vastly different scales to the deduction presented here, which is based on an analysis of Local Group astronomical objects in the weak-field gravitational regime, that DM cosmology is not a correct description of physical reality.

Acknowledgments

The work presented here is based on Kroupa *et al.* (2010) and on contributions by Joerg Dabringhausen (Section 4.3, Figures 4.7–4.8), Manuel Metz (Figure 4.9), Marcel Pawlowski (Figures 4.4, 4.5, Section 4.5), and Benoit Famaey (Section 4.6.3, 4.6.4). This material is partially based on discussions with and contributions by Benoit Famaey, Klaas S. de Boer, Joerg Dabringhausen, Christian Boily, Antonino Del Popolo, Duncan Forbes, Gerhardt Hensler, Helmut Jerjen, Manuel Metz, and Marcel Pawlowski and will appear as a research paper with the above persons as co-authors. I also Iskren Georgiev and Anton Ippendorf for useful inputs, and thank Jelte de Jong for allowing me to use the image from Coleman *et al.* (2007) in Figure 4.6.

REFERENCES

Adén, D., and 10 colleagues. 2009a. A photometric and spectroscopic study of the new dwarf spheroidal galaxy in Hercules. Metallicity, velocities, and a clean list of RGB members. *A&A*, **506**(Nov.), 1147–1168.

Adén, D., Wilkinson, M. I., Read, J. I., Feltzing, S., Koch, A., Gilmore, G. F., Grebel, E. K., and Lundström, I. 2009b. A new low mass for the Hercules dSph: the end of a common mass scale for the dwarfs? *ApJ*, **706**(Nov.), L150–L154.

Angus, G. W. 2008. Dwarf spheroidals in MOND. *MNRAS*, **387**(July), 1481–1488.

Angus, G. W., Famaey, B., and Diaferio, A. 2010. Equilibrium configurations of 11 eV sterile neutrinos in MONDian galaxy clusters. *MNRAS*, **402**(Feb.), 395–408.

Aubert, D., Pichon, C., and Colombi, S. 2004. The origin and implications of dark matter anisotropic cosmic infall on L_* haloes. *MNRAS*, **352**(Aug.), 376–398.

Barazza, F. D., Binggeli, B., and Jerjen, H. 2002. More evidence for hidden spiral and bar features in bright early-type dwarf galaxies. *A&A*, **391**(Sept.), 823–831.

Barnes, J. E. and Hernquist, L. 1992. Formation of dwarf galaxies in tidal tails. *Nature*, **360**(Dec.), 715–717.

Bean, R. 2009. A weak lensing detection of a deviation from General Relativity on cosmic scales. (Sept.), arXiv:0909.3853.

Beasley, M. A., Cenarro, A. J., Strader, J., and Brodie, J. P. 2009. Evidence for the disky origin of luminous Virgo dwarf ellipticals from the kinematics of their globular cluster systems. *AJ*, **137**(June), 5146–5153.

Bekenstein, J. D. 2004. Relativistic gravitation theory for the modified Newtonian dynamics paradigm. *Phys. Rev. D*, **70**(Oct.), 083509.

Bekenstein, J. D. and Milgrom, M. 1984. Does the missing mass problem signal the breakdown of Newtonian gravity?. *ApJ*, **286**(Nov.), 7–14.

Bell, D. L. 1983. The origin of dwarf spheroidal galaxies. *Internal Kinematics and Dynamics of Galaxies*, **100**, 89.

Belokurov, V., and 33 colleagues 2007. Cats and dogs, hair and a hero: a quintet of new Milky Way companions. *ApJ*, **654**(Jan.), 897–906.

Belokurov, V., and 10 colleagues 2010. Big fish, little fish: two new ultrafaint satellites of the Milky Way. *ApJ*, **712**(Mar.), L103–L106.

Bender, R., Burstein, D., and Faber, S. M. 1992. Dynamically hot galaxies. I – Structural properties. *ApJ*, **399**(Nov.), 462–477.

Bertone, G., Hooper, D., and Silk, J. 2005. Particle dark matter: evidence, candidates and constraints. *Phys. Rep.*, **405**(Jan.), 279–390.

Blanchet, L. and Le Tiec, A. 2009. Dipolar dark matter and dark energy. *Phys. Rev. D*, **80**(Sept.), 023524.

Boily, C. M., Nakasato, N., Spurzem, R., and Tsuchiya, T. 2004. Satellite survival in cold dark matter cosmology. *ApJ*, **614**(Oct.), 26–30.

Bosma, A. 1981. 21-cm line studies of spiral galaxies. I - Observations of the galaxies NGC 5033, 3198, 5055, 2841, and 7331. II – The distribution and kinematics of neutral hydrogen in spiral galaxies of various morphological types. *AJ*, **86**(Dec.), 1791–1846.

Bosma, A., Goss, W. M., and Allen, R. J. 1981. The giant spiral galaxy M101. VI – The large scale radial velocity field. *A&A*, **93**(Jan.), 106–112.

Bournaud, F. 2010. Tidal dwarf galaxies and missing baryons. *Advances in Astronomy*, **2010**, 1–7.

Bournaud, F., and 8 colleagues. 2007. Missing mass in collisional debris from galaxies. *Science*, **316**(May), 1166.

Bournaud, F., Duc, P.-A., and Emsellem, E. 2008. High-resolution simulations of galaxy mergers: resolving globular cluster formation. *MNRAS*, **389**(Sept.), L8–L12.

Brada, R. and Milgrom, M. 2000. Dwarf satellite galaxies in the modified dynamics. *ApJ*, **541**(Oct.), 556–564.

Brownstein, J. R. and Moffat, J. W. 2006. Galaxy rotation curves without non-baryonic dark matter. *ApJ*, **636**(Jan.), 721–741.

Bruneton, J.-P., Liberati, S., Sindoni, L., and Famaey, B. 2009. Reconciling MOND and dark matter? *J. Cosmology Astropart. Phys.*, **3**(Mar.), 21.

Bruneton, J.-P. and Esposito-Farèse, G. 2007. Field-theoretical formulations of MOND-like gravity. *Phys. Rev. D*, **76**(Dec.), 124012.

Bullock, J. S., Kolatt, T. S., Sigad, Y., Somerville, R. S., Kravtsov, A. V., Klypin, A. A., Primack, J. R., and Dekel, A. 2001. Profiles of dark haloes: evolution, scatter and environment. *MNRAS*, **321**(Mar.), 559–575.

Busha, M. T., Alvarez, M. A., Wechsler, R. H., Abel, T., and Strigari, L. E. 2010. The Impact of Inhomogeneous Reionization on the satellite galaxy population of the Milky Way. *ApJ*, **710**(Feb.), 408–420.

CDMS II Collaboration, and 60 colleagues. 2010. Dark matter search results from the CDMS II experiment. *Science*, **327**(Mar.), 1619.

Chilingarian, I. V. 2009. Evolution of dwarf early-type galaxies – I. Spatially resolved stellar populations and internal kinematics of Virgo cluster dE/dS0 galaxies. *MNRAS*, **394**(Apr.), 1229–1248.

Ciotti, L., Londrillo, P., and Nipoti, C. 2006. Axisymmetric and triaxial MOND density-potential pairs. *ApJ*, **640**(Apr.), 741–750.

Cole, S., and 30 colleagues. 2005. The 2dF Galaxy Redshift Survey: power-spectrum analysis of the final data set and cosmological implications. *MNRAS*, **362**(Sept.), 505–534.

Coleman, M. G., and 18 colleagues. 2007. The elongated structure of the Hercules Dwarf Spheroidal Galaxy from deep Large Binocular Telescope imaging. *ApJ*, **668**(Oct.), L43–L46.

Combes, F. 2004. Galaxy formation and baryonic dark matter. *Dark Matter in Galaxies*, **220**(July), 219.

Combes, F. 2009. From distances to galaxy evolution and the dark matter problem. Commentary on Tully R. B. and Fisher J. R., 1977, A&A, 54, 661. *A&A*, **500**(June), 119–120.

Combes, F. 2010. Theoretical problems and perspectives. *Galaxies in isolation: exploring nature versus nurture*, **421**(Oct.), 233.

Combes, F. and Tiret, O. 2010. MOND and the galaxies. *American Institute of Physics Conference Series*, **1241**(June), 154–161.

Cooper, A. P., and 11 colleagues. 2010. Galactic stellar haloes in the CDM model. *MNRAS*, **406**(Aug.), 744–766.

Dabringhausen, J., Hilker, M., and Kroupa, P. 2008. From star clusters to dwarf galaxies: the properties of dynamically hot stellar systems. *MNRAS*, **386**(May), 864–886.

Dabringhausen, J., Kroupa, P., and Baumgardt, H. 2009. A top-heavy stellar initial mass function in starbursts as an explanation for the high mass-to-light ratios of ultra-compact dwarf galaxies. *MNRAS*, **394**(Apr.), 1529–1543.

de Blok, W. J. G. and Bosma, A. 2002. High-resolution rotation curves of low surface brightness galaxies. *A&A*, **385**(Apr.), 816–846.

Dekel, A. and Silk, J. 1986. The origin of dwarf galaxies, cold dark matter, and biased galaxy formation. *ApJ*, **303**(Apr.), 39–55.

Dekel, A., and 9 colleagues. 2009. Cold streams in early massive hot haloes as the main mode of galaxy formation. *Nature*, **457**(Jan.), 451–454.

Dekel, A. and Woo, J. 2003. Feedback and the fundamental line of low-luminosity low-surface-brightness/dwarf galaxies. *MNRAS*, **344**(Oct.), 1131–1144.

Del Popolo, A. and Kroupa, P. 2009. Density profiles of dark matter haloes on galactic and cluster scales. *A&A*, **502**(Aug.), 733–747.

Del Popolo, A. and Yesilyurt, I. S. 2007. The cosmological mass function. *Astronomy Reports*, **51**(Sept.), 709–734.

Demleitner, M., Accomazzi, A., Eichhorn, G., Grant, C. S., Kurtz, M. J., and Murray, S. S. 2001. ADS's Dexter Data Extraction Applet. *Astronomical Data Analysis Software and Systems X*, **238**, 321.

Diemand, J., Kuhlen, M., Madau, P., Zemp, M., Moore, B., Potter, D., and Stadel, J. 2008. Clumps and streams in the local dark matter distribution. *Nature*, **454**(Aug.), 735–738.

Disney, M. J., Romano, J. D., Garcia-Appadoo, D. A., West, A. A., Dalcanton, J. J., and Cortese, L. 2008. Galaxies appear simpler than expected. *Nature*, **455**(Oct.), 1082–1084.

D'Onghia, E. and Lake, G. 2008. Small dwarf galaxies within larger dwarfs: why some are luminous while most go dark. *ApJ*, **686**(Oct.), L61–L65.

Einstein, A. 1921. *Festschrift der Kaiser-Wilhelm Gesellschaft zur Föderung der Wissenschaften zu ihrem zehnjährigen Jubiläum dargebracht von ihren Instituten. Berlin: Springer*, 90.

Esposito-Farèse, G. 2011. Motion in alternative theories of gravity. *Mass and Motion in General Relativity*, 461–489.

Famaey, B. and Binney, J. 2005. Modified Newtonian dynamics in the Milky Way. *MNRAS*, **363**(Oct.), 603–608.

Famaey, B., Gentile, G., Bruneton, J.-P., and Zhao, H. 2007. Insight into the baryon-gravity relation in galaxies. *Phys. Rev. D*, **75**(Mar.), 063002.

Ferrarese, L., and 10 colleagues. 2006. The ACS Virgo Cluster Survey. VI. Isophotal analysis and the structure of early-type galaxies. *ApJS*, **164**(June), 334–434.

Forbes, D. A., Lasky, P., Graham, A. W., and Spitler, L. 2008. Uniting old stellar systems: from globular clusters to giant ellipticals. *MNRAS*, **389**(Oct.), 1924–1936.

Füzfa, A. and Alimi, J.-M. 2007. Toward a unified description of dark energy and dark matter from the abnormally weighting energy hypothesis. *Phys. Rev. D*, **75**(June), 123007.

Gao, L., White, S. D. M., Jenkins, A., Stoehr, F., and Springel, V. 2004. The subhalo populations of ΛCDM dark haloes. *MNRAS*, **355**(Dec.), 819–834.

Gavazzi, G. 2009. Downsizing among disk galaxies and the role of the environment. *Revista Mexicana de Astronomia y Astrofisica Conference Series*, **37**(Nov.), 72–78.

Gebhardt, K., and 11 colleagues. 2001. M33: a Galaxy with no supermassive black hole. *AJ*, **122**(Nov.), 2469–2476.

Geha, M., Guhathakurta, P., and van der Marel, R. P. 2003. Internal dynamics, structure, and formation of dwarf elliptical galaxies. II. Rotating versus nonrotating dwarfs. *AJ*, **126**(Oct.), 1794–1810.

Gentile, G., Famaey, B., Combes, F., Kroupa, P., Zhao, H. S., and Tiret, O. 2007. Tidal dwarf galaxies as a test of fundamental physics. *A&A*, **472**(Sept.), L25–L28.

Gentile, G., Famaey, B., Zhao, H., and Salucci, P. 2009. Universality of galactic surface densities within one dark halo scale-length. *Nature*, **461**(Oct.), 627–628.

Gentile, G., Salucci, P., Klein, U., Vergani, D., and Kalberla, P. 2004. The cored distribution of dark matter in spiral galaxies. *MNRAS*, **351**(July), 903–922.

Gibbons, G. W. and Hawking, S. W. 1977. Cosmological event horizons, thermodynamics, and particle creation. *Phys. Rev. D*, **15**(May), 2738–2751.

Gilmore, G., Wilkinson, M. I., Wyse, R. F. G., Kleyna, J. T., Koch, A., Evans, N. W., and Grebel, E. K. 2007. The observed properties of dark matter on small spatial scales. *ApJ*, **663**(July), 948–959.

Graham, A. W. and Worley, C. C. 2008. Inclination- and dust-corrected galaxy parameters: bulge-to-disk ratios and size-luminosity relations. *MNRAS*, **388**(Aug.), 1708–1728.

Grebel, E. K. 1999. Evolutionary histories of dwarf galaxies in the Local Group. *The Stellar Content of Local Group Galaxies*, **192**(Jan.), 17.

Grebel, E. K. 2008. Baryonic properties of the darkest galaxies. *IAU Symposium*, **244**(May), 300–310.

Hartwick, F. D. A. 2000. The structure of the outer halo of the Galaxy and its relationship to nearby large-scale structure. *AJ*, **119**(May), 2248–2253.

Hunter, D. A., Hunsberger, S. D., and Roye, E. W. 2000. Identifying old tidal dwarf irregulars. *ApJ*, **542**(Oct.), 137–142.

Jerjen, H., *et al.*

Jerjen, H., Kalnajs, A., and Binggeli, B. 2000. IC3328: A "dwarf elliptical galaxy" with spiral structure. *A&A*, **358**(June), 845–849.

Kallivayalil, N., van der Marel, R. P., and Alcock, C. 2006. Is the SMC bound to the LMC? The Hubble Space Telescope proper motion of the SMC. *ApJ*, **652**(Dec.), 1213–1229.

Karachentsev, I. D., Karachentseva, V. E., and Sharina, M. E. 2005. Dwarf spheroidal galaxies in nearby groups imaged with HST. *IAU Colloq. 198: Near-fields cosmology with dwarf elliptical galaxies*, 295–302.

Kazantzidis, S., Mayer, L., Mastropietro, C., Diemand, J., Stadel, J., and Moore, B. 2004. Density Profiles of Cold Dark Matter Substructure: Implications for the Missing-Satellites Problem. *ApJ*, **608**(June), 663–679.

Kent, S. M. 1989. An improved bulge model for M31. *AJ*, **97**(June), 1614–1621.

Kirby, E. N., and 11 colleagues 2009. The role of dwarf galaxies in building large stellar halos. *astro2010: The Astronomy and Astrophysics Decadal Survey*, **2010**, 156.

Klessen, R. S. and Kroupa, P. 1998. Dwarf spheroidal satellite galaxies without dark matter: results from two different numerical techniques. *ApJ*, **498**(May), 143.

Klimentowski, J., Łokas, E. L., Knebe, A., Gottlöber, S., Martinez-Vaquero, L. A., Yepes, G., and Hoffman, Y. 2010. The grouping, merging and survival of subhaloes in the simulated Local Group. *MNRAS*, **402**(Mar.), 1899–1910.

Klypin, A., Kravtsov, A. V., Valenzuela, O., and Prada, F. 1999. Where are the missing galactic satellites? *ApJ*, **522**(Sept.), 82–92.

Koch, A. and Grebel, E. K. 2006. The anisotropic distribution of M31 satellite galaxies: a polar great plane of early-type companions. *AJ*, **131**(Mar.), 1405–1415.

Komatsu, E., and 18 colleagues. 2009. Five-year Wilkinson Microwave Anisotropy Probe observations: cosmological interpretation. *ApJS*, **180**(Feb.), 330–376.

Koposov, S. E., Yoo, J., Rix, H.-W., Weinberg, D. H., Macciò, A. V., and Escudé, J. M. 2009. A quantitative explanation of the observed population of Milky Way satellite galaxies. *ApJ*, **696**(May), 2179–2194.

Kroupa, P. 1997. Dwarf spheroidal satellite galaxies without dark matter. *New A*, **2**(July), 139–164.

Kroupa, P., and 9 colleagues. 2010. Local-Group tests of dark matter concordance cosmology. Toward a new paradigm for structure formation. *A&A*, **523**(Nov.), A32.

Kroupa, P., Theis, C., and Boily, C. M. 2005. The great disk of Milky-Way satellites and cosmological sub-structures. *A&A*, **431**(Feb.), 517–521.

Kruijssen, J. M. D. and Lamers, H. J. G. L. M. 2008. The photometric evolution of star clusters and the preferential loss of low-mass bodies – with an application to globular clusters. *A&A*, **490**(Oct.), 151–171.

Kruijssen, J. M. D. and Mieske, S. 2009. Dissolution is the solution: on the reduced mass-to-light ratios of Galactic globular clusters. *A&A*, **500**(June), 785–799.

Kunkel, W. E. 1979. On the origin and dynamics of the Magellanic Stream. *ApJ*, **228**(Mar.), 718–733.

Latronico, L. and for the Fermi LAT Collaboration 2009. Measurement of the Cosmic Ray electron plus positron spectrum from 20 GeV to 1 TeV with the Fermi Large Area Telescope. (Sept.), arXiv:0907.0452.

Law, D. R., Majewski, S. R., and Johnston, K. V. 2009. Evidence for a triaxial Milky Way dark matter halo from the Sagittarius stellar tidal stream. *ApJ*, **703**(Sept.), L67–L71.

Li, Y.-S. and Helmi, A. 2008. Infall of substructures on to a Milky Way-like dark halo. *MNRAS*, **385**(Apr.), 1365–1373.

Li, Y.-S., De Lucia, G., and Helmi, A. 2010. On the nature of the Milky Way satellites. *MNRAS*, **401**(Jan.), 2036–2052.

Libeskind, N. I., Frenk, C. S., Cole, S., Jenkins, A., and Helly, J. C. 2009. How common is the Milky Way-satellite system alignment? *MNRAS*, **399**(Oct.), 550–558.

Lingenfelter, R. E., Higdon, J. C., and Rothschild, R. E. 2009. Is there a dark matter signal in the Galactic positron annihilation radiation? *Physical Review Letters*, **103**(July), 031301.

Lisker, T. 2009. Early-type dwarf galaxies in clusters: a mixed bag with various origins? *Astronomische Nachrichten*, **330**(Dec.), 1043.

Llinares, C., Knebe, A., and Zhao, H. 2008. Cosmological structure formation under MOND: a new numerical solver for Poisson's equation. *MNRAS*, **391**(Dec.), 1778–1790.

Lynden-Bell, D. 1976. Dwarf galaxies and globular clusters in high velocity hydrogen streams. *MNRAS*, **174**(Mar.), 695–710.

Lynden-Bell, D. and Lynden-Bell, R. M. 1995. Ghostly streams from the formation of the Galaxy's halo. *MNRAS*, **275**(July), 429–442.

Macciò, A. V. and Fontanot, F. 2010. How cold is dark matter? Constraints from Milky Way satellites. *MNRAS*, **404**(May), L16–L20.

Macciò, A. V., Dutton, A. A., van den Bosch, F. C., Moore, B., Potter, D., and Stadel, J. 2007. Concentration, spin and shape of dark matter haloes: scatter and the dependence on mass and environment. *MNRAS*, **378**(June), 55–71.

Macciò, A. V., Kang, X., and Moore, B. 2009. Central mass and luminosity of Milky Way satellites in the Λ cold dark matter model. *ApJ*, **692**(Feb.), L109–L112.

Macciò, A. V., Kang, X., Fontanot, F., Somerville, R. S., Koposov, S., and Monaco, P. 2010. Luminosity function and radial distribution of Milky Way satellites in a ΛCDM Universe. *MNRAS*, **402**(Mar.), 1995–2008.

Macciò, A. V., Murante, G., and Bonometto, S. P. 2003. Mass of clusters in simulations. *ApJ*, **588**(May), 35–49.

Majewski, S. R. 1994. The Fornax-Leo-Sculptor stream revisited. *ApJ*, **431**(Aug.), L17–L21.

Martínez-Delgado, D. Private communication.

Mateo, M. L. 1998. Dwarf galaxies of the Local Group. *ARA&A*, **36**, 435–506.

Mateo, M., Olszewski, E. W., Pryor, C., Welch, D. L., and Fischer, P. 1993. The Carina dwarf spheroidal galaxy – how dark is it? *AJ*, **105**(Feb.), 510–526.

McGaugh, S. S. 2004. The mass discrepancy-acceleration relation: disk mass and the dark matter distribution. *ApJ*, **609**(July), 652–666.

McGaugh, S. S. 2005a. Balance of dark and luminous mass in rotating galaxies. *Physical Review Letters*, **95**(Oct.), 171302.

McGaugh, S. S. 2005b. The baryonic Tully-Fisher relation of galaxies with extended rotation curves and the stellar mass of rotating galaxies. *ApJ*, **632**(Oct.), 859–871.

McGaugh, S. S. 2008. Milky Way mass models and MOND. *ApJ*, **683**(Aug.), 137–148.

McGaugh, S. S., de Blok, W. J. G., Schombert, J. M., Kuzio de Naray, R., and Kim, J. H. 2007. The rotation velocity attributable to dark matter at intermediate radii in disk galaxies. *ApJ*, **659**(Apr.), 149–161.

Metz, M. 2008. PhD Thesis. *University of Bonn.*

Metz, M. and Kroupa, P. 2007. Dwarf spheroidal satellites: are they of tidal origin? *MNRAS*, **376**(Mar.), 387–392.

Metz, M., Kroupa, P., and Jerjen, H. 2007. The spatial distribution of the Milky Way and Andromeda satellite galaxies. *MNRAS*, **374**(Jan.), 1125–1145.

Metz, M., Kroupa, P., and Jerjen, H. 2009a. Discs of satellites: the new dwarf spheroidals. *MNRAS*, **394**(Apr.), 2223–2228.

Metz, M., Kroupa, P., and Libeskind, N. I. 2008. The orbital poles of Milky Way satellite galaxies: a rotationally supported disk of satellites. *ApJ*, **680**(June), 287–294.

Metz, M., Kroupa, P., Theis, C., Hensler, G., and Jerjen, H. 2009b. Did the Milky Way dwarf satellites enter the halo as a group? *ApJ*, **697**(May), 269–274.

Mieske, S. and Kroupa, P. 2008. An extreme IMF as an explanation for high M/L ratios in UCDs? The CO index as a tracer of bottom-heavy IMFs. *ApJ*, **677**(Apr.), 276–282.

Milgrom, M. 1983. A modification of the Newtonian dynamics as a possible alternative to the hidden mass hypothesis. *ApJ*, **270**(July), 365–370.

Milgrom, M. 1995. MOND and the seven dwarfs. *ApJ*, **455**(Dec.), 439.

Milgrom, M. 1999. The modified dynamics as a vacuum effect. *Physics Letters A*, **253**(Mar.), 273–279.

Milgrom, M. 2007. MOND and the mass discrepancies in tidal dwarf galaxies. *ApJ*, **667**(Sept.), L45–L48.

Milgrom, M. 2008. Marriage Á-la-MOND: baryonic dark matter in galaxy clusters and the cooling flow puzzle. *New A Rev.*, **51**(May), 906–915.

Milgrom, M. 2009. The central surface density of 'dark haloes' predicted by MOND. *MNRAS*, **398**(Sept.), 1023–1026.

Moffat, J. W. 2006. Scalar tensor vector gravity theory. *J. Cosmology Astropart. Phys.*, **3**(Mar.), 4.

Moffat, J. W. and Toth, V. T. 2008. Testing modified gravity with globular cluster velocity dispersions. *ApJ*, **680**(June), 1158–1161.

Moffat, J. W. and Toth, V. T. 2009a. Modified gravity and the origin of inertia. *MNRAS*, **395**(May), L25–L28.

Moffat, J. W. and Toth, V. T. 2009b. Fundamental parameter-free solutions in modified gravity. *Classical and Quantum Gravity*, **26**(Apr.), 085002.

Moore, B., Diemand, J., Madau, P., Zemp, M., and Stadel, J. 2006. Globular clusters, satellite galaxies and stellar haloes from early dark matter peaks. *MNRAS*, **368**(May), 563–570.

Moore, B., Ghigna, S., Governato, F., Lake, G., Quinn, T., Stadel, J., and Tozzi, P. 1999a. Dark matter substructure within galactic halos. *ApJ*, **524**(Oct.), L19–L22.

Moore, B., Lake, G., Quinn, T., and Stadel, J. 1999b. On the survival and destruction of spiral galaxies in clusters. *MNRAS*, **304**(Apr.), 465–474.

Murray, N. 2009. The sizes and luminosities of massive star clusters. *ApJ*, **691**(Feb.), 946–962.

Navarro, J. F., and 8 colleagues. 2010. The diversity and similarity of simulated cold dark matter haloes. *MNRAS*, **402**(Feb.), 21–34.

Navarro, J. F., Frenk, C. S., and White, S. D. M. 1997. A universal density profile from hierarchical clustering. *ApJ*, **490**(Dec.), 493.

Nipoti, C., Ciotti, L., Binney, J., and Londrillo, P. 2008. Dynamical friction in modified Newtonian dynamics. *MNRAS*, **386**(June), 2194–2198.

Okamoto, T. and Frenk, C. S. 2009. The origin of failed sub-haloes and the common mass scale of the Milky Way satellite galaxies. *MNRAS*, **399**(Oct.), L174–L178.

Okazaki, T. and Taniguchi, Y. 2000. Dwarf galaxy formation Induced by galaxy interactions. *ApJ*, **543**(Nov.), 149–152.

Okamoto, T., Frenk, C. S., Jenkins, A., and Theuns, T. 2010. The properties of satellite galaxies in simulations of galaxy formation. *MNRAS*, **406**(July), 208–222.

Oort, J. H. 1932. The force exerted by the stellar system in the direction perpendicular to the galactic plane and some related problems. *Bull. Astron. Inst. Netherlands*, **6**(Aug.), 249.

Ott, J., Walter, F., and Brinks, E. 2005. A Chandra X-ray survey of nearby dwarf starburst galaxies – II. Starburst properties and outflows. *MNRAS*, **358**(Apr.), 1453–1471.

Palma, C., Majewski, S. R., and Johnston, K. V. 2002. On the distribution of orbital poles of Milky Way satellites. *ApJ*, **564**(Jan.), 736–761.

Parriott, J. R. and Bregman, J. N. 2008. Mass loss from evolved stars in elliptical galaxies. *ApJ*, **681**(July), 1215–1232.

Peñarrubia, J., McConnachie, A. W., and Navarro, J. F. 2008. The cold dark matter halos of Local Group dwarf spheroidals. *ApJ*, **672**(Jan.), 904–913.

Perlmutter, S., and 32 colleagues. 1999. Measurements of Omega and Lambda from 42 high-redshift supernovae. *ApJ*, **517**(June), 565–586.

Pflamm-Altenburg, J. and Hensler, G. 2011. Accretion regulated star formation in late-type galaxies. *UP2010: Have Observations Revealed a Variable Upper End of the Initial Mass Function?* **440**(June), 403.

Pflamm-Altenburg, J. and Kroupa, P. 2009. Recurrent gas accretion by massive star clusters, multiple stellar populations and mass thresholds for spheroidal stellar systems. *MNRAS*, **397**(July), 488–494.

Piontek, F. and Steinmetz, M. 2009. The modeling of feedback processes in cosmological simulations of disk galaxy formation. (Sept.), arXiv:0909.4167.

Pompei, E. and Iovino, A. 2010. Compact groups of galaxies: small, dense, and elusive. *Galaxies in Isolation: Exploring Nature Versus Nurture*, **421**(Oct.), 279.

Press, W. H., Teukolsky, S. A., Vetterling, W. T., and Flannery, B. P. 1992. Numerical recipes in C. The art of scientific computing. 2nd ed. *Cambridge UK: University Press.*

Primack, J. R. 2009. Cosmology: small scale issues. *American Institute of Physics Conference Series*, **1166**(Sept.), 3–9.

Recchi, S., Theis, C., Kroupa, P., and Hensler, G. 2007. The early evolution of tidal dwarf galaxies. *A&A*, **470**(July), L5–L8.

Riess, A. G., and 19 colleagues 1998. Observational evidence from supernovae for an accelerating universe and a cosmological constant. *AJ*, **116**(Sept.), 1009–1038.

Rubin, V. C. and Ford, W. K., Jr. 1970. Rotation of the Andromeda nebula from a spectroscopic survey of emission regions. *ApJ*, **159**(Feb.), 379.

Sanders, R. H. 2005. A tensor-vector-scalar framework for modified dynamics and cosmic dark matter. *MNRAS*, **363**(Oct.), 459–468.

Sanders, R. H. 2008a. Forming galaxies with MOND. *MNRAS*, **386**(May), 1588–1596.

Sanders, R. H. 2008b. From dark matter to MOND. (June), arXiv:0806.2585.

Sanders, R. H. and McGaugh, S. S. 2002. Modified Newtonian dynamics as an alternative to dark matter. *ARA&A*, **40**, 263–317.

Sanders, R. H. and Noordermeer, E. 2007. Confrontation of modified Newtonian dynamics with the rotation curves of early-type disc galaxies. *MNRAS*, **379**(Aug.), 702–710.

Serra, A. L., Angus, G. W., and Diaferio, A. 2010. Implications for dwarf spheroidal mass content from interloper removal. *A&A*, **524**(Dec.), A16.

Shaya, E., and 20 colleagues 2009. Properties of dark matter revealed by astrometric measurements of the Milky Way and local galaxies. *astro2010: The Astronomy and Astrophysics Decadal Survey*, **2010**, 274.

Silk, J. and Norman, C. 1981. Dissipational galaxy formation – confrontation with observations. *ApJ*, **247**(July), 59–76.

Simon, J. D. and Geha, M. 2007. The kinematics of the ultrafaint Milky Way satellites: solving the missing satellite problem. *ApJ*, **670**(Nov.), 313–331.

Skordis, C. 2009. TOPICAL REVIEW: The tensor-vector-scalar theory and its cosmology. *Classical and Quantum Gravity*, **26**(July), 143001.

Spergel, D. N., and 21 colleagues. 2007. Three-year Wilkinson Microwave Anisotropy Probe (WMAP) observations: implications for cosmology. *ApJS*, **170**(June), 377–408.

Strigari, L. E., Bullock, J. S., Kaplinghat, M., Simon, J. D., Geha, M., Willman, B., and Walker, M. G. 2008. A common mass scale for satellite galaxies of the Milky Way. *Nature*, **454**(Aug.), 1096–1097.

Tegmark, M., and 65 colleagues. 2004. The three-dimensional power spectrum of galaxies from the Sloan Digital Sky Survey. *ApJ*, **606**(May), 702–740.

Tiret, O. and Combes, F. 2007. Evolution of spiral galaxies in modified gravity. *A&A*, **464**(Mar.), 517–528.

Tiret, O. and Combes, F. 2008. Interacting galaxies with modified Newtonian dynamics. *Formation and Evolution of Galaxy Disks*, **396**(Oct.), 259.

Tollerud, E. J., Bullock, J. S., Strigari, L. E., and Willman, B. 2008. Hundreds of Milky Way satellites? Luminosity bias in the satellite luminosity function. *ApJ*, **688**(Nov.), 277–289.

Tully, R. B. and Fisher, J. R. 1977. A new method of determining distances to galaxies. *A&A*, **54**(Feb.), 661–673.

Unruh, W. G. 1976. Notes on black-hole evaporation. *Phys. Rev. D*, **14**(Aug.), 870–892.

van den Bergh, S. 2008. Astrophysics: how do galaxies form? *Nature*, **455**(Oct.), 1049–1051.

van der Marel, R. P., Alves, D. R., Hardy, E., and Suntzeff, N. B. 2002. New understanding of Large Magellanic Cloud Structure, dynamics, and orbit from carbon star kinematics. *AJ*, **124**(Nov.), 2639–2663.

Walsh, S. M., Willman, B., and Jerjen, H. 2009. The invisibles: a detection algorithm to trace the faintest Milky Way satellites. *AJ*, **137**(Jan.), 450–469.

Walsh, S. M., Willman, B., Sand, D., Harris, J., Seth, A., Zaritsky, D., and Jerjen, H. 2008. Bootes II reBooted: An MMT/MegaCam study of an ultrafaint Milky Way satellite. *ApJ*, **688**(Nov.), 245–253.

Watkins, L. L., and 10 colleagues. 2009. Substructure revealed by RRLyraes in SDSS Stripe 82. *MNRAS*, **398**(Oct.), 1757–1770.

Weidner, C., Kroupa, P., and Larsen, S. S. 2004. Implications for the formation of star clusters from extragalactic star formation rates. *MNRAS*, **350**(June), 1503–1510.

Weil, M. L., Eke, V. R., and Efstathiou, G. 1998. The formation of disc galaxies. *MNRAS*, **300**(Nov.), 773–789.

Wetzstein, M., Naab, T., and Burkert, A. 2007. Do dwarf galaxies form in tidal tails? *MNRAS*, **375**(Mar.), 805–820.

Wolf, J., Martinez, G. D., Bullock, J. S., Kaplinghat, M., Geha, M., Muñoz, R. R., Simon, J. D., and Avedo, F. F. 2010. Accurate masses for dispersion-supported galaxies. *MNRAS*, **406**(Aug.), 1220–1237.

Zhao, H. 1996. A steady-state dynamical model for the COBE-detected Galactic bar. *MNRAS*, **283**(Nov.), 149–166.

Zhao, H. S. 2008. An ecological approach to problems of dark energy, dark matter, MOND and neutrinos. *Journal of Physics Conference Series*, **140**(Nov.), 012002.

Zlosnik, T. G., Ferreira, P. G., and Starkman, G. D. 2007. Modifying gravity with the aether: an alternative to dark matter. *Phys. Rev. D*, **75**(Feb.), 044017.

Zwicky, F. 1933. Die Rotverschiebung von extragalaktischen Nebeln. *Helvetica Physica Acta*, **6**, 110–127.

5. Stellar tidal streams

R. IBATA

5.1 Introduction

Stellar streams represent the remnants of ancient accretion events into a galaxy, and are thus extremely important as tracers of the galaxy formation process. In recent years, it has been increasingly recognized that many of the clues to the fundamental problem of galaxy formation are preserved in fossil substructures (Freeman and Bland-Hawthorn, 2002), particularly in the outskirts of galaxies. Hierarchical formation models suggest that galaxy outskirts form by accretion of minor satellites, predominantly at early epochs when large disk galaxies were assembling for the first time. The size, metallicity, and amount of substructure in the faint outskirts of presentday galaxies are therefore directly related to issues such as the small-scale properties of the primordial power spectrum of density fluctuations and the suppression of star formation in small halos (Springel *et al.*, 2005).

Remarkable progress has been made in recent years in understanding galaxy formation and evolution. High redshift observations have revealed the star formation history of the universe and the evolution of galaxy morphology (see, e.g., Bell *et al.*, 2005; Ryan *et al.*, 2008). However, the nature of look-back observations does not allow one to study the evolution of individual galaxies, and low-mass or small-scale structures remain out of reach in all but the nearest galaxies. Hence high spatial-resolution observations in nearby galaxies are required to complement the samples at cosmological distances to answer many of the big fundamental questions of galaxy formation such as how did the Milky Way build up, and how typical was this formation history? What fraction of the halo, disk, thick disk, and bulge components was accreted, and how, why, and when? What fraction formed in situ? What was the detailed chemical enrichment history? The study of stellar streams will be central in our attempts to answer these questions, as the fossil evidence we seek is mainly to be found in these structures (or the remnants thereof).

Streams are also sensitive probes of the gravitational field that they inhabit, allowing us to recover the underlying distribution of dark matter, thereby shedding light on the nature of this most enigmatic component of galaxies.

5.1.1 *Streams are a natural outcome of dynamical evolution in a tidal field*

Most stars form in clusters from gas that has collapsed out of the interstellar medium into dense giant molecular clouds (see, e.g., McKee and Ostriker, 2007). After the first generation of stars ignite, their strong radiation field rapidly expels much of the gas from the proto-cluster, changing its gravitational potential. Furthermore, gas removal becomes even more severe once the first supernovae type II explode. What happens when gas is ejected from a star cluster? From an initial equilibrium state where the velocity dispersion of the young stars was in balance with the total mass present,

$$\sigma_{\text{stars,initial}} = \sqrt{\frac{G(M_{\text{stars}} + M_{\text{gas}})}{R}}, \tag{5.1}$$

the stars find themselves in a lower potential,

$$\sqrt{\frac{GM_{\text{stars}}}{R}} < \sigma_{\text{stars,initial}}, \tag{5.2}$$

with too high velocity dispersion after the gas has been expelled. The natural outcome is that many of the higher velocity stars simply escape from the cluster (Geyer and Burkert, 2001).

Once the initial formation of the cluster or satellite is complete, other dynamical processes compete to further deplete the structure (see, e.g., Binney and Tremaine, 1987). These include violent relaxation of the system, where the rapid change of the potential causes some fraction of the structure to collapse into a dense core, while other stars attain large energies and are lost to the system. The timescale of this process is related to the rate of change of the potential $\dot{\Phi}$ as follows:

$$t_{vr} = \frac{3}{4}\left\langle \frac{\dot{\Phi}^2}{\Phi^2} \right\rangle^{1/2}, \tag{5.3}$$

and is a few (~15) dynamical (crossing) times,

$$t_{cr} = \frac{r_h}{\sigma_h} \approx \sqrt{\frac{r_h^3}{GM_h}}, \tag{5.4}$$

(where the h subscript denotes half-mass values). On longer timescales, as stars undergo many distant encounters with other stars in the system, they slowly migrate into regions of phase-space where they are no longer bound; this is especially important for low mass systems. Typically, this evaporation time is,

$$t_{ev} \approx 136 t_{rx}, \tag{5.5}$$

where t_{rx} is the two body relaxation time:

$$t_{rx} = \frac{9}{16\sqrt{\pi}} \frac{\sigma^3}{G^2 m \rho \ln(\gamma N)} \approx \frac{0.1N}{\ln(N)} t_{cr}. \tag{5.6}$$

Here, N is the number of stars in the dynamical system, m is the stellar mass, σ is the velocity dispersion, and ρ is the stellar density. The formation of hard binaries provides an additional means to unbind cluster stars; subsequent three-body interactions result in a further hardening of the binary and (often) the ejection of the third star with high velocity.

Stars lost in any of these ways (initial escape, after violent relaxation, after a three-body encounter, or after slow evaporation) obviously soon find their way into the Galactic tidal field. What happens then?

5.1.2 *Galactic tides*

Let us therefore try to understand how tides work. Consider a host galaxy and a satellite as sketched in Figure 5.1. Here $\vec{R}$ is the position vector from the center of the host galaxy to the center of the satellite, $\vec{X}$ is the position vector of a location under consideration, and $\vec{x} = \vec{X} - \vec{R}$. Tides are a second order effect of the potential, so to explore them, let us expand the potential in a Taylor series about the point of interest. Recall that the usual Taylor series of some function f of a scalar X

$$\begin{aligned} f(X) = {} & f(R) \\ & + (X - R) f'(R) \\ & + \frac{1}{2!}(X - R)^2 f''(R) \\ & + \frac{1}{3!}(X - R)^3 f'''(R) \\ & + \ldots \end{aligned} \tag{5.7}$$

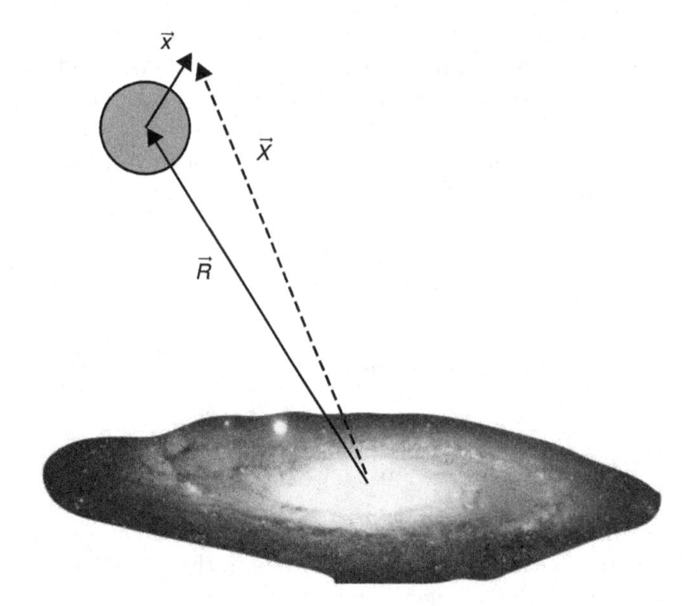

FIG. 5.1. The geometry of the tidal field being investigated. The vector $\vec{R}$ connects the center of the host galaxy and the satellite, $\vec{X}$ is a position vector to the location of interest, and $\vec{x} = \vec{X} - \vec{R}$.

becomes

$$f(\vec{X}) = \sum_{j=0}^{\infty} \left[\frac{1}{j!} (\vec{x} \cdot \nabla_{R'})^j f(R') \right], \tag{5.8}$$

when the function has a vector argument. Expanding the potential ϕ around $\vec{X}$ therefore gives

$$\begin{aligned}
\phi(\vec{X}) &= \phi(\vec{R}) \\
&+ \vec{x} \cdot [\nabla_{\vec{R}'} \phi(\vec{R}')] \Big|_{\vec{R}'=\vec{R}} \\
&+ \frac{1}{2} \vec{x} \cdot [\vec{x} \cdot \nabla_{\vec{R}'} (\nabla_{\vec{R}'} \phi(\vec{R}'))] \Big|_{\vec{R}'=\vec{R}} \\
&+ \frac{1}{6} \vec{x} \cdot [\vec{x} \cdot [\vec{x} \cdot \nabla_{\vec{R}'} (\nabla_{\vec{R}'} (\nabla_{\vec{R}'} \phi(\vec{R}')))]] \Big|_{\vec{R}'=\vec{R}} \\
&+ \ldots .
\end{aligned} \tag{5.9}$$

The first right-hand-side (RHS) term $\phi(\vec{R})$ is simply a constant (hence there are no associated forces) and can be ignored. The second term is the first order gradient of the potential along the direction $\vec{x}$. This is a gravitational acceleration that is uniform in the vicinity of $\vec{R}$ (and because of this is uninteresting in the present discussion). The first interesting term in the expansion is the third, which represents a saddle-point (it is the Hessian of ϕ).

Let us see how this works with some concrete examples. A simple case to examine first is the behavior of a Keplerian potential:

$$\phi(\vec{X}) = \frac{1}{|\vec{X}|}. \tag{5.10}$$

The second RHS term in Equation 5.9 becomes then

$$\vec{x} \cdot [\nabla_{\vec{R}'} \phi(\vec{R}')]\Big|_{\vec{R}'=\vec{R}} = \frac{\vec{x} \cdot \vec{R}}{|\vec{R}|^3}, \tag{5.11}$$

while the third term becomes

$$\frac{1}{2}\vec{x} \cdot [\vec{x} \cdot \nabla_{\vec{R}'}(\nabla_{\vec{R}'} \phi(\vec{R}'))]\Big|_{\vec{R}'=\vec{R}} = -\frac{1}{2}\frac{|\vec{x}|^2}{|\vec{R}|^3} + \frac{3}{2}\frac{(\vec{x} \cdot \vec{R})^2}{|\vec{R}|^5}. \tag{5.12}$$

Likewise, for a Keplerian potential, the acceleration field

$$g(\vec{X}) = -\nabla \phi(\vec{X}) \tag{5.13}$$

becomes

$$\begin{aligned} g(\vec{X}) = &-\frac{\hat{R}}{|\vec{R}|^2} \\ &-\frac{\vec{x}}{|\vec{R}|^3} + 3\frac{(\vec{x} \cdot \hat{R})\hat{R}}{|\vec{R}|^3}. \end{aligned} \tag{5.14}$$

The first RHS term of Equation 5.14 can be immediately recognized as the Newtonian gravity term, while the second line is the tidal term of interest to us here. When $\vec{x}$ is parallel to $\vec{R}$ this term becomes $2|x|/|\vec{R}|^3$, whereas when $\vec{x}$ is perpendicular to $\vec{R}$ it becomes $-|x|/|\vec{R}|^3$; this means tidal stretching occurs along $\vec{R}$ and tidal compression occurs perpendicular to that direction.

Another simple potential that is of relevance to galaxies is the logarithmic halo,

$$\phi(\vec{X}) = \frac{V_c^2}{2} \ln \left(R_c^2 + \vec{X}^2\right), \tag{5.15}$$

where V_c is a scale velocity and R_c is a core radius. A similar expansion of the acceleration field gives

$$\begin{aligned} \frac{g(\vec{X})}{V_c^2} = &-\frac{\vec{R}}{R_c^2 + \vec{R}^2} \\ &-\frac{\vec{x}}{\left(R_c^2 + \vec{R}^2\right)^{3/2}} + 2\frac{(\vec{x} \cdot \vec{R})\hat{R}}{\left(R_c^2 + \vec{R}^2\right)^2}. \end{aligned} \tag{5.16}$$

Figures 5.2a and 5.2b compare the acceleration field of this potential to the Keplerian case considered above.

What effect do the tides have on the density of the satellite? To examine this, let us consider the tidal acceleration $\vec{h}$, which from Equation 5.9, is

$$\begin{aligned} \vec{h}(\vec{X}) &= -\nabla\left(\frac{1}{2}\vec{x} \cdot [\vec{x} \cdot \nabla_{\vec{R}'}(\nabla_{\vec{R}'} \phi(\vec{R}'))]\Big|_{\vec{R}'=\vec{R}}\right) \\ &= -\vec{x} \cdot \nabla_{\vec{R}'}(\nabla_{\vec{R}'} \phi(\vec{R}'))\Big|_{\vec{R}'=\vec{R}}. \end{aligned} \tag{5.17}$$

If we define $\vec{n}$ to be a unit vector normal to the surface of the satellite under consideration, the average outward component of $\vec{h}$ is

$$\langle h_{out} \rangle = -\int \vec{n} \cdot [\vec{x} \cdot \nabla_{\vec{R}'}(\nabla_{\vec{R}'} \phi(\vec{R}'))]\Big|_{\vec{R}'=\vec{R}} dS, \tag{5.18}$$

which due to the divergence theorem is

$$\langle h_{out} \rangle = -\int \nabla \cdot [\vec{x} \cdot \nabla_{\vec{R}'}(\nabla_{\vec{R}'} \phi(\vec{R}'))]\Big|_{\vec{R}'=\vec{R}} dV. \tag{5.19}$$

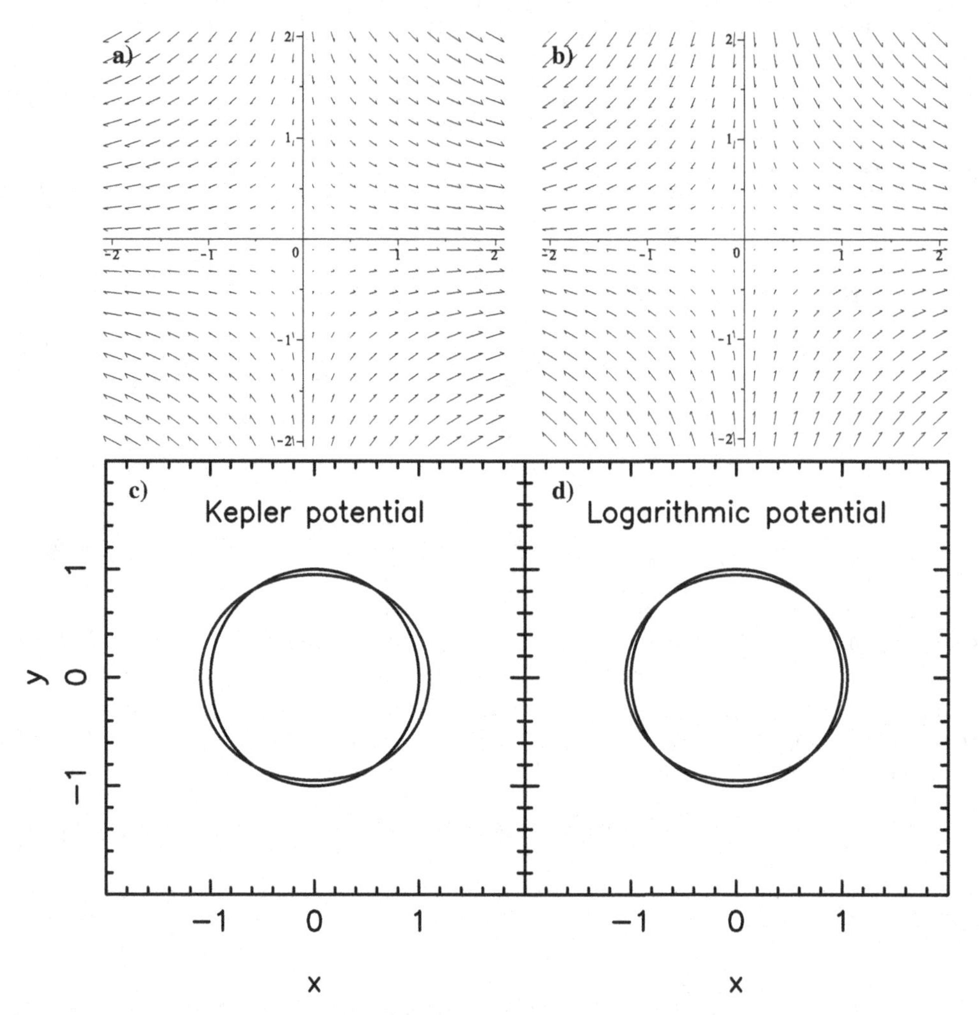

FIG. 5.2. (a) Acceleration field for a Keplerian potential; (b) and a logarithmic potential. The horizontal axis points toward the host galaxy. Note that in the logarithmic potential the acceleration is considerably stronger along the vertical axis than in the Keplerian case. (c-d) show the corresponding deformation of a spherical body.

If no sources of gravity are present $\nabla \cdot \nabla_{\vec{R}'}\phi(\vec{R}') = 0$ and hence $\langle h_{out} \rangle = 0$. This means that tidal deformation maintains the satellite at a constant density in a volume that does not contain any sources of gravity. This is indeed the case in a Keplerian potential, though it is not true in a logarithmic potential. The bottom panels in Figure 5.2 show how an initially spherical body would become deformed in these two potentials. In the logarithmic halo the tidal deformation results in a net compression (note that the third axis in Figure 5.2 is not depicted!).

How large can a satellite be in a given potential? The criterion that the gravitational acceleration must exceed the tidal force for the body to remain bound results in the classical Roche Limit for the easy case of a rigid sphere,

$$x < R\left(\frac{2m}{M}\right)^{1/3}, \tag{5.20}$$

where x is the maximum size of a satellite of mass m in a circular orbit of radius R about a host of mass M. Similar calculations for a non-rigid satellite are unfortunately quite messy, as one needs to consider the potential of nested homeoids (involves elliptic integrals).

However, by running a library of numerical N-body models of satellites, Oh *et al.* (1995) were able to show that satellites with King-model profiles survive as bound entities for a Hubble time if their limiting radius $r_{\rm lim}$ is smaller than $2r_t$, where

$$r_t = \begin{cases} \left[\frac{\alpha}{3+e}\right]^{1/3}(1-e)a & \text{Keplerian potential} \\ \left[\frac{\alpha(1-e)^2}{[(1+e)^2/2e]\ln[(1+e)/(1-e)]+1}\right]^{1/3} a & \text{Logarithmic potential} \end{cases} \quad (5.21)$$

Here α is the satellite to host mass ratio, a is the semi-major axis of the orbit, and e is the eccentricity of the orbit. This gives us a handy criterion to determine whether a realistic satellite model remains bound or not.

5.1.3 *Aside – orbital games and N-body calculations*

In the next section we see how stars are lost from the system and what happens to them, but first it will be instructive to attempt to integrate some orbits. One of the simplest accurate schemes that one can use is Runge-Kutta (see, e.g., Press *et al.*, 1992). To advance from the current position of the system x_n, y_n in a small step h to the next position $x_{n+1}(= x_n + h)$, y_{n+1} can be achieved by applying the following algorithm:

$$y_{n+1} = y_n + \frac{1}{6}k_1 + \frac{1}{3}k_2 + \frac{1}{3}k_3 + \frac{1}{6}k_4 + O(h^5)\,, \quad (5.22)$$

with

$$\begin{aligned} k_1 &= hf(x_n, y_n) \\ k_2 &= hf\left(x_n + \frac{1}{2}h, y_n + \frac{1}{2}k_1\right) \\ k_3 &= hf\left(x_n + \frac{1}{2}h, y_n + \frac{1}{2}k_2\right) \\ k_4 &= hf(x_n + h, y_n + k_3)\,. \end{aligned} \quad (5.23)$$

The function f here is the first order differential equation (or the set of first order differential equations) that govern the system. A very simple FORTRAN95 program listed at the end of this chapter implements this algorithm for a Plummer potential (it is very straightforward to alter the program to use other potentials).

Investigating the more interesting problem of the dynamical evolution of a satellite requires modeling the satellite as a collection of particles and integrating the motions of these particles under the influence of each other and of the host. I highly recommend working through the excellent on line book *The Art of Computational Science* by Hut and Makino (`http://www.artcompsci.org/`) to gain a thorough understanding of how to design and construct an N-body integrator. However, running the full-scale simulations is best undertaken with state-of-the-art algorithms such as "gyrfalcon" (Dehnen, 2002) which is conveniently an element of the "NEMO" stellar dynamics tool box (available at `http://bima.astro.umd.edu/nemo/`) (Teuben, 1995).

In Figure 5.3 I have used NEMO to run N-body simulations of the evolution of the Sagittarius dwarf galaxy. I began by integrating the orbit of a point-mass in the realistic Galactic potential of Dehnen and Binney (1998) to provide the starting point for a King-model realization of the dwarf galaxy, and then used "gyrfalcon" to evolve that King model within the same Galactic potential. Although it is not as good as a movie, the sequence in Figure 5.3 shows how escaping stars from a tidally perturbed satellite give rise to star streams. But how do the stars escape? Let us next take a closer look at how this works.

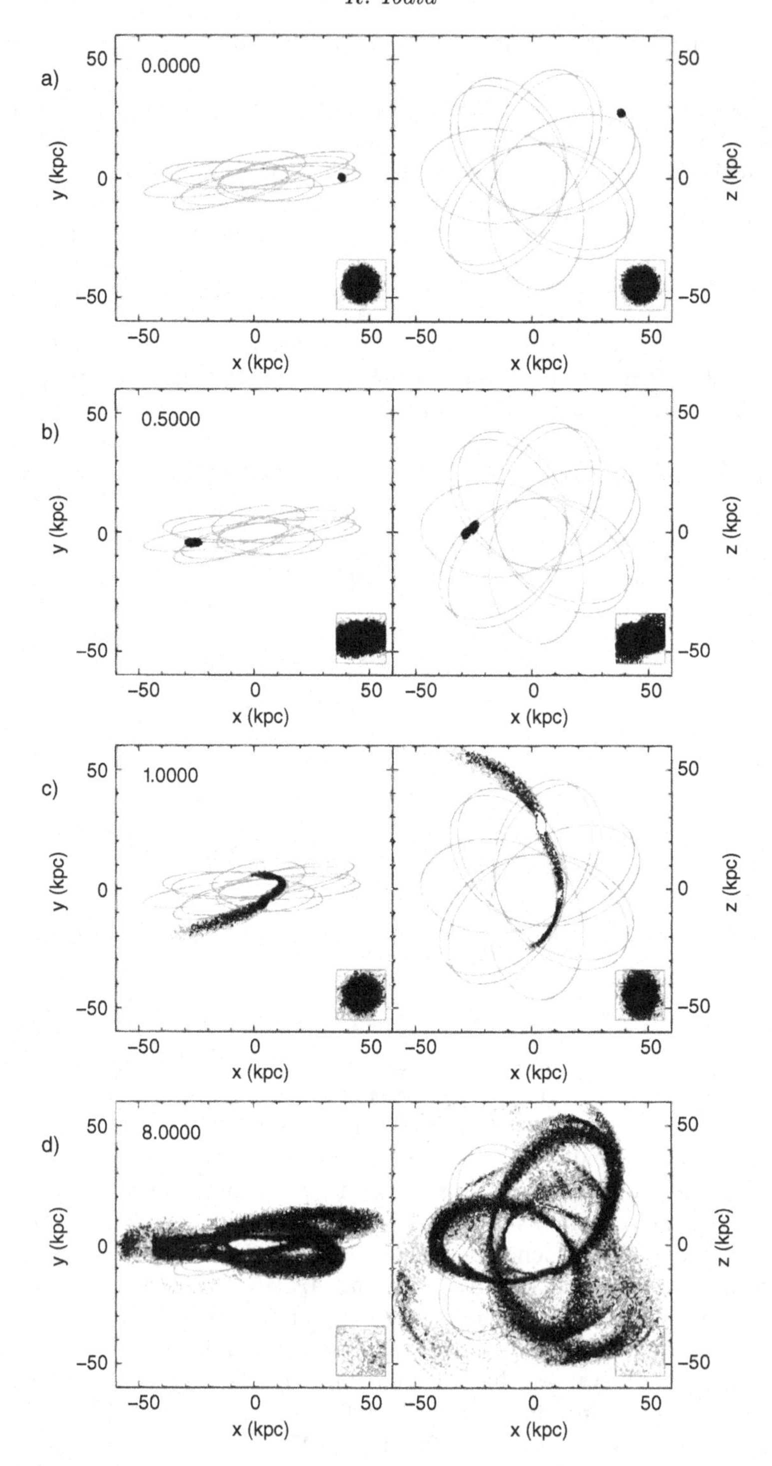

FIG. 5.3. N-body simulation of the dynamical evolution of the Sagittarius dwarf galaxy in the Milky Way potential of Dehnen and Binney (1998). From (a) to (d) the panels show the x-y and x-z distributions of particles at $T = 0, 0.5, 1.0$, and 8 Gyr. The small insert in each panel is a 4×4 kpc zoom of the remnant structure. The simulations were set up within "NEMO" and run with "gyrfalcon."

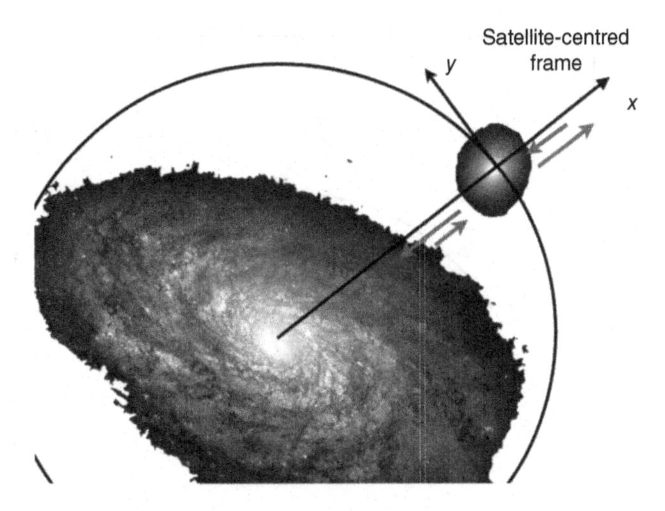

FIG. 5.4. The geometry of a satellite orbiting a host galaxy on a circular orbit. The x and y coordinates are centerd on the satellite. In this frame, the dynamics of a star around the satellite is governed by the interplay of the gravitational attraction of the host galaxy and the centrifugal force.

5.1.4 *The simplest case – satellite on a circular orbit*

Consider a satellite orbiting a host galaxy in a circular orbit in the manner depicted in Figure 5.4. If the galaxies are approximated as point-masses, the motion of a body in this system is governed by the following set of coupled equations:

$$\begin{aligned}
\ddot{x} &= 2\dot{y} + x - \frac{(1-\mu)(x+\mu)}{((x+\mu)^2 + y^2 + z^2)^{3/2}} \\
&\quad - \frac{\mu(x+\mu-1.0)}{((x+\mu-1)^2 + y^2 + z^2)^{3/2}} \\
\ddot{y} &= 2\dot{x} + x - \frac{(1-\mu)y}{((x+\mu)^2 + y^2 + z^2)^{3/2}} \\
&\quad - \frac{\mu y}{((x+\mu-1)^2 + y^2 + z^2)^{3/2}} \\
\ddot{z} &= - \frac{(1-\mu)z}{((x+\mu)^2 + y^2 + z^2)^{3/2}} \\
&\quad - \frac{\mu z}{((x+\mu-1)^2 + y^2 + z^2)^{3/2}},
\end{aligned} \tag{5.24}$$

where μ is the mass ratio of the two galaxies. The potential in the rotating frame is of course simply the classical Lagrange case, with the Lagrange point L1 (nearest the host) on the negative x axis and L2 on the positive side of the x axis. It is amusing to integrate orbits within this system, and I highly encourage the reader to modify the previous orbit program to use these equations of motion (Equation 5.24).

To escape from the satellite, it seems reasonable to expect that stars need only have energy greater than some critical value. However, this expectation turns out to be fallacious (Fukushige and Heggie, 2000). The structure of the potential around the satellite is essentially like that of a steep-sided container with two narrow exits (the Lagrange points L1 and L2). If these exit points are targeted sufficiently well, the star may escape as in Figure 5.5; otherwise, near misses such as those shown in Figure 5.6 will occur. It also transpires that some unbound stars remain close in phase-space around the satellite (Ross *et al.*, 1997) so that the satellite may be accompanied by an entourage of unbound

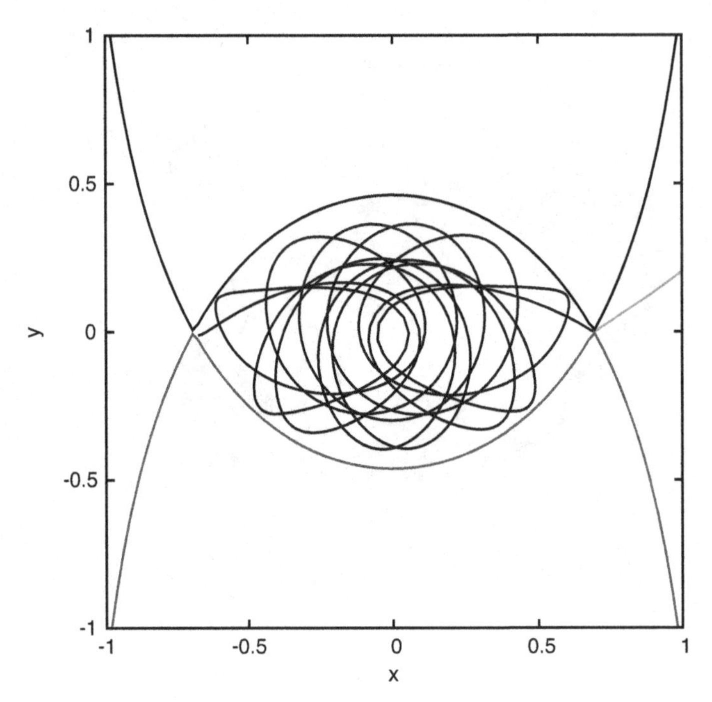

FIG. 5.5. The orbit of an escaping star in the system depicted in Figure 5.4, taking a point-mass potential. After many revolutions around the satellite, the star eventually manages to escape via the L2 point. (Credit: Douglas Heggie).

stars (Johnston *et al.*, 1999a), complicating the analysis of tidal debris, and even the identification of the bound extent of the satellite.

5.1.5 *Dynamical friction*

Even though the satellite may initially form far from the gravitational influence of its host, the dynamical interaction between the satellite and the wake it induces on the halo of the host will cause it to lose energy over time, resulting in the orbital decay of the satellite. In Figure 5.7, I show the results of two N-body simulations of a point-mass satellite orbiting within a realistic Milky Way model. In this simulation, the satellite begins its evolution close to the virial radius of the host on a nearly radial orbit. The left-hand right-hand panels show the evolution of a $10^9\,M_\odot$ and $10^{10}\,M_\odot$ satellite, respectively. Although the $10^9\,M_\odot$ satellite retails its initial orbit over the age of the Universe, clearly the more massive case (similar in mass to the Large Magellanic Cloud) undergoes severe orbital decay.

An approximation by Chandrasekhar can be used to derive an analytical estimate of this "dynamical friction" (see, e.g., Binney and Tremaine, 1987), according to which a satellite of mass M and velocity $\vec{v_M}$ traversing a medium of density ρ suffers a drag:

$$\frac{d\vec{v}_M}{dt} = -\frac{4\pi \ln \Lambda G^2 \rho M}{v_M^3} \left[\mathrm{erf}(X) - \frac{2X}{\sqrt{\pi}} e^{-X^2} \right] \vec{v}_M \, , \tag{5.25}$$

where $\ln \Lambda$ is the Coulomb logarithm and $X \equiv \frac{v_M}{\sqrt{2}\sigma}$. If we restrict ourselves to the case of a satellite on a circular orbit of initial radius r_i, the dynamical friction timescale becomes

$$t_{fric} = \frac{1.17}{\ln \Lambda} \frac{r_i^2 v_c}{GM} \, . \tag{5.26}$$

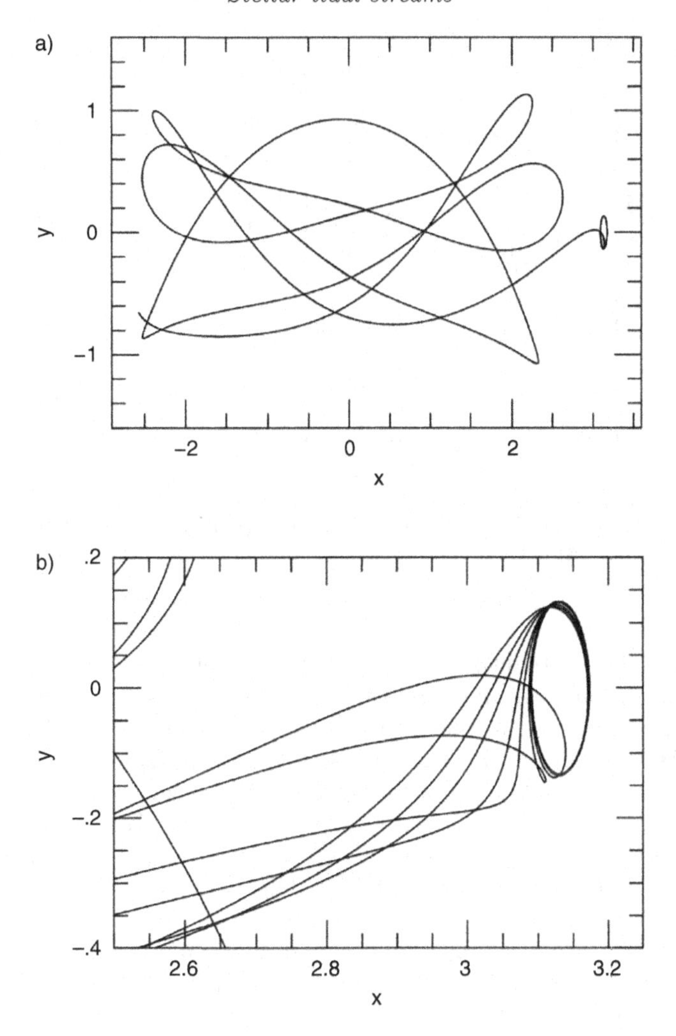

FIG. 5.6. From Fukushige and Heggie (2000). (a) The orbit of a star with energy above the escape energy. The star attempts to leave the system close to the L2 point (located at $x \sim 3.1$, $y = 0$), but "bounces" off it. (b) shows a close-up of this region, with several orbits of similar energy also failing to escape.

Using this equation, we see that dynamical friction is completely unimportant for globular clusters beyond the bulge; it is also relatively unimportant for the dwarf spheroidal galaxy satellites beyond ~20 kpc. However, applying this equation to the case of the Large Magellanic Cloud orbiting our Galaxy (see Binney and Tremaine, 1987), yields

$$t_{fric} = \frac{10^{10}}{\ln \Lambda}\left(\frac{r_i}{60\,\mathrm{kpc}}\right)^2 \left(\frac{v_c}{220\,\mathrm{km\,s^{-1}}}\right)\left(\frac{2\times 10^{10}\,\mathrm{M_\odot}}{M}\right)\mathrm{yr}, \tag{5.27}$$

and since $\ln \Lambda \sim 3$ for the LMC, the frictional timescale is approximately a quarter of the age of the Universe. So dynamical friction is essential for the buildup of the galaxy from more massive accretions.

To summarize, baryons in a dark matter satellite collapse, forming stars; alternatively, gas collapses without dark matter, most likely with the aid of tides. Stars are necessarily lost from these birth sites and mass-loss continues throughout the life of the satellite, even in the absence of tides, due to violent relaxation, energy equipartition, interactions with hard binaries, and evaporation. These stars are lost into a dynamic tidal environment. Tides themselves are a second order effect of a force field (given by the matrix of second

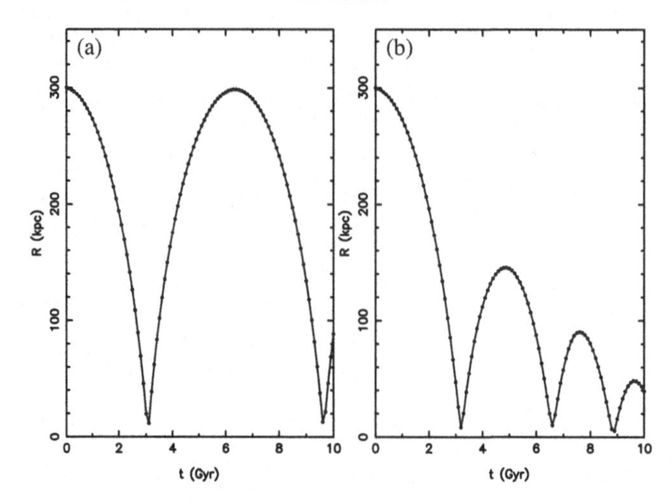

FIG. 5.7. N-body simulation of the orbital evolution of a point-mass satellite of mass: (a) $10^9\,M_\odot$ and (b) $10^{10}\,M_\odot$ in a realistic Milky Way potential. The more massive model suffers from severe orbital decay.

order derivatives of the potential). They perturb the intrinsic potential of the satellite giving rise to an elongation in the direction parallel to the line connecting the body to its perturber, with compression in the perpendicular direction. Outside sources of gravity, constant density is retained; otherwise the satellite suffers tidal compression.

The orbit of the satellite will decay via dynamical friction. Tidal sculpting, then tidal disruption, will take place simultaneously with the internal dynamical evolution of the satellite. So stars are lost into the host galaxy. In the slow case, these stars leave thorough the Lagrange points L1 and L2 of the satellite. This causes two tidal arms to be formed (this can be readily seen in the simulations on the second row of Figure 5.3). Since the L1 point is deeper in the potential, stars escape from it with a larger velocity, giving rise to the leading arm. However, as we have seen, not all unbound stars are completely lost to the system. I note in passing that phase-mixing will decrease the velocity dispersion of the stream with time (see, e.g., Helmi and White, 1999).

5.2 The utility of star streams

5.2.1 *Streams as tracers of satellite accretion*

It is now recognized (e.g., Johnston *et al.*, 1996) that star streams are long-lived features in the halos of galaxies, as long as they do not stray too close to the dense inner regions where dynamical times are short and the related precession and phase-mixing very fast. If the halo potential is approximately spherical, the stream debris should not precess significantly and will therefore be aligned in a certain orbital plane. As viewed from Earth, which is close to the center of the Milky Way as far as outer halo structures are concerned, such material will be seen to be aligned (approximately) along great circles in the sky. Indeed, the stream of the Sagittarius dwarf galaxy was first discovered by searching for stellar alignments in great-circle bands (Ibata *et al.*, 2001).

Thus the number of observed streams should give us a handle on the number of accretion events that a giant galaxy has experienced. Johnston *et al.* (1999a) used the physical argument that stars pulled off a satellite at a break radius radius $r_{\rm break}$ at a distance R from the host should change in orbital frequency by an amount

$$\Delta\Omega \sim r_{\rm break}\frac{d\Omega}{dR} \tag{5.28}$$

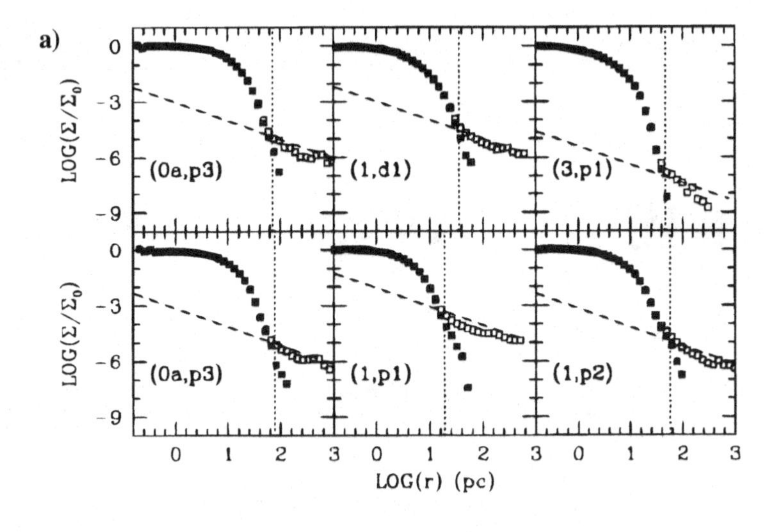

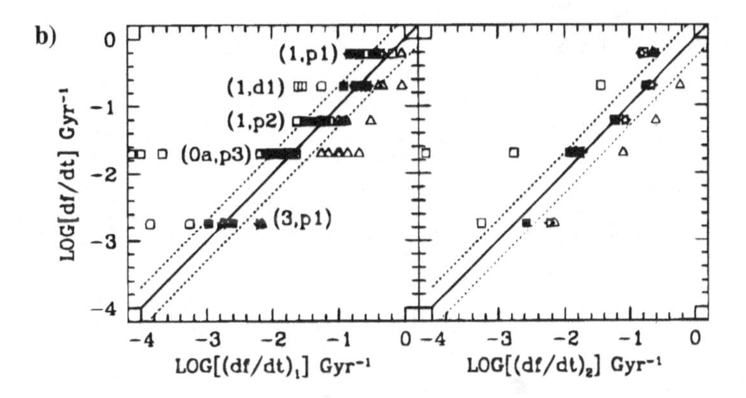

FIG. 5.8. (a) Simulations by Johnston *et al.* (1999a) showing the surface density of particles along the major axis of the satellite model and into the stellar stream. At the radius marked by the vertical dashed line, the satellite surface density profile breaks into the stream population. (b) These panels compare the measured mass loss rate df/dt with two estimates $(df/dt)_1$ and $(df/dt)_2$ derived from the break radius position. The fact that these estimates correlate well with the measured values means that it is not necessary to undertake a full dynamical analysis of a stream (which is not possible in most real situations given our lack of kinematic knowledge) to know its mass-loss rate.

(where $\Omega(R)$ is the frequency of a circular orbit around the host) to make an estimate of the mass-loss rate from an accreted satellite. Figure 5.8 shows that their estimate gives a good proxy of the mass loss rate measured from N-body simulations, and opens up the possibility of using streams to measure the mass accretion rate into the halos of giant galaxies.

Recently, Johnston *et al.* (2008) have used the expectations for satellite properties in standard cosmology to show that the observed morphology of stream debris in the halo of a galaxy can be used to understand the properties of the dwarf galaxies that were accreted. For instance, the observation of a large fraction of stars in a halo being confined to streams implies that many recent accretions have taken place; large streams should be associated with high luminosity progenitors; a large number of streams would (rather obviously) imply a large number of accretions; debris on clouds plumes or shells (see Figure 5.9) imply that the progenitors arrived on radial orbits whereas great-circle streams are due to satellites on more circular orbits; high metallicity of the debris should

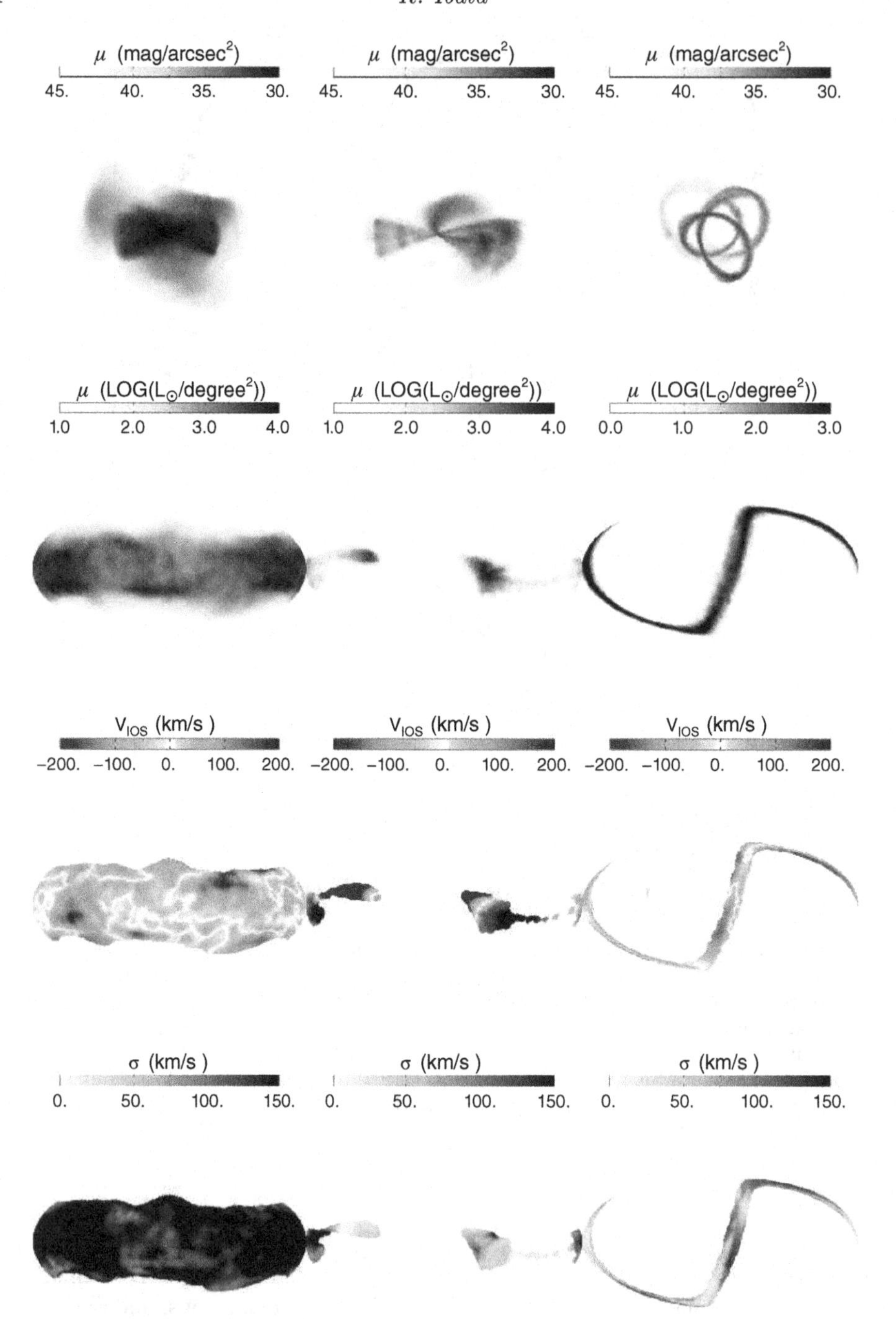

FIG. 5.9. (From Johnston *et al.*, 2008). Example of the properties of different stream types seen in simulations: mixed (left-hand panels), shells/plumes/clouds (middle panels) and great circles (right-hand panels). The top panels are views of external systems. The lower rows show a view from inside the simulation (i.e., for comparison to Galactic structures).

imply a high luminosity satellite; finally debris rich in α elements would imply that the satellite was accreted early in the formation of the giant galaxy.

Although this is promising, getting useful cosmological constraints from an analysis of observations of streams in external galaxies seems a long way off. Currently, it is very hard to detect structures of surface brightness fainter than $\Sigma_V \sim 28\,\text{mag arcsec}^{-2}$

unless one resorts to counting individual resolved stars. This means that only very bright (i.e., recent) accretions can currently be studied through surface brightness techniques (but in any case one cannot probe very far out into the distant Universe due to the problem of cosmological dimming). Resolved stars can be used to detect very low surface brightness structures, but it is not possible to probe much beyond $\lesssim 10\,\mathrm{Mpc}$, and even at that distance the currently best instrument available (the ACS camera on board the Hubble Space Telescope) has a very small field of view, making it impractical to survey a significant fraction of the halo of any galaxy. Understanding galaxy buildup from such observables will probably only be realistic using the next generation of telescopes.

5.2.2 *Streams as probes of the host potential*

Currently, the most reliable means to determine the potential of a spiral galaxy is by measuring the kinematics of the H I gas, and deriving the rotation curve. Generally, such data are of very high quality but rather limited spatially to the confines of the disk (normally they probe slightly beyond the optical disk, with virtually no vertical extent). As such, this method does not lend itself to the analysis of the properties of the dark matter beyond the inner regions of a galaxy.

The outer regions of galaxies are the most dark matter dominated and where baryons have had the least influence on the distribution of dark matter, so they are arguably the most interesting to study. Unfortunately, however, the paucity of baryonic (visible) tracers makes this endeavor very challenging. One means that is used to probe the large-scale distribution of mass is to measure the line-of-sight kinematics of halo satellites (dwarf galaxies and globular clusters) and to analyze their velocities using the Jeans equation. Even under the assumption that the halo is spherically symmetric and stationary the equation is complex:

$$\frac{GM(r)}{r} = -\sigma_r^2 \left[\frac{d \ln \rho}{d \ln r} + \frac{d \ln \sigma_r^2}{d \ln r} + 2\left(1 - \frac{\sigma_\theta^2}{\sigma_r^2}\right) \right], \tag{5.29}$$

where $M(r)$ is the cumulative mass inside radius r, ρ is the tracer density, and σ_r and σ_θ are the radial and tangential velocity dispersions, respectively. Typically, less than 100 tracers will be available at interesting (large) radii in a galaxy (e.g., Battaglia *et al.*, 2005), and since Equation 5.29 involves the derivatives in both density and kinematics, it can be appreciated that the method gives very uncertain results.

In contrast, streams are potentially much more powerful. The reason for this is that a low-mass stream follows the guiding center orbit to high accuracy, so the stars along such a stream trace out an orbit in the galaxy. Since energy

$$E = \text{constant} = \frac{1}{2}\left(v_x^2 + v_y^2 + v_z^2\right) + \Phi \tag{5.30}$$

along an orbit, it follows that contours of $\vec{v}$ along the stream are simply contours of the potential. This is of course a means to measure the potential in the spatial region inhabited by the stream. Having access to several streams would allow one to build up a complete picture of the mass distribution in the galaxy and would be much easier and more reliable than solving the Jeans's equation. In practize however, one does not yet have the luxury of knowing the full three-dimensional velocities and positions of the stars in the Milky Way, although this is set to improve dramatically with GAIA data over the next decade. Indeed, as shown by Johnston *et al.* (1999b), only 100 accurately measured stream stars are needed to constrain the Galactic potential to a few percent, and the analysis appears to be tractable even if the Galactic potential has changed substantially over time (Zhao *et al.*, 1999). While the prospects look good for uncovering the dark matter distribution in the halo of the Milky Way (and other nearby galaxies),

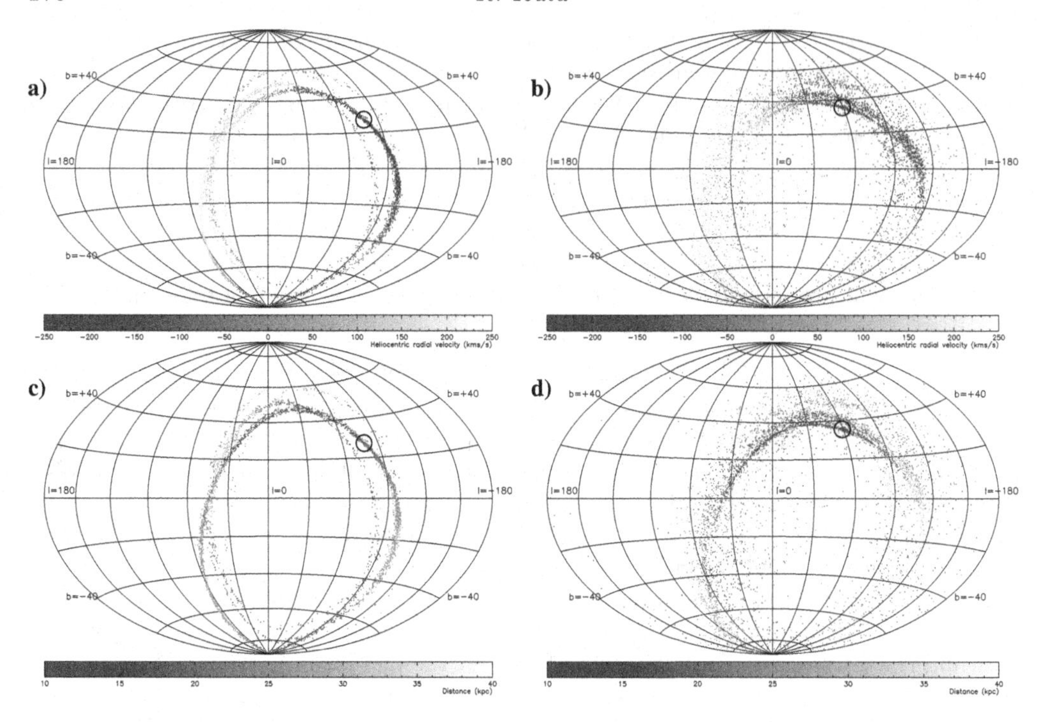

FIG. 5.10. (From Ibata *et al.*, 2002b). The effect of dark matter halo substructure on low-mass star streams. In a smooth halo (a and c), low-mass streams follow narrow paths on the sky, confined in a narrow range of velocity (a and b), and distance (c and d). In the presence of dark matter substructure (b and d), the potential becomes uneven and the streams become messy.

this method is unlikely to be useful for understanding any but the youngest of streams in the disk of a galaxy, because of the efficient radial mixing (Roškar *et al.*, 2008) there.

5.2.3 *Streams as Galactic seismometers*

Cold dark matter cosmology predicts that galaxies contain hundreds of dark matter substructures (Klypin *et al.*, 1999) with masses similar to dwarf satellite galaxies. Only a small fraction of these dark satellites can be identified with the observed population of satellites, however, which begs the question of where the missing satellites are, and even whether they exist or not. One possibility to address this issue is via the influence massive dark satellites have on dynamically fragile structures such as streams. As shown by Ibata *et al.* (2002b), the halo substructures change the galaxy from a smooth calm force field (left-hand panels of Figure 5.10) into a "choppy sea" where the stream and its progenitor are tossed hither and thither (right-hand panels of Figure 5.10). The effect of this is that the stream becomes dynamically heated, with a significantly larger line of sight dispersion, velocity dispersion, and width. Indeed, this is even more striking in energy and angular momentum space (Figure 5.11). Surveys are currently under way to test this heating on the globular cluster Palomar 5; finding a single unheated ancient globular cluster stream would disprove the existence of CDM structures of mass $< 10^7\,\mathrm{M}_\odot$, and would be a great challenge to this cosmology.

5.2.4 *Testing the nature of gravity*

Some theories posit that a scalar field is responsible for dark energy, and it has been proposed that this scalar field may interact with dark matter. The scalar field may mediate additional long-range forces between dark matter particles of comparable strength to the canonical gravitational force, although the theory currently makes no predictions about

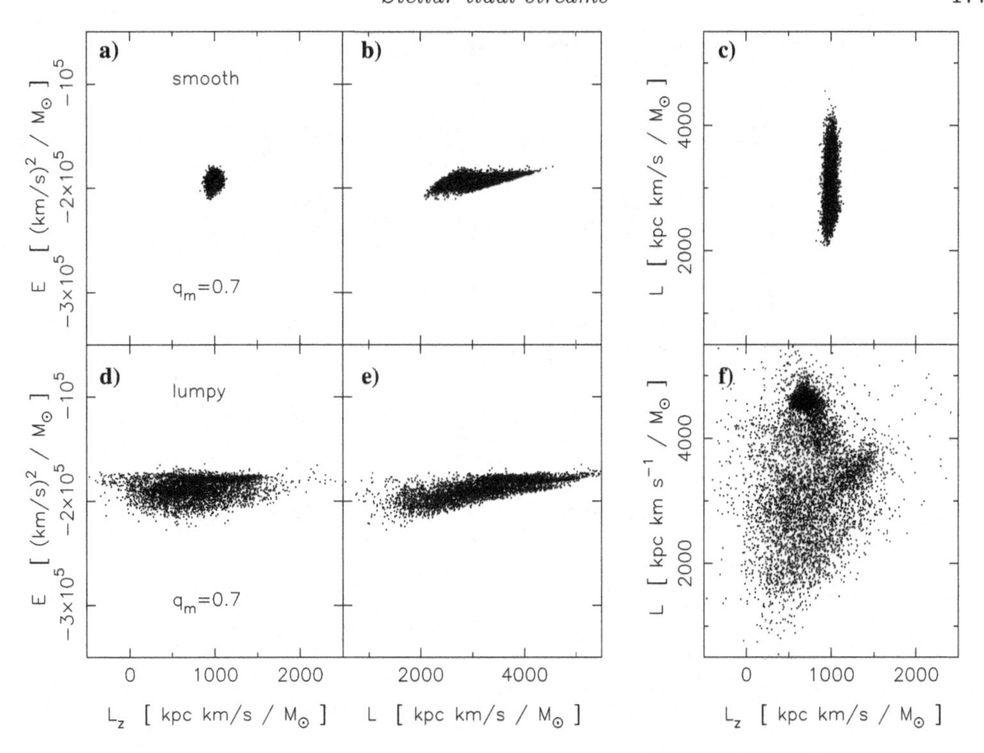

FIG. 5.11. The heating effect of dark matter substructure is readily discernible in energy and angular momentum. (a), (b), and (c) show the final structure of a globular cluster star stream in a smooth oblate halo (of flattening $q_m = 0.7$), while (d), (e), and (f) show the same stream simulated with the addition of the expected dark matter substructures. In the latter, the stream has been substantially heated by the substructures. From Ibata *et al.* (2002b).

the strength of the extra force. So does dark matter accelerate differently from baryons in reality? We would essentially like to be able to conduct Galileo's famous "Tower of Pisa" thought experiment comparing the motions of baryons and dark matter particles, and for this we need to find a very tall tower!

It turns out to be very difficult to test this idea. However, Kesden and Kamionkowski (2006a,b) showed that disrupting dwarf satellite galaxies, which are expected to contain dark matter in addition to their stars, creates an ideal laboratory to study this effect. They undertook a series of simulations of the Sagittarius dwarf galaxy (see Figure 5.12) in which they studied the influence of changing the strength of gravity for the dark matter. If the additional force is stronger than Newtonian gravity, the dark matter particles accelerate faster into the Milky Way potential, leaving the stars slightly behind. Because of this, any stars that do leave the system during the tidal disruption process are more likely to leave through the L2 Lagrange point, so that the resulting star stream appears asymmetric, with a leading arm less populated than the trailing arm. With current observations of the Sagittarius dwarf stream, Kesden and Kamionkowski (2006a,b) were able to rule out a 9% higher acceleration for the dark matter. It should be possible to improve these constraints substantially with deeper observations of the Sagittarius stream.

5.3 Unresolved substructure in galaxies

5.3.1 *Case study – Milky Way*

We have seen that stellar streams are the natural consequence of the tidal dissolution of dwarf satellite galaxies orbiting a massive companion. Over time, more and more such

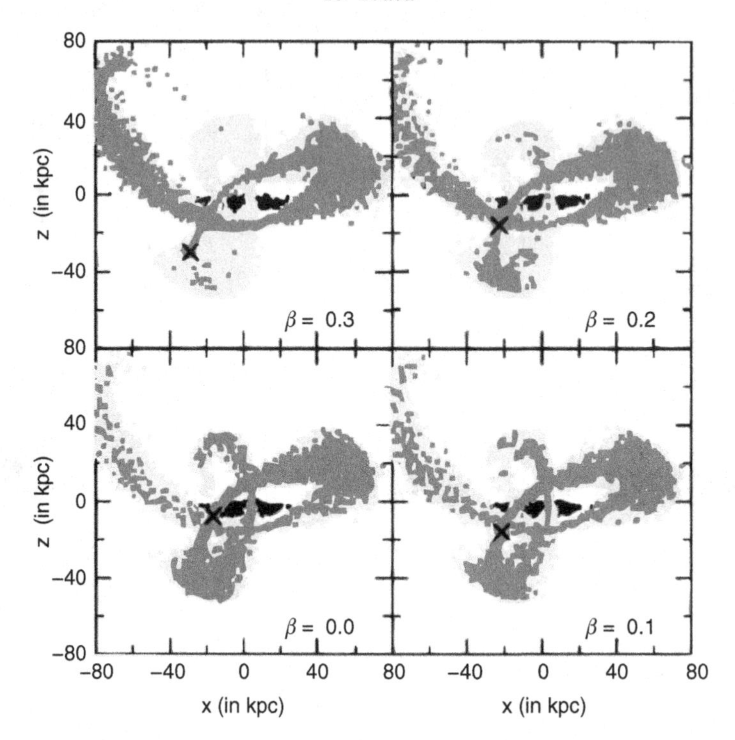

FIG. 5.12. Simulations of the tidal disruption of the Sagittarius dwarf galaxy under different assumptions for the gravitational force that acts upon dark matter. Stars are shown as dark gray, dark matter particles with light gray. The case $\beta = 0$ corresponds to the classical Newtonian case, with positive β signifying stronger acceleration on dark matter than on baryons, in which case the leading stream becomes poorly populated with stars. From Kesden and Kamionkowski (2006a).

systems are accreted and assimilated, while individual streams stretch out, becoming phase mixed and so less readily identifiable. Thus, we expect to see a progression from satellites to satellite streams, to mixed streams, and finally to a homogenous stellar halo (although the Universe is not old enough for any Milky Way–like galaxy to have reached this final stage). While satellite streams are now more or less routinely detected, until very recently it had not been possible to detect the more ancient phased-mixed accretions.

The pioneering study by Bell *et al.* (2008) addressed this question, using the Sloan Digital Sky Survey to probe the inhomogeneities in the halo of the Milky Way. Figure 5.13 provides a striking visual demonstration of the inherent asymmetry of our Galaxy, showing that lines of sight through the halo of the Milky Way that should have identical color-magnitude structure (and normalization) are in reality very different. Indeed, selecting stars that are consistent with being in various distance slices (Figure 5.14) reveals numerous coherent density substructures, many of which were identified in earlier studies (e.g., Ibata *et al.*, 2001; Belokurov *et al.*, 2006; Duffau *et al.*, 2006)). To quantify the incidence of substructure, Bell *et al.* (2008) applied the following statistic to the SDSS survey:

$$\sigma/\text{total} = \frac{\sqrt{\frac{1}{n}\sum_i (D_i - M_i)^2 - \frac{1}{n}\sum_i (M_i' - M_i)^2}}{\frac{1}{n}\sum_i D_i} \tag{5.31}$$

where D_i are the observed stars in spatial bin i, M_i are the corresponding counts in a smooth underlying halo model, and M_i' are Poisson realizations of that model. Thus the σ/total statistic expresses the fractional rms deviation of density substructures in the survey over and above the expected Poisson variations in a smoothed parametrized model. This turns out to be a very convenient statistic, as it is independent of the

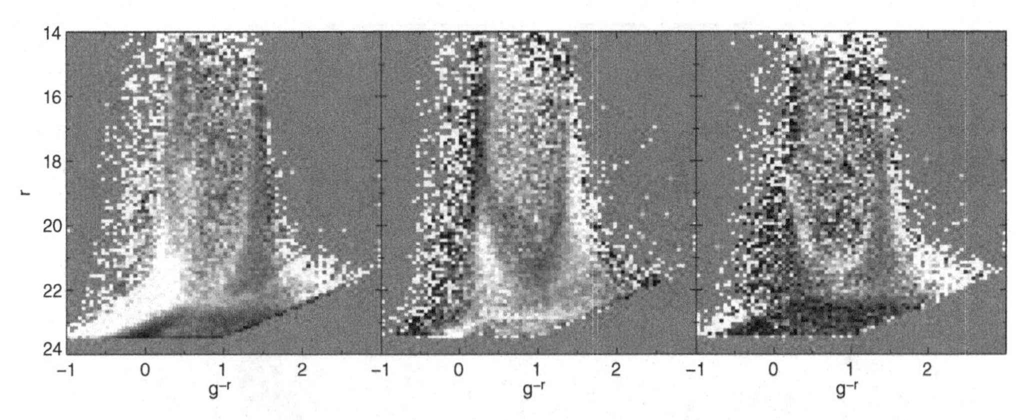

FIG. 5.13. These Hess diagrams (from Bell *et al.*, 2008) show the difference between 3 pairs of lines of sight that should have almost identical color-magnitude structure. In a symmetric Galactic halo, the Hess diagrams should be close to zero in all three panels, yet obvious residuals are present, implying significant departures from axial symmetry.

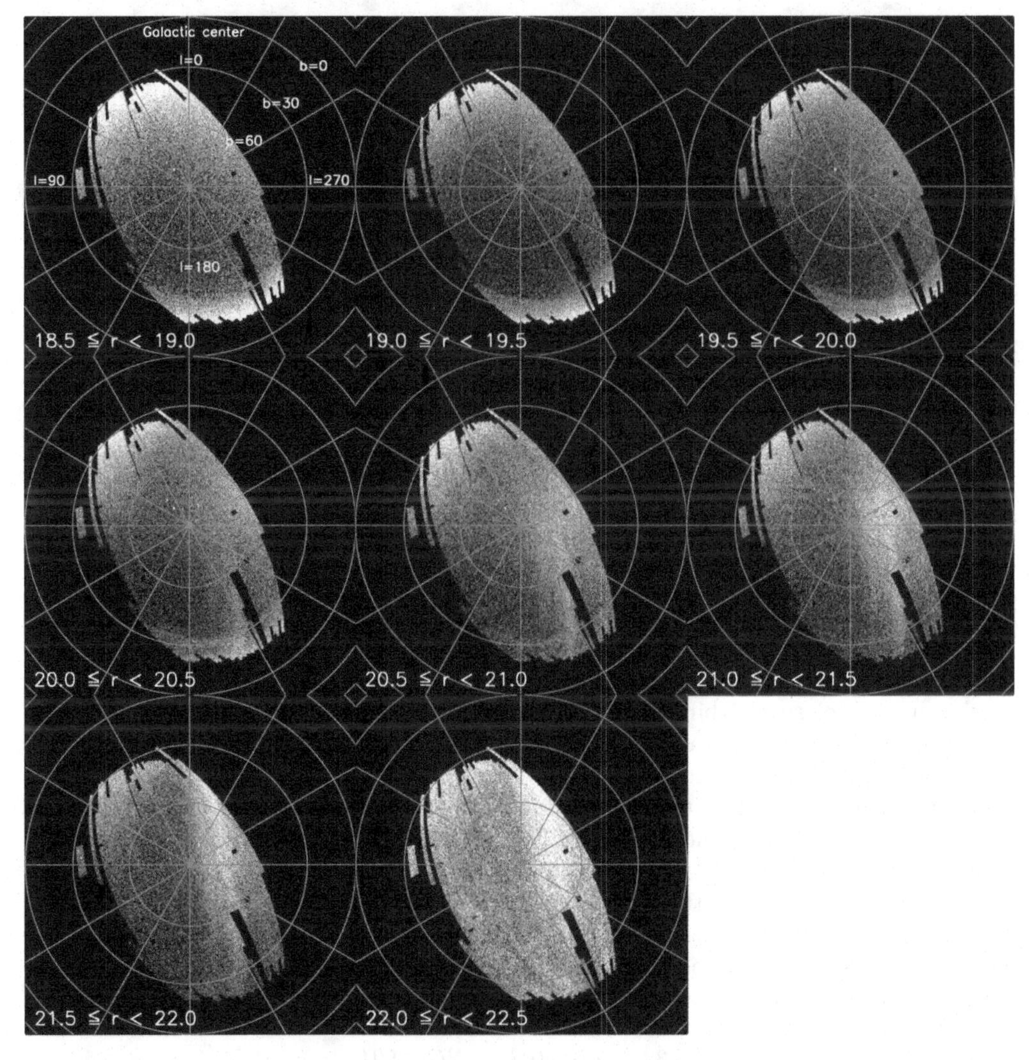

FIG. 5.14. SDSS star-counts (from Bell *et al.*, 2008) in slices in magnitude (i.e., distance) through the Galactic halo, corresponding to distances of approximately 7 to 35 kpc. Several asymmetric features are immediately apparent in these data.

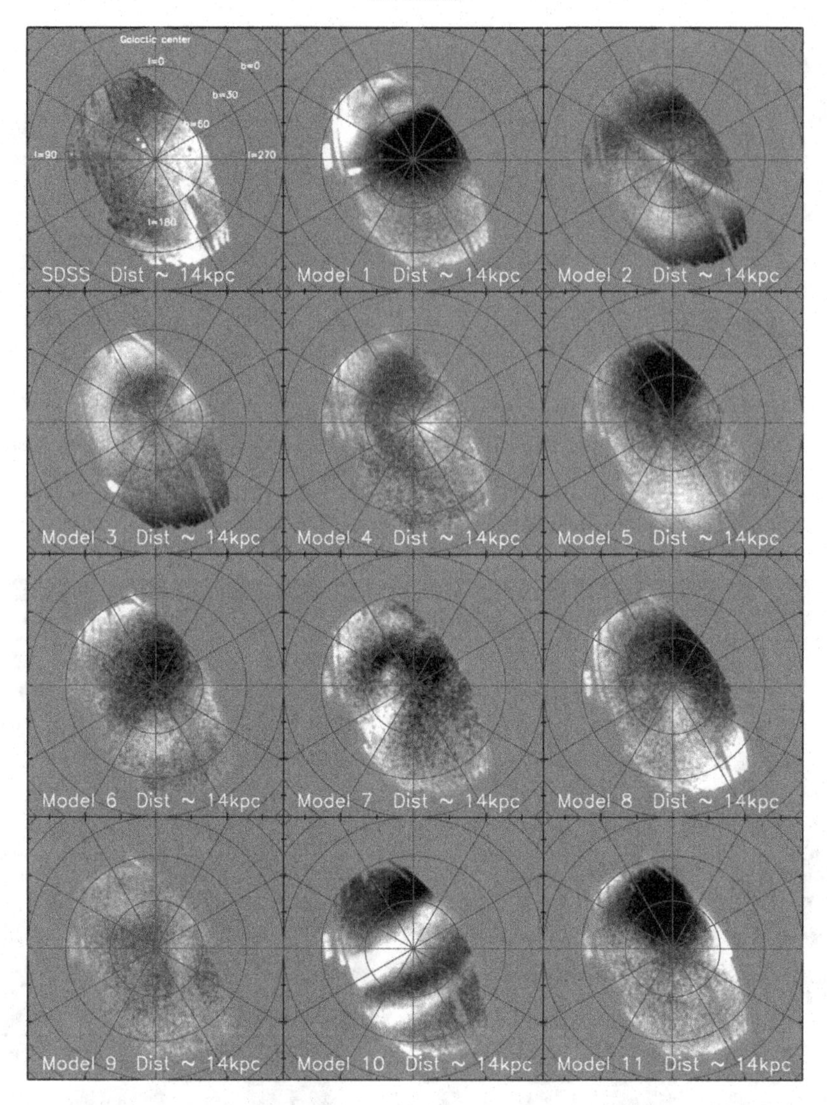

FIG. 5.15. The top left panel shows the residuals of the counts of blue SDSS stars ($0.2 \leq g - r \leq 0.4$) consistent with being at a distance of ~14 kpc from a smoothed model. The other panels show the particle counts from 11 simulations of the halo formation models of Bullock and Johnston (2005).

chosen number of spatial bins and of the total number of data points, and it is also corrected for Poisson uncertainties. Bell *et al.* (2008) calculated σ/total in the distance bins shown in Figure 5.14 and found a fractional rms deviation of ~40% beyond 14 kpc, which interestingly is similar to the value found by applying this substructure statistic to the halo models of Bullock and Johnston (2005). Figure 5.15 compares the 14 kpc distance slice in the SDSS with those halo models treated in a similar manner. The visual similarity is quite striking, lending further support for hierarchical formation models. Thus it has become clear that the Galactic halo is substantially lumpy and about as much so as the most extreme cosmological models predict. Is this a peculiarity of the Milky Way, or can it be confirmed in other systems?

5.3.2 *Case study - NGC 891*

To address this question, we undertook a study (Mouhcine *et al.*, 2007; Rejkuba *et al.*, 2009; Ibata *et al.*, 2009) of 3 deep HST/ACS fields in the halo of the edge-on galaxy NGC 891, considered to be the nearest ($D = 9.7$ Mpc) edge-on analog of the Milky Way.

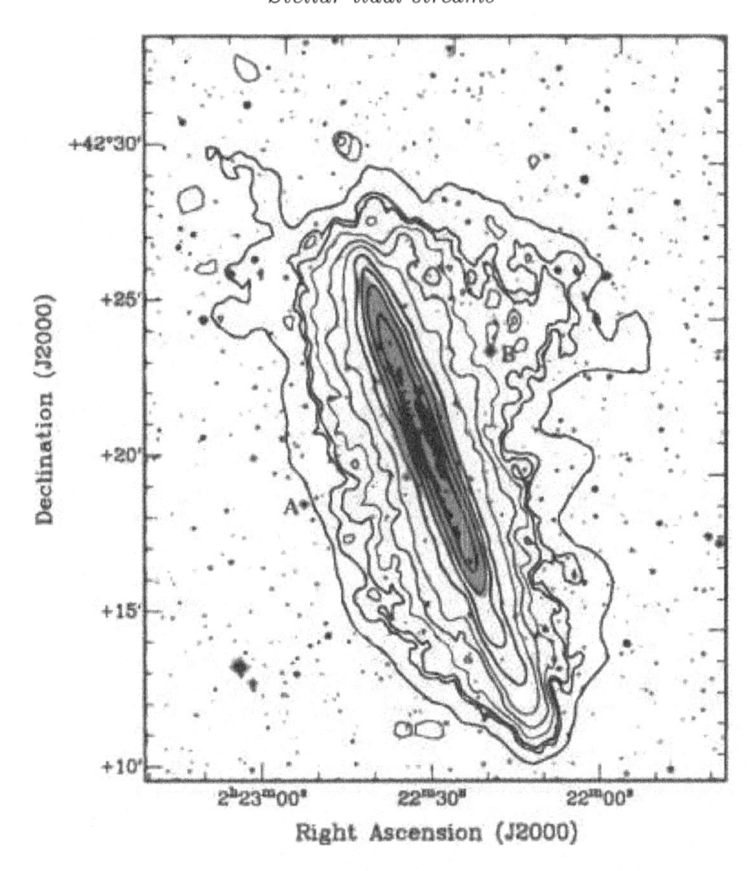

FIG. 5.16. (From Oosterloo *et al.*, 2007) Total H I image of NGC 891 from observations with the Westerbork Synthesis Radio Telescope. The contours mark levels at 0.01, 0.02, 0.05, 0.1, 0.2, 0.5, 1.0, 2.0, and 5.0 $\times 10^{21}$ cm^{-2}.

NGC 891 has similar morphology and mass as the Galaxy, previous studies have presented no indications of disturbance in the stellar component, and it is almost perfectly edge-on ($i = 89.8° \pm 0.5°$ – Kregel and van der Kruit, 2005), essential to disentangle the halo and disk in the absence of kinematic information.

One feature of NGC 891 that sets it apart from the Milky Way is that it possesses a substantial H I halo (Sancisi and Allen, 1979; Oosterloo *et al.*, 2007) (see Figure 5.16), which close to the plane of the galaxy is due primarily to Galactic fountain phenomena, although ≈10% may be of accretion origin (Fraternali and Binney, 2008). It appears that this gaseous component is linked to active ongoing star formation, and indeed the accretion rate matches closely the star-formation rate of 2.9 $M_{\odot}$ yr^{-1} (Fraternali and Binney, 2008). An ionized gas "halo" component, almost certainly related to the star formation activity, has been detected as an Hα envelope surrounding the disk of the galaxy (Dettmar, 1990; Rand *et al.*, 1990; Rossa and Dettmar, 2003). Star formation in the Milky Way is known to be episodic, with numerous short-term peaks in the last 2 Gyr during which the star-formation rate reached 3–5 times the steady rate (de la Fuente Marcos and de la Fuente Marcos, 2004). Given the supernova-driven galactic fountain, it is likely that NGC 891 is currently undergoing such an episode. Since there is no reason to expect a causal connection between this halo gas and the halo or thick disk stellar populations, which in the Milky Way and in simulations are exclusively composed of ancient stars, we believe that NGC 891 can be examined as a plausible analog of a galaxy formed in a similar manner as the Milky Way.

The three fields indicated in Figure 5.17 were observed with the Wide-Field Camera of the Advanced Camera for Surveys (ACS) on board the Hubble Space Telescope. These

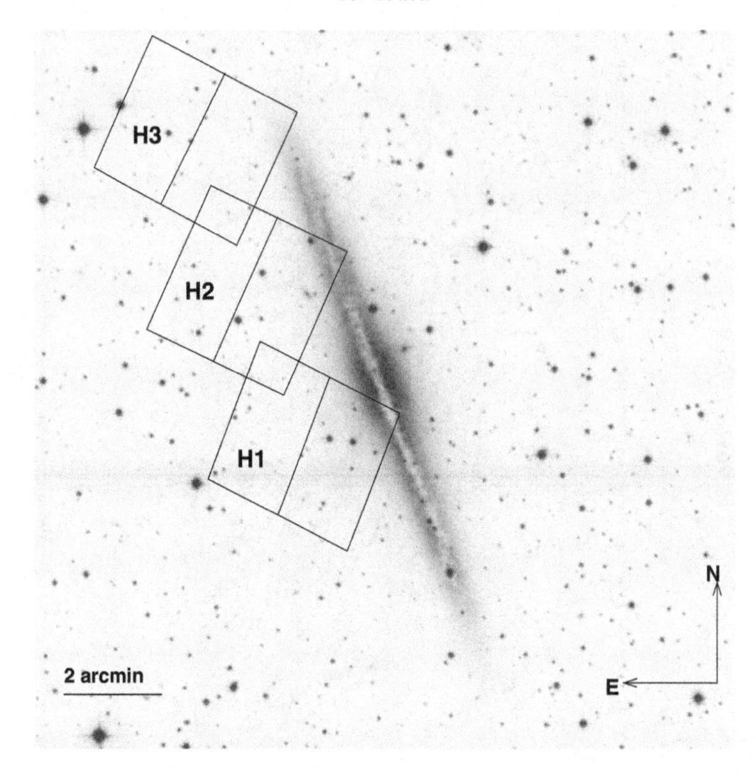

FIG. 5.17. The locations of the three ACS fields in NGC 891 studied by Mouhcine *et al.* (2007); Rejkuba *et al.* (2009), and Ibata *et al.* (2009), superimposed on photographic data from the Digitized Sky Survey.

fields probe both the disk and thick disk of the galaxy over a wide radial range as well as the inner halo, and have slight overlaps that are very convenient as a means to ensure photometric consistency and to check the reliability of the photometric uncertainties. The photometry reaches a limiting magnitude of $I \sim 29$, approximately 3 magnitudes below the tip of the red giant branch (RGB).

Figure 5.18 shows our analysis of the star-count distribution in the halo of NGC 891: a two-dimensional model (panel a) subtracted from the data leaves the lumpy residuals shown in panel b; the significance of these residuals is indicated in the lowest two panels for two different choices for the spatial binning. Interestingly, just as in the Milky Way, we see large pixel-to-pixel variations, well beyond Poisson counting uncertainties. The region at $X > 10$ kpc and $Z < -4$ kpc appears devoid of any very significant localized peaks (which could be satellite galaxies). For this region we calculate $\sigma/\text{total} = 0.14$, lower than the value obtained by Bell *et al.* (2008) for the Milky Way. The probability of finding such a large value of this statistic by chance is 0.8%, assuming the smooth model of panel a. We view a projection of NGC 891, whereas the Bell *et al.* (2008) resolved the Milky Way into various shells in distance. Thus one would expect a weaker substructure signal in NGC 891, consistent with what is seen.

However, the real power of the ACS survey of NGC 891 is that we can estimate a metallicity for each star from its color and magnitude by comparison to stellar isochrone models. Figure 5.19a shows the resulting map of the median metallicity over the survey region. (The Voronoi bins were chosen to have a signal-to-noise ratio of at least 10.) It is immediately obvious from this map that there is a gradient in metallicity from the disk out into the halo. The properties of these extra-planar stars in NGC 891 question the classical view of what a stellar halo should be. The contiguous ACS pointings show that the metallicity distribution at $\sim$10 kpc perpendicular to the disk of the galaxy is dominated by stars with [Fe/H] ~ -0.8 (Mouhcine *et al.*, 2007). This is a full $\sim$0.5 dex

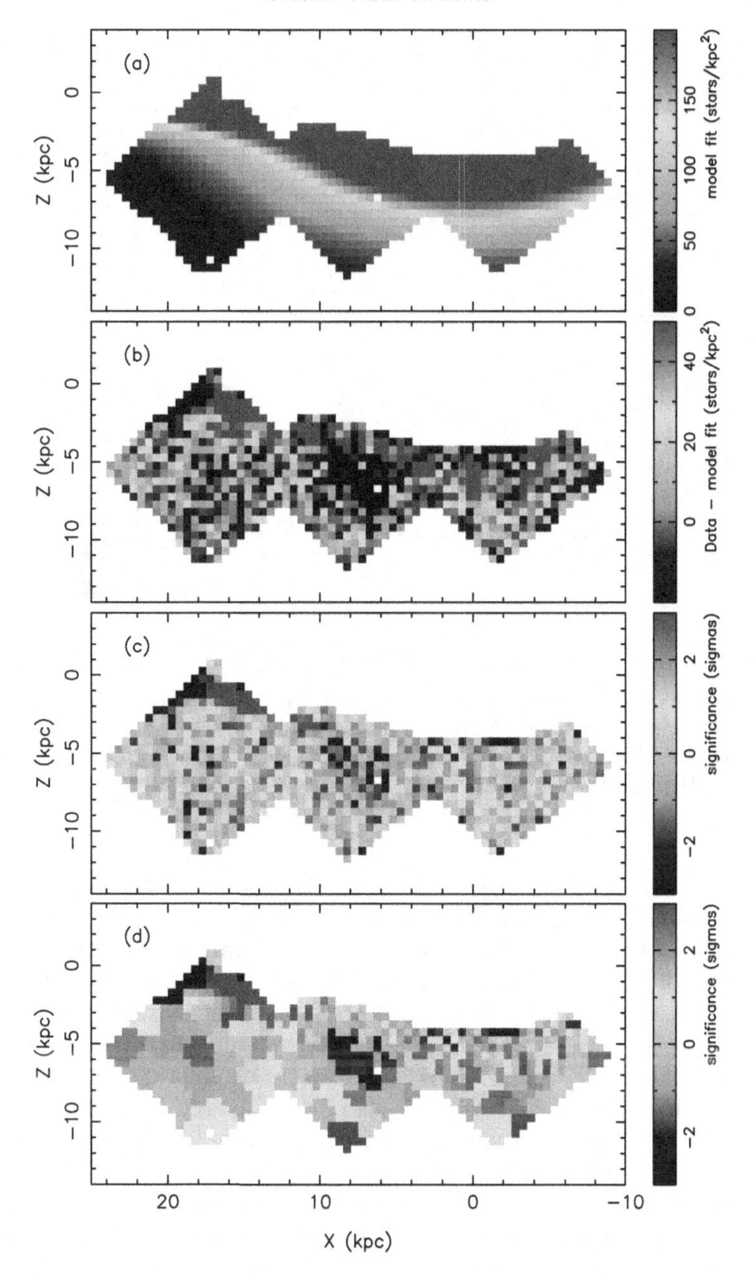

FIG. 5.18. (From Ibata *et al.*, 2009). Panel (a) displays a two-component (disk and spheroid) model fit to the ACS star-counts distribution in NGC 891; the corresponding residuals are shown in panel (b). Panel (c) displays the corresponding significance map, revealing several isolated pixels containing significant counts above the model; the same data are reproduced in panel (d), where we have summed over the large Voronoi super-pixels used later for the metallicity analysis in Figure 5.19.

more metal-rich than what is measured for the Galaxy halo stars at a comparable height (see Figure 5.20), and suggests that the metal-poor stars observed in the Galaxy may not be the dominant population in galactic halos in general (Mouhcine, 2006).

However, perhaps the most striking aspect of this data set is that it shows that the spatial metallicity distribution is not smooth. By subtracting a smooth model from the metallicity map and dividing by the uncertainties (Figure 5.21), we discover highly significant pixel-to-pixel variations in the median metallicity, reminiscent of the variations seen previously in the density. This appears to indicate that the halo of NGC 891 is

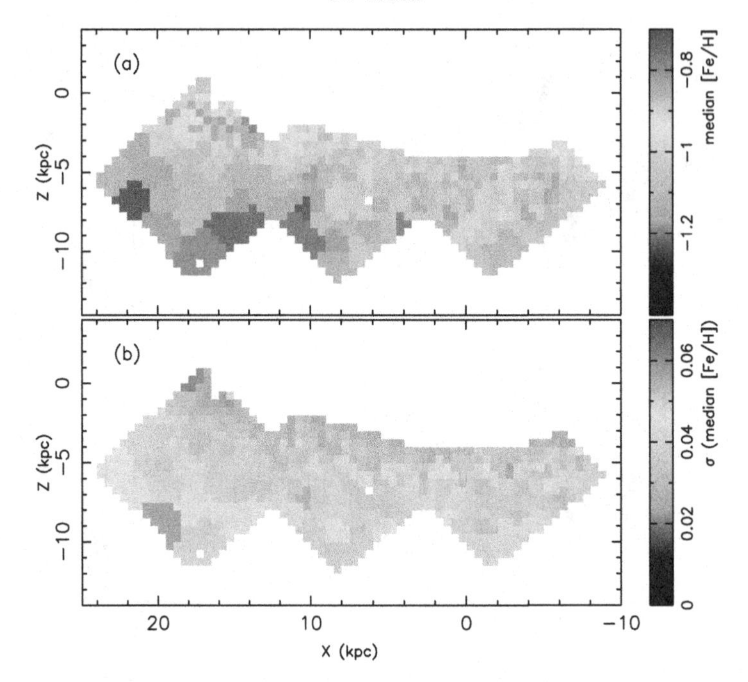

FIG. 5.19. (From Ibata *et al.*, 2009). (a): map of the median metallicity in NGC 891, color-coded with the values shown on the wedge. A large-scale gradient in the halo region is noticeable, as well as a more metal-rich "thick disk". (b): map of the uncertainties on the median metallicity, derived from simulating the effect of the photometric measurement uncertainties on the metallicity interpolation.

composed of a multitude of small accretions that have not yet been fully blended into a "normal" smooth halo. This result based on metallicity variations, is in broad agreement with the findings of Bell *et al.* (2008), based on density variations in the Milky Way.

Thus the halo of NGC 891 shows evidence for substantial amounts of substructure. These clumps show similar statistics in the scatter of stellar density as the Milky Way, a finding that supports the most extreme models of hierarchical halo formation. These substructures are also detected from the small-scale statistical fluctuations in the metallicity distribution, the first time it has been possible in any galaxy, and which gives complementary information on their nature. Evidently, these arise from independent accretions each with a different history of star formation. For the future, it will be interesting to extend this analysis to other extra-galactic systems, and in particular to assess the level of substructure as a function of galaxy morphology, type, mass, and accretion history.

5.4 Finding streams

As we have seen, star streams can be extremely valuable tools for undercovering the details of galaxy formation, for probing the distribution of dark mass in a galaxy, and even for probing the nature of gravity itself. I conclude this chapter by summarising some of the techniques that have been used to find these structures.

The main difficulty is that star streams tend to have extremely low surface brightness and they tend to be superimposed on a foreground (or background) of contaminating "normal" Galactic stars that vastly outnumber them. We are therefore faced with the problem of finding a "needle in a haystack." One of the techniques that is often employed in such circumstances is the so-called matched filter method, which is nicely explained by Rockosi *et al.* (2002) in the context of finding spatial overdensities that have a characteristic feature in color-magnitude space.

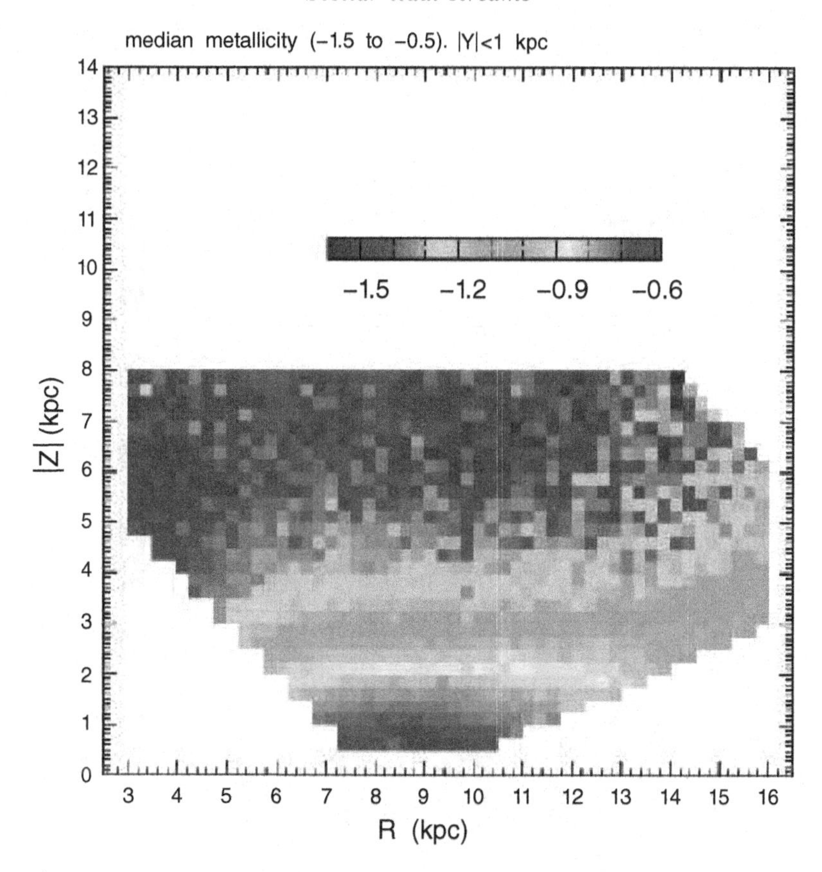

FIG. 5.20. (From Ivezić *et al.*, 2008). The metallicity distribution of stars in the Milky Way, derived from the SDSS as a function of Galactocentric radius R and height above the plane z.

Suppose that the number of stars of a stream feature we wish to detect is a factor α times, say, a globular cluster template f_{cl}:

$$n_{cl}(\text{color}, \text{mag}) = \alpha f_{cl}(\text{color}, \text{mag})\,; \tag{5.32}$$

here we are considering n and f as counts in a Hess diagram (i.e., binned in color and magnitude). But what is actually observed is a number of stars n_{stars}, made up of those target stream stars plus a contaminating background:

$$n_{stars,(i,j)} = \alpha f_{cl,(i,j)} + n_{bg,(i,j)}\,. \tag{5.33}$$

When the background counts dominate (i.e., in almost all situations of interest!), the signal uncertainty becomes simply the usual $\sqrt{n}$ Poisson counting uncertainty, so

$$\chi^2 = \sum_{i,j} \frac{(n_{stars,(i,j)} - [\alpha f_{cl,(i,j)} + n_{bg,(i,j)}])^2}{n_{bg,(i,j)}}\,, \tag{5.34}$$

from which we can derive a relation for the factor α:

$$\alpha = \frac{\sum_{i,j}[f_{cl,(i,j)}/n_{bg,(i,j)}]\delta_{stars} - \int f_{cl} d(\text{color}, \text{mag})}{\int (f_{cl}^2/n_{bg}) d(\text{color}, \text{mag})}\,. \tag{5.35}$$

The weight applied to each survey star is simply

$$h(\text{color}, \text{mag}) = \frac{f_{cl}(\text{color}, \text{mag})}{n_{bg}(\text{color}, \text{mag})}\,. \tag{5.36}$$

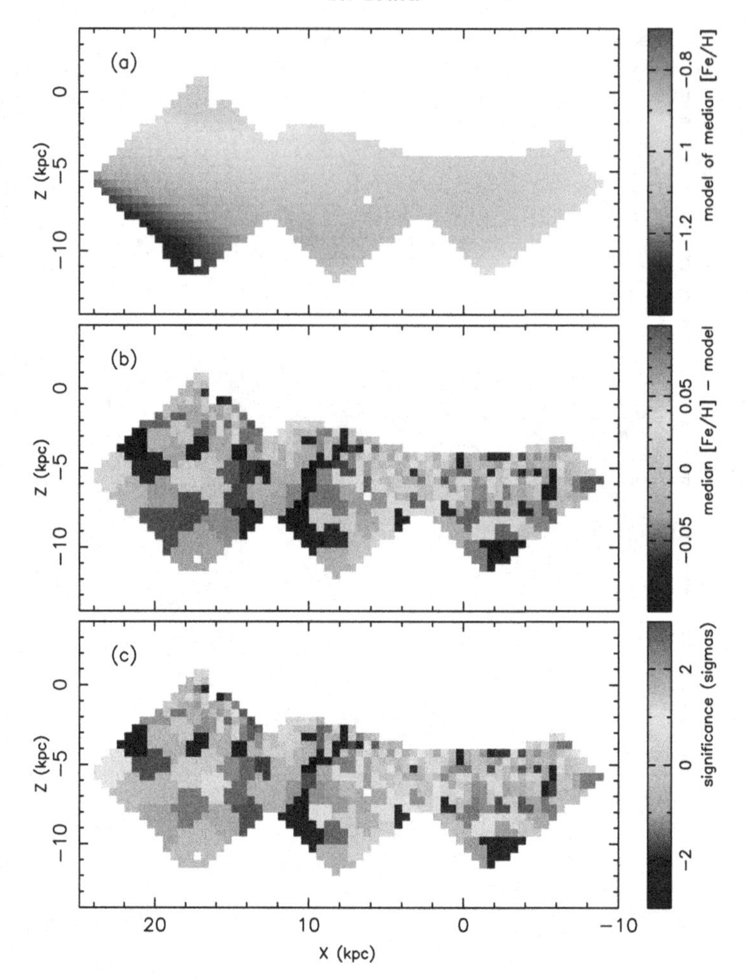

FIG. 5.21. (From Ibata *et al.* 2009). Inhomogeneities in the median metallicity over NGC 891. The upper panel shows a Legendre polynomial fit to the median metallicity map. The middle panel displays the difference between the median metallicity map and this fit, while the lower panel reports the significance level of the differences. These maps show that the halo has highly significant local variations in the median metallicity.

Figure 5.22 shows schematically how the method works. Many of the low-contrast streams detected in the SDSS (e.g., Grillmair, 2006; Grillmair and Dionatos, 2006) were discovered by using this technique.

Another technique that has been used to to find star streams is the so-called great circle counts method (Johnston *et al.*, 1996). Stars of some all-sky survey are summed in a band around a great circle on the sky, and the measured counts are assigned to the corresponding pole. This is repeated for all poles (binned to a resolution appropriate to the chosen width of the band). In this way, structures that follow great circle paths are easily detected as point-like overdensities at their poles. The method was successfully used to discover the stream of the Sagittarius dwarf galaxy (Ibata *et al.*, 2001, 2002a).

Over the next years the most promising prospect for this field of study is the arrival of the GAIA astrometric data set, which will provide an unparalleled leap forward in the mapping of our Galaxy. Given the kinematic and distance information that this survey will reveal, we will be able to search for streams directly in phase-space. As shown by Helmi and de Zeeuw (2000), energy, angular momentum, and z-component of angular momentum are conserved in stellar streams that disrupt sufficiently far away in the Galactic halo (see Figure 5.23). Adding in detailed element abundance information

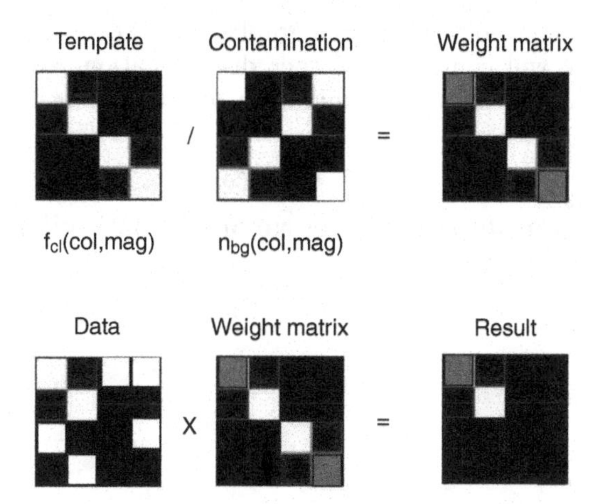

FIG. 5.22. Schematic working of the Matched Filter technique. The template matrix is divided by the contamination matrix to yield the weight matrix. (This weight matrix has high weight where the template is high and contamination is low, intermediate weight where the template is high and contamination high, and zero weight where the template is zero). The filtered result is found by multiplying the data matrix (i.e., in our case, the Hess diagram) by that weight matrix.

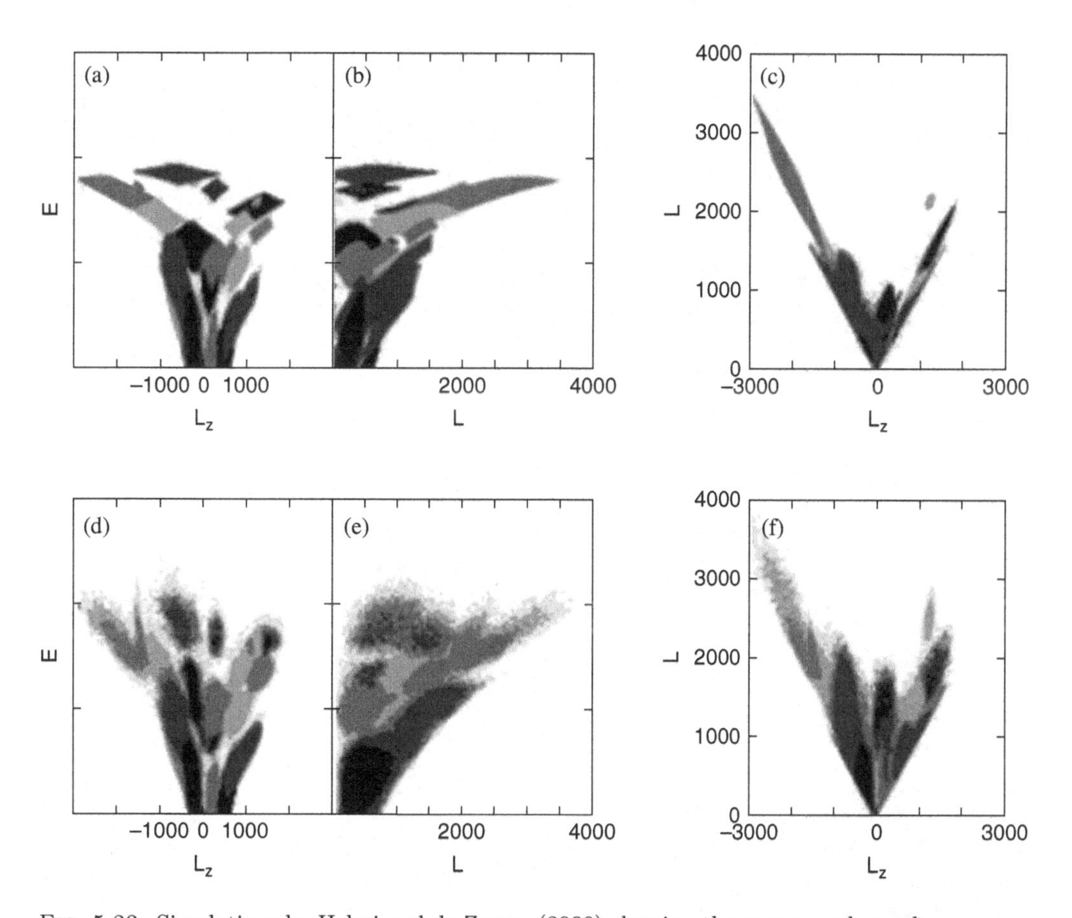

FIG. 5.23. Simulations by Helmi and de Zeeuw (2000) showing the energy and angular momentum of a set of satellite galaxies accreted into a model galaxy. (a), (b), and (c) show the initial properties of the structures; (d), (e), and (f) show the phase-space structure of the debris after 12 Gyr.

from next-generation multi-object spectrographs (as suggested by Freeman and Bland-Hawthorn, 2002) will contribute further discrimination power between populations, allowing us to uncover most if not all of the accretions that made up this ancient component.

5.5 Appendix – a simple orbit integrator to play with

```
 program orbit
  implicit none

! plotting variables:
  integer :: i,j
  integer, parameter :: n = 10000                       ! number of steps in orbit

! integration variables
  integer, parameter :: np = 6                          ! dimensions of phase-space vector
  real    :: x(np),dxdt(np)                             ! phase-space vector and derivatives
  real    :: dt
  real    :: x1(np),x2(np),x3(np),x4(np)                ! intermediate variables
  real    :: k1(np),k2(np),k3(np),k4(np)                !           "


  x=0.0 ; x(2)=4.0 ; x(4)=0.2                           ! chosen starting position
  dt=1.e-2                                              ! integration timestep

  open(1,file='orbit.out',status='unknown')             ! output file

  do j=1,n
     x1=x        ; call pot(x1,dxdt) ; k1=dt*dxdt
     x2=x+k1*0.5 ; call pot(x2,dxdt) ; k2=dt*dxdt
     x3=x+k2*0.5 ; call pot(x3,dxdt) ; k3=dt*dxdt
     x4=x+k3     ; call pot(x4,dxdt) ; k4=dt*dxdt
     x = x + k1/6.0 + k2/3.0 + k3/3.0 + k4/6.0          ! Runge-Kutta

     write(1,*) x                                       ! output x,y,z,vx,vy,vz
  end do

  close(1)

end program orbit


subroutine pot(x,dxdt)
  implicit none

  integer, parameter :: np = 6
  real*4 x(np),dxdt(np)

  call Plummer(x,dxdt)

end subroutine pot
```

```
subroutine Plummer(x,dxdt)
  implicit none

  integer :: i
  integer, parameter :: np = 6
  real, parameter    :: M = 1.0, a = 1.0
  real    :: x(np),dxdt(np),dPhidx(np/2),r,s

!            G*M
!  Phi = - ---------------
!          sqrt(a^2 + r^2)

! For simplicity, we choose units in which G=1

  r=sqrt(x(1)**2+x(2)**2+x(3)**2)
  s=sqrt(a**2 + r**2)

  do i=1,np/2
     dPhidx(i)     = M * x(i) / s**3
  end do

  do i=1,np/2
     dxdt(i)   = x(i+3)          ! change in position = vel
     dxdt(i+3) = -dPhidx(i)      ! change in vel
  end do

end subroutine Plummer
```

REFERENCES

Battaglia, G., and 8 colleagues. 2005. The radial velocity dispersion profile of the Galactic halo: constraining the density profile of the dark halo of the Milky Way. *MNRAS*, **364**(Dec.), 433–442.

Bell, E. F., and 13 colleagues. 2005. Toward an understanding of the rapid decline of the cosmic star formation rate. *ApJ*, **625**(May), 23–36.

Bell, E. F., and 17 colleagues. 2008. The accretion origin of the Milky Way's stellar halo. *ApJ*, **680**(June), 295–311.

Belokurov, V., and 20 colleagues. 2006. The field of streams: Sagittarius and its siblings. *ApJ*, **642**(May), L137–L140.

Binney, J. and Tremaine, S. 1987. Galactic dynamics. Princeton, NJ: Princeton University Press.

Bullock, J. S. and Johnston, K. V. 2005. Tracing galaxy formation with stellar halos. I. Methods. *ApJ*, **635**(Dec.), 931–949.

de la Fuente Marcos, R. and de la Fuente Marcos, C. 2004. On the recent star formation history of the Milky Way disk. *New A*, **9**(July), 475–502.

Dehnen, W. 2002. A hierarchical $\mathcal{O}(N)$ force calculation algorithm. *Journal of Computational Physics*, **179**(June), 27–42.

Dehnen, W. and Binney, J. 1998. Mass models of the Milky Way. *MNRAS*, **294**(Mar.), 429.

Dettmar, R.-J. 1990. The distribution of the diffuse ionized interstellar medium perpendicular to the disk of the edge-on galaxy NGC 891. *A&A*, **232**(June), L15–L18.

Duffau, S., Zinn, R., Vivas, A. K., Carraro, G., Méndez, R. A., Winnick, R., and Gallart, C. 2006. Spectroscopy of QUEST RR Lyrae variables: The new Virgo stellar stream. *ApJ*, **636**(Jan.), L97–L100.

Fraternali, F. and Binney, J. J. 2008. Accretion of gas on to nearby spiral galaxies. *MNRAS*, **386**(May), 935–944.

Freeman, K. and Bland-Hawthorn, J. 2002. The new galaxy: signatures of its formation. *ARA&A*, **40**, 487–537.

Fukushige, T. and Heggie, D. C. 2000. The time-scale of escape from star clusters. *MNRAS*, **318**(Nov.), 753–761.

Geyer, M. P. and Burkert, A. 2001. The effect of gas loss on the formation of bound stellar clusters. *MNRAS*, **323**(May), 988–994.

Grillmair, C. J. 2006. Substructure in tidal streams: tributaries in the anticenter stream. *ApJ*, **651**(Nov.), L29–L32.

Grillmair, C. J. and Dionatos, O. 2006. A 22° tidal tail for Palomar 5. *ApJ*, **641**(Apr.), L37–L39.

Helmi, A. and White, S. D. M. 1999. Building up the stellar halo of the Galaxy. *MNRAS*, **307**(Aug.), 495–517.

Helmi, A. and de Zeeuw, P. T. 2000. Mapping the substructure in the Galactic halo with the next generation of astrometric satellites. *MNRAS*, **319**(Dec.), 657–665.

Ibata, R. A., Lewis, G. F., Irwin, M. J., and Cambrésy, L. 2002a. Substructure of the outer Galactic halo from the 2-Micron All-Sky Survey. *MNRAS*, **332**(June), 921–927.

Ibata, R. A., Lewis, G. F., Irwin, M. J., and Quinn, T. 2002b. Uncovering cold dark matter halo substructure with tidal streams. *MNRAS*, **332**(June), 915–920.

Ibata, R.A., Lewis, G. F., Irwin, M., Totten, E., and Quinn, T. 2001. Great circle tidal streams: evidence for a nearly spherical massive dark halo around the Milky Way. *ApJ*, **551**(Apr.), 294–311.

Ibata, R. A., Mouhcine, M., and Rejkuba, M. 2009. An HST/ACS investigation of the spatial and chemical structure and substructure of NGC 891, a Milky Way analog. *MNRAS*, **395**(May), 126–143.

Ivezić, Ž., and 52 colleagues. 2008. The Milky Way tomography with SDSS. II. Stellar metallicity. *ApJ*, **684**(Sept.), 287–325.

Johnston, K. V., Bullock, J. S., Sharma, S., Font, A., Robertson, B. E., and Leitner, S. N. 2008. Tracing galaxy formation with stellar halos. II. Relating substructure in phase and abundance space to accretion histories. *ApJ*, **689**(Dec.), 936–957.

Johnston, K. V., Hernquist, L., and Bolte, M. 1996. Fossil signatures of ancient accretion events in the halo. *ApJ*, **465**(July), 278.

Johnston, K. V., Sigurdsson, S., and Hernquist, L. 1999a. Measuring mass-loss rates from Galactic satellites. *MNRAS*, **302**(Feb.), 771–789.

Johnston, K. V., Zhao, H., Spergel, D. N., and Hernquist, L. 1999b. Tidal streams as probes of the Galactic potential. *ApJ*, **512**(Feb.), L109–L112.

Kesden, M. and Kamionkowski, M. 2006a. Galilean equivalence for Galactic dark matter. *Physical Review Letters*, **97**(Sept.), 131303.

Kesden, M. and Kamionkowski, M. 2006b. Tidal tails test the equivalence principle in the dark-matter sector. *Phys. Rev. D*, **74**(Oct.), 083007.

Klypin, A., Kravtsov, A. V., Valenzuela, O., and Prada, F. 1999. Where are the missing Galactic satellites? *ApJ*, **522**(Sept.), 82–92.

Kregel, M. and van der Kruit, P. C. 2005. Structure and kinematics of edge-on galaxy discs – IV. The kinematics of the stellar discs. *MNRAS*, **358**(Apr.), 481–502.

McKee, C. F. and Ostriker, E. C. 2007. Theory of star formation. *ARA&A*, **45**(Sept.), 565–687.

Mouhcine, M. 2006. The outskirts of spiral galaxies: evidence for multiple stellar populations. *ApJ*, **652**(Nov.), 277–282.

Mouhcine, M., Rejkuba, M., and Ibata, R. 2007. The stellar halo of the edge-on galaxy NGC 891. *MNRAS*, **381**(Oct.), 873–880.

Oh, K. S., Lin, D. N. C., and Aarseth, S. J. 1995. On the tidal disruption of dwarf spheroidal galaxies around the galaxy. *ApJ*, **442**(Mar.), 142–158.

Oosterloo, T., Fraternali, F., and Sancisi, R. 2007. The cold gaseous halo of NGC 891. *AJ*, **134**(Sept.), 1019.

Press, W. H., Teukolsky, S. A., Vetterling, W. T., and Flannery, B. P. 1992. *Numerical recipes in FORTRAN. The art of scientific computing.* 2nd ed. Cambridge UK: Cambridge University Press.

Rand, R. J., Kulkarni, S. R., and Hester, J. J. 1990. The distribution of warm ionized gas in NGC 891. *ApJ*, **352**(Mar.), L1–L4.

Rejkuba, M., Mouhcine, M., and Ibata, R. 2009. The stellar population content of the thick disc and halo of the Milky Way analog NGC 891. *MNRAS*, **396**(July), 1231–1246.

Rockosi, C. M., and 9 colleagues. 2002. A matched-filter analysis of the tidal tails of the globular cluster Palomar 5. *AJ*, **124**(July), 349–363.

Roškar, R., Debattista, V. P., Quinn, T. R., Stinson, G. S., and Wadsley, J. 2008. Riding the spiral waves: implications of stellar migration for the properties of galactic disks. *ApJ*, **684**(Sept.), L79–L82.

Ross, D. J., Mennim, A., and Heggie, D. C. 1997. Escape from a tidally limited star cluster. *MNRAS*, **284**(Feb.), 811–814.

Rossa, J. and Dettmar, R.-J. 2003. An Hα survey aiming at the detection of extraplanar diffuse ionized gas in halos of edge-on spiral galaxies. II. The Hα survey atlas and catalog. *A&A*, **406**(Aug.), 505–525.

Ryan, R. E., Jr., Cohen, S. H., Windhorst, R. A., and Silk, J. 2008. Galaxy mergers at z>1 in the HUDF: evidence for a peak in the major merger rate of massive galaxies. *ApJ*, **678**(May), 751–757.

Sancisi, R. and Allen, R. J. 1979. Neutral hydrogen observations of the edge-on disk galaxy NGC 891. *A&A*, **74**(Apr.), 73–84.

Springel, V., and 16 colleagues. 2005. Simulations of the formation, evolution and clustering of galaxies and quasars. *Nature*, **435**(June), 629–636.

Teuben, P. 1995. The stellar dynamics tool box NEMO. *Astronomical Data Analysis Software and Systems IV*, **77**, 398.

Zhao, H., Johnston, K. V., Spergel, D. N., and Hernquist, L. 1999. Studying evolution of the Galactic potential and halo streamers with future astrometric satellites. *The Third Stromlo Symposium: The Galactic Halo*, **165**, 130.

6. Tutorial: The analysis of color-magnitude diagrams

D. VALLS-GABAUD

6.1 Introduction

> The plotting of the colors (or spectra) of stars as abscissae against their absolute magnitudes (total magnitudes) has become one of the most lucrative adventures in the study of star light.
>
> Shapley (1960)

It is appropriate to recall, in the context of this volume, that just over a century ago the first color-magnitude diagram (CMD) was published. The author of this landmark paper was not Ejnar Hertzsprung nor Henry N. Russell, but Hans O. Rosenberg, a colleague of Karl Schwarzschild at Göttingen. Rosenberg had been working since 1907 on getting spectral properties of stars by measuring plates obtained with the Zeiss objective prism camera (Hermann, 1994). To maximize the number of spectra per plate, he observed the Pleiades cluster and obtained spectra for about 60 of them, over 1907–1909, noting that their inferred effective temperatures correlated with their apparent magnitudes in the first ever published CMD (Rosenberg, 1910).[1] His goal was to "make the most accurate determination of the spectral types of stars in the Pleiades" by using a "physiological blend" of the depth and width of the Ca II K line (393.37 nm) with the Balmer Hδ and Hζ lines. He excluded the Ca II H line at 396.9 nm as it was blended with Hϵ in the very low dispersion spectra he used (1.9 mm from Hγ to Hζ). With an exposure time of 90 minutes he could measure spectra down to the 10th photographic magnitude, finding that for the actual members of the Pleiades "there is a strict relation between the brightness and the spectral type, with no exception in the interval from the 3rd to the 9th magnitude." Hertzsprung's diagrams (magnitude *vs.* color) of the Pleiades and the Hyades would appear a year later (Hertzsprung, 1911) whereas Russell's version for field stars with parallaxes (with absolute magnitude *vs.* spectral type) would only appear in 1914 (Russell, 1914a,b), although the correlation between luminosity and spectral type was noted by Hertzsprung in 1905 and in more detail by Russell (1912). The key difference is that while Russell required parallaxes to ascertain the distances to the stars, Rosenberg carefully checked the membership of stars in the Pleiades, rejecting outliers and non-members.

Henry Russell was obsessed by priority, (self-)attribution and promotion (Devorkin, 2000). For example, the famous Vogt(–Russell) theorem on stellar structure first appeared in 1926 (Vogt, 1926) and in his influential textbook Russell does give full credit to Vogt (Russell *et al.*, 1927), yet he will later claim (Russell, 1931) that he had found it independently. In the case of the CMD, Russell called it in private the "Russell diagram," but this was not accepted in public, as the contribution by Hertzsprung (unlike Rosenberg's) was well and widely known. Russell pored over the astronomical journals and was well aware of Hertzsprung's results. We know he read the *Astronomischen Nachrichten* systematically, as one of the leading journals of the time, and that he was well aware of Hertzsprung's papers, just as was his mentor, E. C. Pickering, who received them and wrote to Hertzsprung discussing several issues in spectral classification. Russell wrote to Hertzsprung on September 27, 1910, thanking him for sending copies of his papers (Hearnshaw, 1986). Hertzsprung's (1911) paper contained the CMDs of the Hyades and the Pleiades, citing explicitly the previous -and pioneering-work by Rosenberg (1910).

[1] A translation into English is available at Leos Ondra's website www.leosondra.cz/en/first-hr-diagram.

With the rising influence of European (mostly Dutch) astronomers in the United States, the issue of the proper acknowledgment became very serious and created frictions and debate within the community. After two decades, the "Russell diagram" became known as the Hertzsprung-Russell diagram, thanks in part to the influential conference delivered in 1933 by B. Strömgren at the meeting of the Astronomische Gesellschaft, but much to the irritation of many, including Russell himself who even refused to acknowledge that Hertzsprung had found (and coined the terms) "giant" and "dwarf" stars (Smith, 1977). The (proper) renaming of the diagram was a long battle that lasted till the late 1940s, when S. Chandraskhar, advising the *Astrophysical Journal* and tired of the controversy, decided that the standard nomenclature would be the "H–R Diagram" (Devorkin, 2000). Rosenberg's pioneering contribution has been unfairly forgotten from the history describing the elaboration of the first CMDs (see, e.g., Waterfield, 1956; Nielsen, 1969; Devorkin, 2000).

Hans Rosenberg was born in Berlin on May 18, 1879, and studied first in Berlin, under W. Foerster, and then in Strasbourg, with E. Becker, obtaining his Dr. Phil. with a thesis on the period changes that χ Cygni underwent from 1686 to 1901. Interested in both instrumentation and astrophysics, he moved to Göttingen to work with Karl Schwarzschild where he was to produce the first large survey of stellar temperatures estimated with objective prism spectra. For years, the Rosenberg temperature scale would set the standard and be widely used, as well as his review on photoelectric photometry in the *Handbuch der Astrophysik* (Rosenberg, 1929). His *Habilitation* thesis from Tübingen in 1910 was titled "*The relation between brightness and spectral type in the Pleiades*," whose results were published in (Rosenberg, 1910). His instrumental expertise allowed him to get spectra of comets Daniel (1907 IV) and Morehouse (1908 III). From 1910 on, he worked at Tübingen first as Privatdozent, where he founded its observatory in Österberg, becoming its director in 1912 and professor at the university in 1916 while serving in the army during World War I. With P. Goetz he made the first photometric map of the Moon, and developed the use of photoelectric cells as astronomical detectors in 1913. Moving to Kiel in 1925 he worked on solar eclipses, developed direct measures of the colors of stars, and carried heavy teaching duties. In spite of his position as professor at Kiel university, on April 1, 1933, three uniformed members of the feared SA Nazi paramilitary group came to his apartment and forced him to resign (Duerbeck, 2006) even so, he would have lost his position because of the racial laws enacted in 1935. He was immediately invited in 1934 by the University of Chicago to work at Yerkes Observatory where he stayed three years, working on measures of the limb darkening and color indices in eclipsing binaries (e.g., Rosenberg, 1936). In 1938 he was appointed director of Istanbul Observatory, where he reorganized the teaching of astronomy at the University of Istanbul and set up new priorities for the observing campaigns. He passed away there on July 26, 1940, a week after suffering a heat stroke (Gleissberg, 1940).

An even earlier relationship between color and magnitude was found by Charlier (1889), who studied the correlation between magnitude and a color-like quantity (the difference between the magnitude he measured in a photographic plate and the visual magnitude) as determined by Max Wolf in Heidelberg. However, Charlier interpreted the correlation found (the visually fainter stars had larger differences) as a systematic error in the visual magnitudes. As Rosenberg correctly pointed out, this could also be produced by selective reddening, hence the importance of getting spectra. The spectral type-magnitude correlation found in the Pleiades could not be produced by dust, hence "the plausible color differences among the stars in the Pleiades – the fainter the star, the redder it is – following from the optical and photographic brightness measurements are confirmed by the spectral properties." This groundbreaking result would be put to good use to lay the foundations of stellar physics (see, e.g., Salaris and Cassisi, 2005), but the true pioneer has unfairly been forgotten, thus, it would be a fitting tribute to rename the diagram the

Rosenberg-Hertzsprung-Russell diagram (RHR). Ironically, HR can also stand for Hans Rosenberg's initials.

The final word may come from Hertzsprung himself. His modesty made him avoid talking about his own contributions to astronomy, and, as Strand (1968) reminds us, he remarked on the controverted issue of the naming of the diagram

> "Why not call it the color–magnitude diagram? Then we know what it is all about."

The use of CMDs to constrain the physical properties of the stars was noticed very quickly. In fact, Russell (1912) was the very first to use the correlation between absolute magnitude and spectral type for (dwarf) stars with measured parallaxes to infer a distance to the Pleiades of 500 light-years.[2] In this chapter a brief summary is presented of the uses of CMDs to infer properties of single stars (Section 6.1), of detached binary stars whose components are assumed to have evolved independently of each other (Section 6.2), and coeval stellar populations such as (presumably) those in clusters (Section 6.3). The thorny issue of transforming isochrones from/to the theoretical diagram to/from observed CMDs will not be dealt with here. Although many efforts have been made to find the best transformations, the differences observed cannot (yet?) be ascribed to either systematics in the observations or in missing/wrong physics in the stellar evolutionary calculations. For instance, the recent analysis by VandenBerg *et al.* (2010) of some globular clusters, and by An *et al.* (2007) of open clusters shows that while some CMDs can be well fitted, other color-magnitude combinations of the *same* clusters show anomalies that go well beyond the corrections for systematics, rotation, activity, transformations, metallicity, etc. Empirical bolometric corrections are another source of uncertainty (Torres, 2010) as are the systematics in the determination of effective temperatures (e.g., Ramírez and Meléndez, 2005). It makes sense to adopt a standard set of values (Harmanec and Prša, 2011) with nominal values to avoid some of the systematics arising with the adoption of different key values (solar radius, mass, etc.). Section 6.4 deals with the general problem of inverting the CMDs of a mixture of resolved stellar populations to infer their chemical and star formation rate histories, and Section 6.5 is a very brief discussion on CMDs of pixels in unresolved stellar populations. Section 6.6 closes the chapter with a discussion of some statistical issues in the interpretation of CMDs.

SectionectionSingle stars Fitting isochrones to a set of stars is the main method to constrain their physical properties; other techniques, such as stellar oscillations, are limited to nearby stars. The set of basic properties includes the age t, Helium abundance Y, metallicity Z, α elements over abundance, distance modulus $\mu = m - M$, convection mixing length, etc., that is, all the quantities that determine the position of stars in a CMD. In contrast with its importance for stellar evolution, relatively little work has been done to formulate mathematically the problem to go beyond the "fit-by-eye" approach that has characterized this field (and unfortunately still does!). Perhaps the first attempt to find the distance of a star from an isochrone comes from Schaltenbrand (1974) who developed a simple method to estimate the nearest point of the zero-age main sequence to a given star in a two-color diagram under the assumption of Gaussian errors. This approach was further formalized in a proper probabilistic framework by Luri *et al.* (1992) as the so-called "proximity parameter", similar to the deterministic "near point" estimator by Flannery and Johnson (1982). Here one computes the sum of minimum distances from a set of N_* stars to a given isochrone with given properties (age t, Helium abundance Y, metal content Z, etc.) along with properties of the set of stars, which can also

[2] The spectra of the fainter members of the Pleiades was noted by EC. Pickering and A. Cannon, who made no mention of Rosenberg's (1910) work nor the comprehensive survey by Hertzsprung (1911).

be applied to the theoretical isochrone, such as the distance modulus μ or the extinction. We represent this set as the vector of parameters $\vec{\vartheta} = (t, Y, Z, \mu, \cdots)$. One can form the statistic Ψ^2 which is calculated as the sum of minimal (squared) distances:

$$\Psi^2(\vec{\vartheta}) = \sum_{i=1}^{N_*} \min_m \left(d_{im}^2\right), \tag{6.1}$$

where d_{im} is the geometrical distance between star number i and the point m on the given isochrone:

$$d_{im}^2 = \left(\frac{m_i - m_m}{\sigma(m_i)}\right)^2 + \left(\frac{c_i - c_m}{\sigma(c_i)}\right)^2, \tag{6.2}$$

and m and c are magnitudes and colors, respectively, or gravities and temperatures, or any two observables that can be predicted with the models, whereas $\sigma(m_i)$ and $\sigma(c_i)$ are the errors in these quantities for the given star labeled i. This distance can obviously be generalized to any n-dimensional space. If (and this is a major assumption, discussed later) the model stars were uniformly distributed along the isochrone, then the probability that one observed star comes from that isochrone would be

$$\mathcal{P}_i(\vec{\vartheta}) = \frac{1}{2\,\pi\,\sigma(m_i)\,\sigma(c_i)} \sum_{m=1}^{M} \exp\left(-d_{im}^2/2\right), \tag{6.3}$$

where M is the total number of points the isochrone has been divided into. The probability that an *ensemble* of N_* stars comes from an isochrone with properties $\vec{\vartheta}$ is just the product of the individual probabilities

$$\mathcal{P}_{total}(\vec{\vartheta}) = \prod_{i=1}^{N_*} \mathcal{P}_i(\vec{\vartheta}). \tag{6.4}$$

One can then use the standard maximum likelihood technique to find the set of parameters $\vec{\vartheta}$ that maximizes this probability. The maximum likelihood statistic (MLS) would, for instance, be

$$\mathrm{MLS}(\vec{\vartheta}) = -\log \mathcal{P}_{total}(\vec{\vartheta}) = \sum_{i=1}^{N_*} \log \mathcal{P}_i(\vec{\vartheta}). \tag{6.5}$$

For, say, $m = 3$ parameters (distance modulus, age, and metallicity) this would be an optimization problem: find the point in this three-dimensional space for which MLS reaches a maximum.

The procedure is best illustrated for the case of a single star ($N_* = 1$), as shown in Figure 6.1 where a star (and its errors, which determine the elliptical confidence region) is represented in a CMD along with two possible isochrones of very different properties but which lie at the same normalized distance d_{1m} of unity. Clearly many more isochrones can have the same normalized distance or even smaller ones, but they both show that one cannot decide which isochrone fits best, as both contribute the same amount to the Ψ^2 statistic or to the MLS one. The problem, thus formulated, is intrinsically degenerate, as there are multiple solutions: many different combinations of the basic parameters $\vec{\vartheta}$ can give the same maximum in the likelihood.

The reason for this degeneracy stems from the key assumption made in deriving the probability that the star was sampled from the isochrone (Equation 6.3): it was assumed that the number of stars *along* the isochrone didn't change. That is, we were only concerned about the *geometrical shape* of the isochrone, and not about the *density* of stars

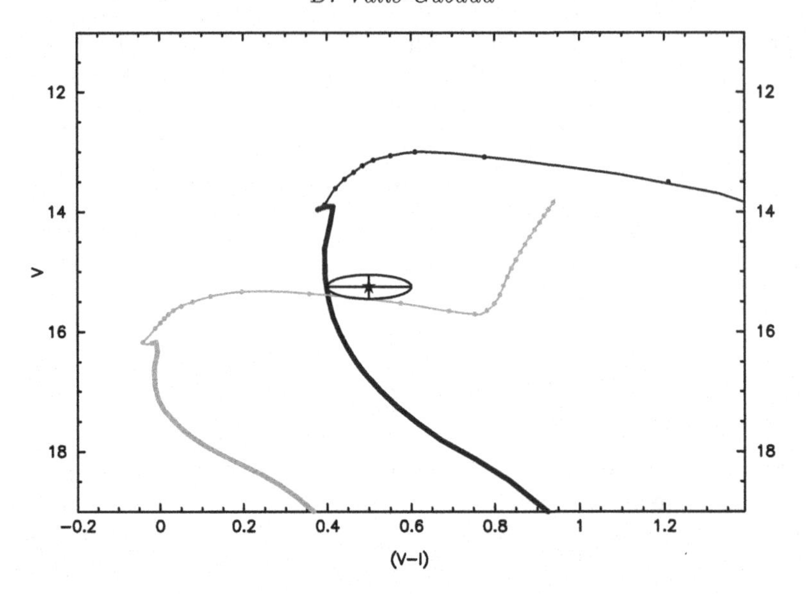

FIG. 6.1. The observation of a single star in a given CMD. Which of the two isochrones is more likely to be the correct one? Only two isochrones are plotted, for illustrative purposes, with different properties (age, distance, metallicity, etc.), and the points along each isochrone are such that the difference in stellar mass is the same (namely $\Delta M = 0.01\mathrm{M}_\odot$). In their main sequence the density of stars is very large (slow evolution) whereas after the turn-off, the much faster evolutionary speed makes the stars become more widely separated for this fixed ΔM. Many different isochrones can go through the same observed point (star with error bars at 1σ and with the corresponding probability level given by the elliptical curve), and hence there are, in principle, many possible solutions: there is a huge degeneracy. For clarity, in this figure only two isochrones are marked whose near point lie at a normalized distance of 1σ from the observed position. The geometrical term of their likelihoods is identical (same distance in probability or σ units), yet the lower (light gray) isochrone has an evolutionary term much smaller than that for the left (dark gray) isochrone: the density of points from the lower (light gray) isochrone is much smaller when arriving close to the observed star. Hence it is much less probable that the observed star is drawn from it. The purely geometrical degeneracy can be lifted using the prior information provided by stellar evolution in terms of the evolutionary speed of each track.

along it. This *geometrical method*, while useful in selecting shapes of isochrones that come close to the observed position, is highly degenerate (Figure 6.1).

Figure 6.2 illustrates one aspect of this degeneracy: isochrones of widely different ages and metallicities have a very similar geometrical shape. It is therefore hardly surprising that geometrical methods which rely on the proximity of a star to these isochrones yield huge degeneracies. Adding the uncertainties in distance modulus and dust extinction makes these methods unsuitable for any quantitative analysis.

While extensions of this geometrical method are discussed in their contexts later on, it is worth understanding in more detail the underlying reason for which the distribution of stars is *not* uniform along an isochrone.

Let us consider a curvilinear coordinate s along an isochrone of given parameters. Stellar evolution theory predicts that for a given abundance the number of stars on the isochrone depends only on the age t of that isochrone and on the mass m of the star at that precise locus, so that $s = s(m,t)$ only. An offset in age dt is reflected only through changes in mass m and position s along the isochrone of age t as

$$dt(m,s) = \left.\frac{\partial t}{\partial m}\right|_s dm + \left.\frac{\partial t}{\partial s}\right|_m ds. \tag{6.6}$$

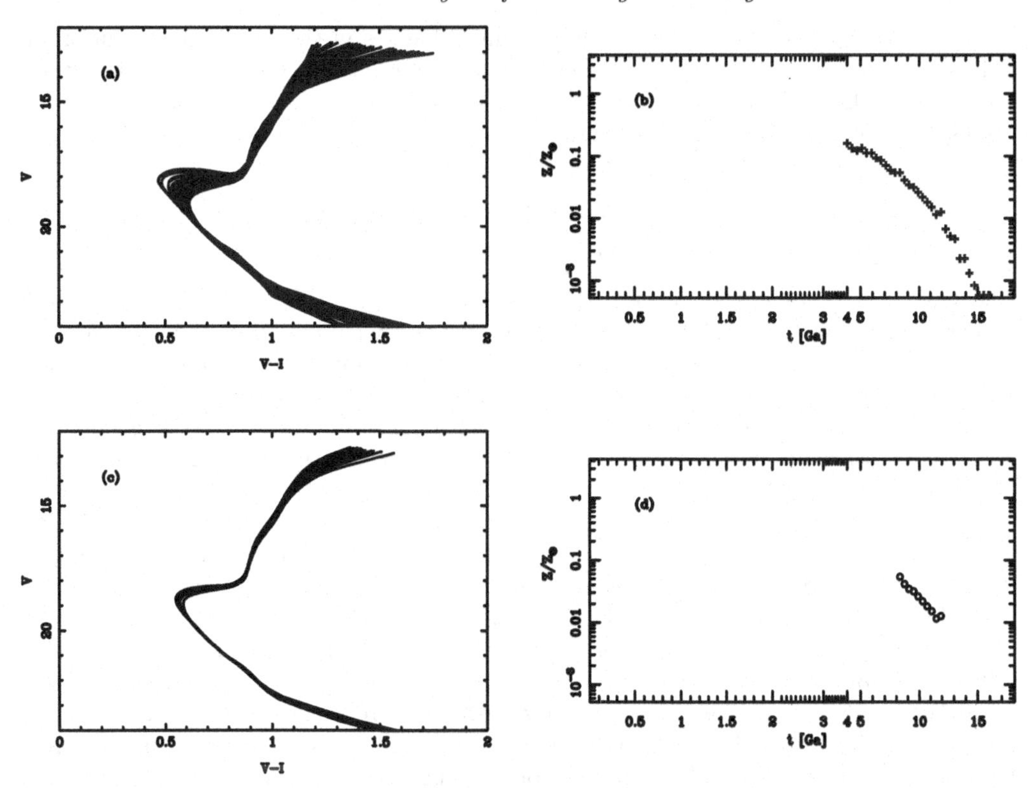

FIG. 6.2. An illustration of the *geometrical* degeneracy between age and metallicity. In panel (a) a series of isochrones of roughly similar shapes is indicated, whose ages and metallicities are given in panel (b). The well-defined locus in the Z-age plane means that many CMD inversions with sparse populations, differential reddening, or binaries will tend to produce age-metallicity "relations" that only reflect this geometrical degeneracy. This is further illustrated in panel (c) where the geometrical shapes are even closer. Yet, as panel (d) shows, any isochrone changing its metallicity by nearly an order of magnitude can produce another isochrone of the very same geometrical shape provided the age changes by about a factor of two. Physically, there is no such degeneracy, as a change of (atmospheric) metallicity reflects a change in the (core) abundance, and hence a different evolutionary speed, as reflected by the density of stars along the isochrone.

For a star in the isochrone, the offset is, by definition, $dt = 0$, hence

$$\left.\frac{\partial m}{\partial s}\right|_t = -\left.\frac{\partial m}{\partial t}\right|_s \times \left(\left.\frac{\partial s}{\partial t}\right|_m\right)^{-1} . \tag{6.7}$$

The first term is always finite. The second term is the evolutionary speed, the rate of change for a given mass m of its coordinate along the isochrone when the age changes by some small amount. One can think of s as representing some evolutionary phase, and so this term will be large when the phase is short-lived: a small variation in age yields a very large change in position along the isochrone. Alternatively, for a given age, and since the first term is always finite, a wide variation in position implies a narrow range in mass. This is the case of the red giant branch or the white dwarf cooling sequence, for instance. On the other hand, slowly evolving phases such as the main sequence have small evolutionary speeds and wide ranges in mass for a given interval along an isochrone. Clearly, the most important phases to discriminate between alternative ages and metallicities will be the post main-sequence ones, where the range of mass is small (and hence insensitive to the details of the stellar mass function), and at the same time where evolutionary speeds are large. If we consider the mid- to lower main sequence, at fixed metallicity, isochrones of all ages trace the same locus, with only marginal changes in the density distribution of points among them. In this sense, one of the parameters, the age t, is to a large extent absent from the main sequence, whereas phases beyond it are always substantially a function of (at least) both age and metallicity.

The density of stars along an isochrone is therefore

$$\frac{dN}{ds} = -\left(\frac{dN}{dm}\right) \times \left(\left.\frac{\partial m}{\partial t}\right|_s\right) \times \left(\left.\frac{\partial t}{\partial s}\right|_m\right). \tag{6.8}$$

The first term is related to the initial stellar mass function (IMF), and, as discussed earlier, the second term is finite and the third is a strong function of the evolutionary speed. If the mass after the turn-off is assumed to be roughly constant, this implies that the *ratio* in the number of stars in two different evolutionary stages after the turn-off will only depend on the ratio of their evolutionary timescales. In the context of stellar population synthesis, this is known as the fuel consumption theorem (Renzini and Buzzoni, 1983).

The way to infer the properties of star, given some observables and our knowledge of stellar evolution, is obviously the Bayesian method, where both the errors in the observables and our prior information (stellar evolution) can be handled properly, even in the case of one single star. Good reviews of Bayesian inference in physics are provided by Cousin (1995), Dose (2003), and Trotta (2008), along with the monographs by Gregory (2005) and Hobson *et al.* (2010). The Bayesian method allows us to answer the question we are interested in: what is the probability of the occurrence of the estimated parameters $\vec{\boldsymbol{\vartheta}}$ given the observed data set $\vec{\mathbf{D}} = (V, V-I, B-R, \cdots)$ and our prior information provided by stellar evolution? The prior information can be formulated as the probability distribution function expected for the parameters, $\pi(\vec{\boldsymbol{\vartheta}})$, on the basis of stellar evolution or other prior knowledge (say, a measure of Z), and is normalized to one, $\int \pi(\vec{\boldsymbol{\vartheta}})\, d\vec{\boldsymbol{\vartheta}} = 1$. The probability of observing the data set $\vec{\mathbf{D}}$, given some parameters $\vec{\boldsymbol{\vartheta}}$ is the likelihood $\mathcal{L}(\vec{\mathbf{D}}|\vec{\boldsymbol{\vartheta}})$. An important quantity is the *evidence* (sometimes also termed marginal likelihood) which is just $E(\vec{\mathbf{D}}) = \int \pi(\vec{\boldsymbol{\vartheta}})\, \mathcal{L}(\vec{\mathbf{D}}|\vec{\boldsymbol{\vartheta}})\, d\vec{\boldsymbol{\vartheta}}$. Bayes (1983)'s theorem[3] then states that

$$\mathcal{P}(\vec{\boldsymbol{\vartheta}}|\vec{\mathbf{D}}) = \pi(\vec{\boldsymbol{\vartheta}})\, \frac{\mathcal{L}(\vec{\mathbf{D}}|\vec{\boldsymbol{\vartheta}})}{E(\vec{\mathbf{D}})}. \tag{6.9}$$

In other words, our prior information of the parameters is modified into our posterior probability distribution function by the ratio of the likelihood over the evidence. We will see that this formulation of the problem also allows us to discriminate among models through the model selection technique. In the case of an observed star associated with an isochrone with n observables $\vec{\mathbf{D}}$ and m parameters $\vec{\boldsymbol{\vartheta}}$, we can define a simple *geometrical* likelihood as

$$\mathcal{L}_{geom}(\vec{\mathbf{D}}|\vec{\boldsymbol{\vartheta}}) = \frac{1}{(2\,\pi)^{n/2} \prod_{k=1}^{n} \sigma_k} \prod_{k=1}^{n} \exp\left(-D_k^2/2\right), \tag{6.10}$$

where D_k is again the normalized distance between the observed quantity S_k and the one predicted by the model M_k, given an observed error σ_k in that quantity

$$D_k^2 = \left(\frac{S_k - M_k}{\sigma_k}\right)^2. \tag{6.11}$$

[3] Dale (1982) explores the issue of whether Laplace (1812) should been given credit for the actual use, proof, and development of the theorem.

For example, we can take $n = 3$ observables such as $\vec{\mathbf{D}} = (m_V^{obs}, T_{\rm eff}^{obs}, Z^{obs})$, which depend on, say, $m = 5$ theoretical parameters such as $\vec{\vartheta} = (m, t, d, Z, \alpha)$ and we would have

$$\mathcal{L}_{geom}(m_V^{obs}, T_{\rm eff}^{obs}, Z^{obs}|m, t, d, Z, \alpha) = \frac{\exp\left(-\chi^2/2\right)}{(2\,\pi)^{3/2}\,\sigma(m_V^{obs})\,\sigma(T_{\rm eff}^{obs})\,\sigma(Z^{obs})}, \qquad (6.12)$$

with, under the assumption that they are uncorrelated,

$$\chi^2 = \left(\frac{m_V^{obs} - m_V^{theo}(m, t, d, Z, \alpha)}{\sigma(m_V^{obs})}\right)^2 + \left(\frac{T_{\rm eff}^{obs} - T_{\rm eff}^{theo}(m, t, d, Z, \alpha)}{\sigma(T_{\rm eff}^{obs})}\right)^2 + \left(\frac{Z^{obs} - Z^{theo}(m, t, d, Z, \alpha)}{\sigma(Z^{obs})}\right)^2. \qquad (6.13)$$

If quantities are correlated, one has to form the pairs

$$\mathcal{L}_{corr}(A, B) = \frac{\exp\left(-\left[D_A^2 + D_B^2 - 2\rho D_A D_B\right] / \left[2(1-\rho^2)\right]\right)}{2\,\pi\,\sigma(A)\,\sigma(B)\,\sqrt{1-\rho^2}}, \qquad (6.14)$$

where ρ is the correlation coefficient between the two quantities A and B, or, in terms of their covariance, $cov(A, B) = \rho\,\sigma(A)\,\sigma(B)$.

We can illustrate this case with an example where we assume to have a prior knowledge of the distance of the star and its bolometric correction so that we can, for instance, use the bolometric luminosity rather than its apparent magnitude in some photometric band.

Figure 6.3 shows as a test case five stars with the same effective temperature but different luminosities so as to sample different evolutionary régimes (top panel). There is no prior information on age or metallicity nor mass. If we are only interested in, say, the ages of these stars, we can consider the mass, the metallicity, the mixing length parameter, etc., as nuisance parameters with flat probability distributions (say, between 0.2 and 200 $M_\odot$, 0.0001 and 0.2, and 0.5 to 2.0, respectively) so that the probability distribution function (PDF) of the age is given by

$$\mathcal{P}(t) = \int\!\!\int\!\!\int dm\, dZ\, d\alpha\, \mathcal{L}_{geom}(t, Z, m, \alpha). \qquad (6.15)$$

Likewise, if we are interested in estimating the mass of that star, we would marginalise over the other (nuisance) parameters to get

$$\mathcal{P}(m) = \int\!\!\int\!\!\int dt\, dZ\, d\alpha\, \mathcal{L}_{geom}(t, Z, m, \alpha). \qquad (6.16)$$

We illustrate the technique here for the ages of the stars, given their importance in the context of both CMDs and stellar evolution (e.g., Lebreton, 2000; Soderblom, 2010), and because they have been widely used, even though they are not physically justified (e.g Lachaume *et al.*, 1999; Reddy *et al.*, 2003). The resulting age PDFs are given in the lower panel of Figure 6.3 and show the widely different distributions depending on the location of the test stars in the diagram (all share the same errors in bolometric luminosity and effective temperature, for the sake of the argument).

More important, however, this formulation of a *geometrical* likelihood (Equation 6.10) does *not* include any information on the evolutionary speed: we have only used the geometrical *shape* of the isochrone, the only quantity that matters when computing the distance from the observed star to a point on the isochrone (Equation 6.11). To incorporate

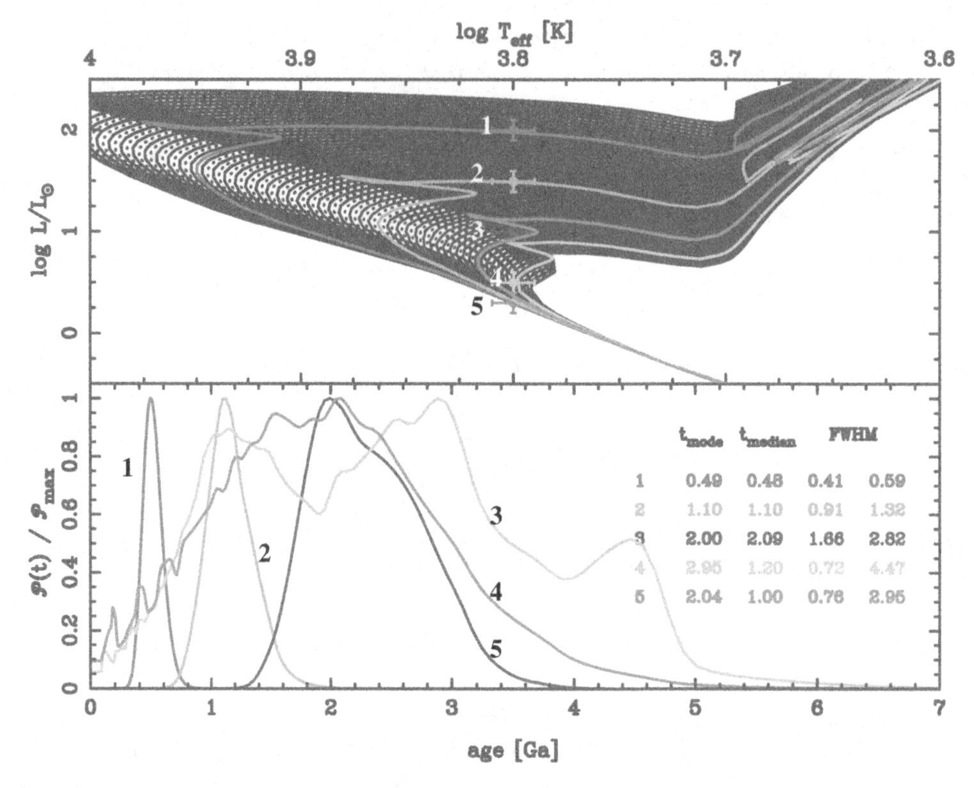

FIG. 6.3. A test case with five stars at fixed effective temperature ($\log T_{\rm eff} = 3.8$, $\sigma(\log T_{\rm eff}) = 0.01$) and 5 different bolometric luminosities sampling the HRD in regions where (*i–ii*) evolution is fast (top two stars, coded 1 and 2); (*iii*) near the TAMS, where multiple isochrones cross (coded 3 star ($\log L/{\rm L}_\odot = 1.0$); (*iv*) a slightly evolved star away from the MS (coded 4); and (*v*) a star very close to the ZAMS traced by very young stars. The lower panel shows the posterior probability distribution function for the age of each star, assuming flat priors for the other parameters. In this figure, the likelihood is computed using the geometrical term *only*, and although the upper two stars appear to have well-defined ages (an FWHM range from 0.41 to 0.59 Gyr for the first star, and from 0.91 to 1.32 Gyr for the second one), the star located between the red and the blue loops has an FWHM range of 1.5 Gyr, and it gets worse for the two lower stars. This is just due to the large density of isochrones in these areas, and the geometrical term of the likelihood cannot disentangle, per se, which set is more appropriate, since no prior information on the evolutionary speed is used.

the *physics* of stellar evolution we need to account for the *density* of stars along the isochrone (Equation 6.8) so that we have a proper *physical* likelihood. To do this, we need to integrate along all possible masses in the isochrone, but noting that (unlike the geometrical case) not all masses are equally probable. Let the density of stars of mass m along an isochrone be $\rho(m)$, then

$$\mathcal{L}_{phys}(\vec{\mathbf{D}}|\vec{\vartheta}) = K^{-1} \int_{m_{lim}}^{m_{top}} dm \, \frac{\rho(m)}{(2\pi)^{n/2} \prod_k^n \sigma_k} \prod_k^n \exp\left(-D_k^2/2\right), \tag{6.17}$$

where, for a proper normalization, we require

$$K = \int_{m_{lim}}^{m_{top}} dm \, \rho(m), \tag{6.18}$$

and m_{lim} and m_{top} are the lower and upper mass limits of the isochrone considered. It is useful to write this in terms of the curvilinear coordinate s along the isochrone, since the mapping of the mass m to a position in the CMD is highly non-linear (cf. Equation 6.8): the red giant branch will be poorly sampled if the mass interval is too large. We can thus re-write Equation 6.17 as

$$\mathcal{L}_{phys}(\vec{\mathbf{D}}|\vec{\vartheta}) = \int_{s=0}^{s=1} \underbrace{ds}_{curvilinear} \underbrace{\frac{dN(m)}{dm}}_{\phi(m)} \underbrace{\frac{dm}{ds}}_{speed} \underbrace{\frac{\prod_k^n \exp\left(-D_k^2/2\right)}{(2\pi)^{n/2}\prod_k^n \sigma_k}}_{geometry}, \tag{6.19}$$

where the lower and upper limits of the curvilinear coordinate s have been set to 0 and 1, respectively, and we can identify, besides the geometrical term, the initial mass function $\phi(m)$ and the evolutionary speed dm/ds along the isochrone. These two functions encapsulate the prior information provided by stellar evolution in a way that the geometrical approximation cannot possibly handle. In reference to previous works (Hernandez *et al.*, 1999; Jørgensen and Lindegren, 2005), we will refer this method as the `BayesGM` method.[4]

The posterior PDF then becomes

$$\mathcal{P}_{post}(\vec{\vartheta}|\vec{\mathbf{D}}) = \prod_{k=1}^{n} \pi(\vec{\vartheta})_k \prod_{i=1}^{N_*} \mathcal{L}_{phys}(\vec{\mathbf{D}}|\vec{\vartheta})_i, \tag{6.20}$$

where $\pi(\vec{\vartheta})_k$ is the prior probability distribution function for the parameter indexed k of the parameter vector $\vec{\vartheta}$. The key point of this formulation is that the prior on mass, the IMF $\phi(m)$, must be *included* in the likelihood, for a proper physical weighting. As the data set is fixed, we do not need to compute the evidence which only acts, in this context, as a normalization constant.

For example, consider a star on the red giant branch: for a long interval along the nearly vertical isochrone, the mass hardly changes, and hence the effective weight of the IMF in the integrand will be very small and the geometrical distance will, in proportion, be more important. In the main sequence the reverse is true. The effects of including this prior information explicitly in the likelihood are dramatic, and are shown in Figure 6.4. There is little difference for the two brightest stars, as isochrones run almost parallel to the effective temperature axis and hence map the plane in a well-behaved way: the only difference with the geometrical method is that our prior information on both evolutionary speed and the IMF will favor tracks with *smaller* masses (they evolve more slowly and are more abundant, two factors that cannot be handled by the geometrical method). The middle star (color-coded in blue) is more interesting as it lies in the area where multiple isochrones cross each other, hence providing a huge geometrical degeneracy as reflected by the age range from 1.66 to 2.82 Gyr (Figure 6.3). In this particular case, large ages appear to be penalized, and the new FWHM range is restricted to 1.66 to 2.29 Gyr (Figure 6.4). The next star illustrates this effect even more clearly, with a reduction of the interval 0.72–4.47 Gyr to 1.48–3.55 Gyr, a contraction of 1.75 Gyr. The faintest star, which appeared deceptively well behaved when using the geometrical method, now reveals that much younger ages are more than twice as, probable, an this is true as well for older ones. The full prior information on the way stars evolve is properly used.

In Figures 6.3 and 6.4, the full posterior age distributions are given. Rather than using the full PDF of the parameters, it is customary to encapsulate the information in quantities such as the (posterior) modes (the values of the parameters where the posterior

[4] In Hernandez *et al.* (1999), the physical likelihood was referred to as the G matrix.

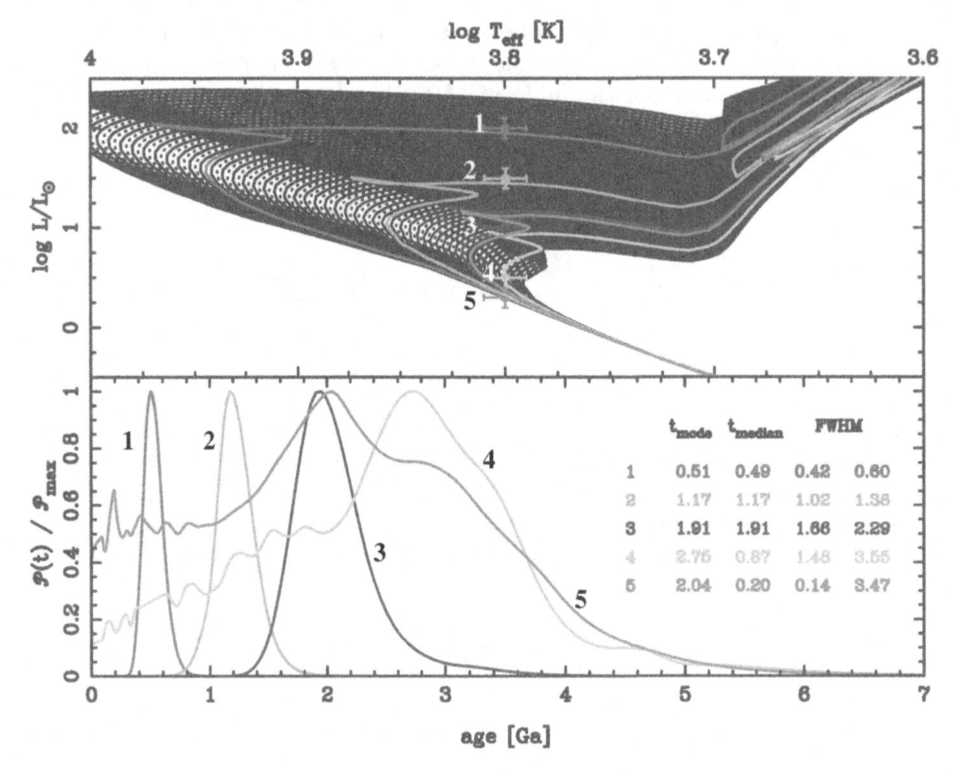

FIG. 6.4. The same set of stars as in Figure 6.3. This time the full likelihood function (that is, with the evolutionary terms) and the prior on the mass function is used. There is a tendency to increase the mode and the median age for the two top stars, although still within the same FWHM range. This slight increase is due to the fact that the probability of observing a less massive star is larger, and hence the ages increase. This effect helps reduce the range for star # 3 by almost 0.5 Gyr, and produces well-defined peaks for the two lower stars, reducing the range of the posterior PDF. The effect is even stronger when a wide range of metallicities is considered: the posterior PDF peaks at the proper ages and metallicities, thereby lifting the purely geometrical degeneracy illustrated in Figure 6.2.

PDF reaches a maximum), the means or expectation values, or even the medians. The frequentist confidence intervals become, within the Bayesian framework, the credible regions (CRs) such that they are the (closed but not necessarily connected) volumes that contain a fraction α of the total volume under the posterior:

$$\int_{CR(\alpha)} d\vec{\vartheta}\ \mathcal{P}_{post}(\vec{\vartheta}|\vec{\mathbf{D}}) = \alpha. \tag{6.21}$$

There are many different possible CRs. The central credible interval (CCI) is defined such that the intervals $(-\infty, \theta_{low})$ and $(\theta_{high}, +\infty)$ each contains $(1-\alpha)/2$ of the posterior volume, and always contains the median. The minimum credible interval (MCI) is built in such way that the posterior PDF is always larger inside the MCI than outside. It contains the mode, obviously, but may not be connected. In Figure 6.4 the FWHM (that is, the values for which the PDF reaches half its maximum value) are indicated, but note that they do not correspond to any fixed probability measure. The figure also illustrates the poor performance of the median when the PDF is wide. Similarly, and contrary to the claims by Burnett and Binney (2010), the mean can be highly biased (although easy to compute). We hence prefer to use the mode of the distribution, as a robust point estimate of the quantity of interest.

Pont and Eyer (2004) have also developed a Bayesian technique to explore the distribution of ages for the particular case of G dwarf stars to revise the age-metallicity relation in

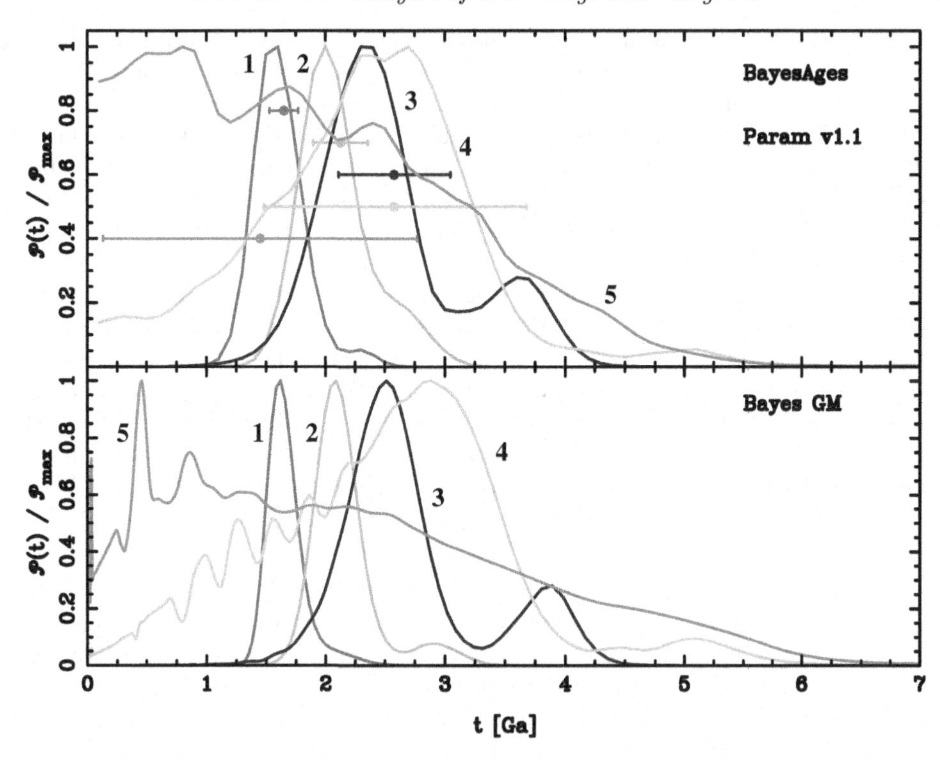

FIG. 6.5. Comparing the posterior PDF in ages for the 5 stars indicated in the top panels of Figs. 6.3 and 6.4, with uncertainties of $\sigma(\log T_{\rm eff}) = 0.01$ and $\sigma(M_V) = 0.1$, this time at $M_V = 2.0, 2.5, 3.0, 3.5$ and 4.0. `BayesAges` (Pont and Eyer, 2004) tends to produce slightly wider PDFs than our `BayesGM` technique, while `PARAM v1.1` (da Silva *et al.*, 2006) yields point estimates slightly older (1σ bars have been offset vertically for clarity). The well-defined peak at $t \approx 0.45$ Gyr predicted by `BayesGM`, and which is not present in `BayesAges` is robust: decreasing the uncertainties in the two observables produces a mode at this age.

the Galactic disc. They use a *geometrical* likelihood in the three-dimensional parameter space of effective temperature, absolute magnitude, and metallicity, and include stellar evolution in terms of priors (for instance, taking the variation of the luminosity as a function of age at fixed temperature). Figure 6.5 compares the results of their method, `BayesAge`, with `BayesGM` for five test stars. While, as expected, the results are very similar for the brightest stars, for the fainter ones the bias they observe toward younger ages is not present in `BayesGM`. For the faintest test star, the peak predicted by `BayesGM` is confirmed when decreasing the error bars in the observables, confirming the robustness of the method. da Silva *et al.* (2006) use a physical likelihood but take the IMF as describing the number of stars along a given isochrone, which is not correct (except on the ZAMS, see Equation 6.8). Nevertheless, the estimates obtained agree with `BayesGM` while they seem to be systematically larger than `BayesAges`, as indicated in Figure 6.5. The code `PARAM v1.1` is available at stev.oapd.inaf.it/cgi-bin/param. Remarkably, Valenti and Fischer (2005) did use what we could call a "empirical" bayesian technique to weigh geometrically estimated ages through a variety of "probabilities" in a purely empirical way. In contrast, Takeda *et al.* (2007) used a proper Bayesian formalism which includes the possible variation of $\Delta Y/\Delta Z$ in stellar tracks, and limits the integration of the posterior PDF to the hyper-box in the space of parameters.

Breddels *et al.* (2010) apply a maximum likelihood technique again using a purely geometrical criterion, while Burnett and Binney (2010) develop a Bayesian method to infer the first two moments of the posterior PDFs, again with a purely geometrical likelihood but properly weighted by priors. In contrast, Casagrande *et al.* (2011), building upon their previous work (Casagrande *et al.*, 2010), include explicitly priors that attempt to correct for biases known to exist in their sample and a proper physical likelihood. Bailer-Jones (2011) also uses a Bayesian framework to estimate extinction and stellar parameters from measures of parallaxes and multi-band photometry. The prior information from stellar evolution is included explicitly by a distribution obtained with 200,000 stars of solar metallicity sampled from a Salpeter IMF and a flat star formation history. The (smoothed) density (obtained via a Gaussian kernel) is used as a prior in absolute magnitude and temperature.

Isochrone ages now have the statistical framework to be inferred properly, and which can be used to calibrate gyrochronological ages (e.g., Chanamé and Ramírez, 2012) and be compared with the independent measures obtained through asteroseismology (e.g., Stello *et al.*, 2009). The field has evolved dramatically since the first attempts (e.g., Perrin *et al.*, 1977) with purely empirical fits to massive surveys with proper statistical methods (e.g., Nordström *et al.*, 2004). The advent of large-scale surveys with measures of fundamental parameters, such as RAVE (e.g., Zwitter *et al.*, 2010) and GAIA will show the usefulness of these Bayesian techniques.

6.1.1 *Caveats and (some) systematics*

There are two important caveats worth keeping in mind, besides the ones noted in the Introduction (photometric corrections, bolometric corrections, and standard values) as there are two underlying assumptions that have been made: first, that the light received in the detector actually corresponds to the star, and second, that the position of the star in the CMDs is only determined by its mass, for a given age and metallicity. The first assumption could be wrong if the detected star is, in fact, an unresolved binary or multiple system. In this case, we are detecting the combined light from the components of the system, and both the magnitude and the color are shifted by an amount that depends on the mass ratio(s). In the theoretical diagram, the luminosity and effective temperature of the combined system is related to the individual properties as

$$L_{A+B+C+\cdots} = L_A + L_B + L_C + \cdots, \tag{6.22}$$

$$T^4_{\mathrm{eff}(A+B+C+\cdots)} = \frac{L_A + L_B + L_C + \cdots}{4\pi\sigma\left(R_A^2 + R_B^2 + R_C^2 + \cdots\right)}, \tag{6.23}$$

and not, as some authors have wrongly claimed (Siess *et al.*, 1997), as the luminosity-weighted mean of the effective temperatures. In the CMDs this effect gives rise to well-known offsets from single star sequences (e.g., Haffner and Heckmann, 1937; Maeder, 1974; Lastennet and Valls-Gabaud, 1996; Hurley and Tout, 1998) and unless one has good reasons to assume the star in consideration is truly single, the effect will produce a systematic bias in the inferred properties.

The second assumption covers several effects. The first one is just a consequence of Vogt's (1926) theorem, which is not valid, for instance, in cooling white dwarfs, where the complicated details of the cooling sequence no longer depend only on the mass. In principle, the parameters controlling the locus of the white dwarf in the cooling sequence could be included in the multivariate set $\vec{\vartheta}$, and the same formalism can be applied (von Hippel, 2005; von Hippel *et al.*, 2006; Jeffery *et al.*, 2007; van Dyk *et al.*, 2009; Jeffery *et al.*, 2011). The same applies to pre-main-sequence stars, whose tracks depend to a large extent on their mass accretion history, which introduces a major complication, with, in principle, a functional degree of freedom that is difficult to constrain (Mayne and Naylor, 2008; Naylor, 2009). Attempts are currently being made to use a Bayesian formalism to tackle this problem (Gennaro *et al.*, 2012).

An equally serious case where the assumption is known to be wrong is provided by massive stars; their fast rotations not only bring fusion products from the core to their surface during core hydrogen burning, and hence affect the abundances, but also their location in a CMD depends at least both on the inclination and on the equatorial velocity, none of which are measurable (only $v \sin i$ can be inferred from the line profiles). The effects of rotation can reach some 0.1 mag or more in both color and magnitude (Maeder and Peytremann, 1970; Collins and Sonneborn, 1977). While statistical techniques (e.g., Collins and Smith, 1985; Lastennet and Valls-Gabaud, 1996) could be used incorporating further parameters into the $\vec{\vartheta}$ vector, the availability of state-of-the-art models of (massive) rotating stars (e.g., Brott *et al.*, 2011; Maeder and Meynet, 2012) may provide another way of dealing with stars in the upper main sequence or beyond.

The third effect is the assumed enrichment $\Delta Y/\Delta Z$ which is built in in the evolutionary tracks. Clearly, different assumptions on both the helium abundance Y and the enrichment ratio have consequences on the stellar speed, and thus far only tests with two different sets have been carried out (Casagrande *et al.*, 2011), but many more tests need to be done.

Last is the thorny issue of the calibration of the mixing length convection theory (MLT): all tracks/isochrones are normalized to the putative solar case, and so stars with widely different masses and metallicities are still assumed to have the very same properties as the solar convective layers. Although there is some mild evidence for a possible variation of the MLT parameter (which describes the mixing length in units of the local pressure scale) with mass in some binary systems (e.g., Lastennet *et al.*, 2003; Yıldız *et al.*, 2006; Yıldız, 2007), no isochrones have so far been computed for different scalings of the MLT parameter with mass and/or metallicity, and yet it is clear that stellar convection must depend on stellar parameters, as three-dimensional simulations are indicating (Ludwig *et al.*, 1999). Similarly, the amount of overshooting is yet another degree of freedom that is rarely taken into account even if tracks are computed for a variety of possible values (e.g., VandenBerg *et al.*, 2006).

All these provisos are worth keeping in mind before any interpretation of the resulting PDFs is carried out.

6.2 Detached binary stars

Binary stars can provide, in some circumstances, the only direct and reliable way to measure stellar masses and hence constitute a benchmark for stellar evolution. They can also provide accurate measures of many other quantities, and in some cases the full orbital and physical parameters. Astrometric (or interferometric) orbits combined with radial velocities, or detached eclipsing binaries that are also double-lined spectroscopic ones yield a wealth of precise measures reaching sometimes 1% in masses and radii (see Torres *et al.*, 2010, for a comprehensive review). A catalog of over 130 well-measured binary systems is mantained at www.astro.keele.ac.uk/jkt/debcat.

It is therefore quite natural to check how their own color-magnitude diagrams can test stellar evolution, when the assumption is made that both components evolved independently of each other (i.e. there were no mass transfer episodes).

There have been many attempts at using classical frequentist statistical methods (e.g., Young *et al.*, 2001; Lastennet and Valls-Gabaud, 2002; Malkov *et al.*, 2010, and references therein), but the Bayesian framework presented in the previous section allows one to infer more robustly the distribution functions of the parameters we are interested in. Although a detailed analysis is beyond the scope of this chapter, we can easily check, as an example, the effects of different observables on the inferred ages of the components.

We use the `BayesGM` formalism to infer the posterior distribution function of age for each component of the well-studied binary system β Aur (Torres *et al.*, 2010). Figure 6.6 shows the resulting PDFs, with no prior assumption on their possible coevality. Their posterior modes coincide at an age of 0.41 Gyr when using effective temperatures combined with absolute magnitudes in the V band. Figure 6.7 shows the same posterior age

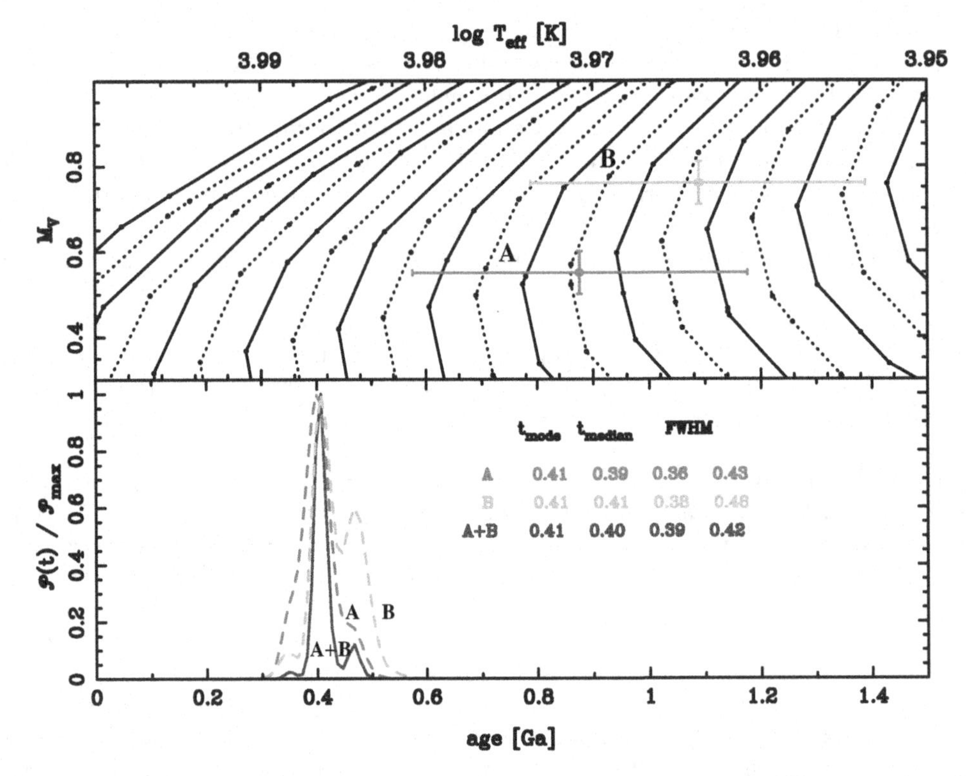

FIG. 6.6. Testing Bayesian ages in well-studied detached binary systems. Here we selected β Aur to minimize the uncertainties in the bolometric corrections. The values for the parameters are given by Torres *et al.* (2010). The posterior PDF in age are coded A and B. In spite of a secondary peak in the latter, they have the same mode, and their medians are very similar. If we impose the constraint that they ought to have the same age, the prior is then $\pi(t) = \delta(t_A - t_B)$, resulting in a joint PDF given by the dark gray solid curve, with a much reduced FWHM range.

distributions when another pair of observables is used: gravity and effective temperature. In this case, the posterior PDFs are more concentrated, but their modes are significantly different from the ones using absolute magnitudes and temperatures. In spite of small bolometric corrections, using absolute luminosities rather than absolute magnitudes in V yields wider PDFs (Figure 6.8), with ages consistently larger than the ones inferred from the $(\log g, \log T_{\rm eff})$ pairs. Remarkably, in all cases the PDFs for each component overlap, arguing strongly for coevality (but note this was not assumed a priori). If we impose the condition of coevality, the likelihoods of each component are combined and multiplied by a Dirac distribution $\delta(t_A - t_B)$. The resulting PDFs are shown with solid lines in these three figures. As expected in the Bayesian framework, the more information included results in posterior distributions that are much narrower.

Whereas there are no other independent constraints on age, one could use the isochrones to estimate the posterior PDFs on the masses, which are measured independently, or else use the measures as priors. This can very easily be done in the definition of the physical likelihood.

One can also combine prior information when binaries are known to be members of a cluster: as their share the same distance, one can impose this condition in the likelihood to get better constraints on the other parameters (e.g., Lastennet *et al.*, 1999), even though in the case of very nearby clusters the depth or extent along the radial direction may be an issue.

Systematics are likely to become the major source of uncertainties in these analyses-limb darkening "laws" appropriate for the stars in consideration, the amount of "third

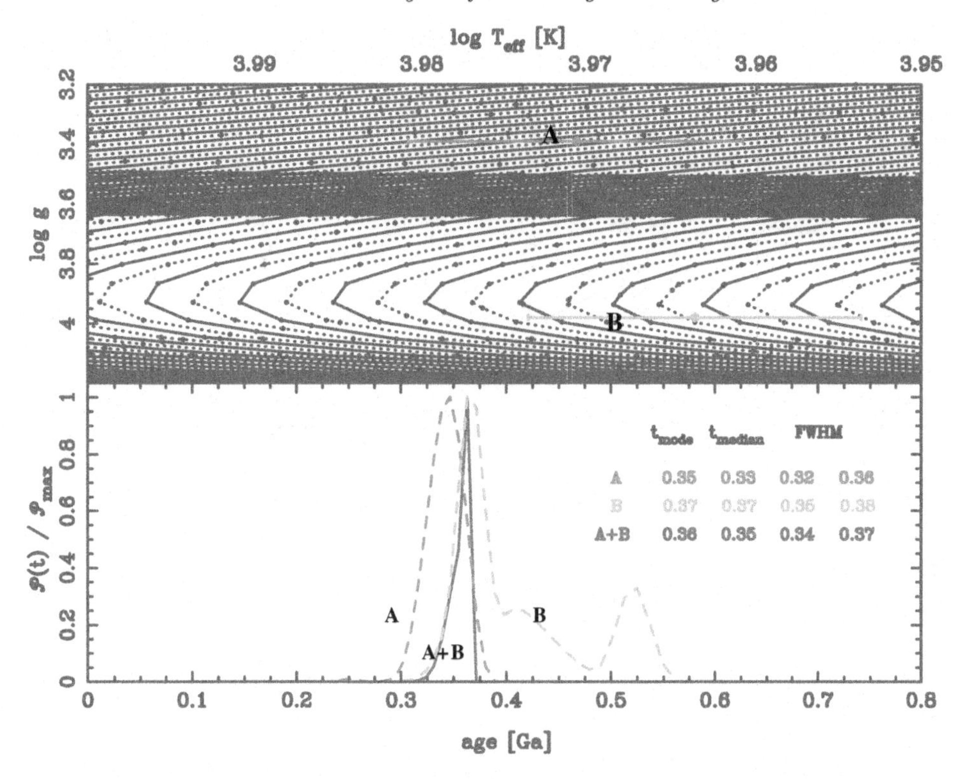

FIG. 6.7. The double-lined eclipsing binary β Aur (as in Figure 6.6) this time using effective temperatures and gravities, whose measures are assumed to be uncorrelated. The error bars $\sigma(\log g) \sim 0.005$ are too small to be seen at the scale of this diagram. While the posterior modes are different, the ranges are very similar. The joint distribution assuming coevality yields a modal age significantly younger than the one inferred using absolute magnitudes and effective temperatures (Figure 6.6).

light," the sources of noise, etc. (Southworth, 2011). Yet, the advent of both massive variability surveys and space missions is providing light curves of such quality (Bruntt and Southworth, 2008) that empirical methods can no longer be reasonably used.

6.3 Coeval stellar populations

The *quantitative* comparison of synthetic CMDs with observations relied initially on either post-MS phases, or the ratio of giants to main-sequence stars, the tips of various loops, etc. (see, e.g., Meyer-Hofmeister, 1969; Robertson, 1974; Becker and Mathews, 1983, for some early attempts). The advent of high-quality CCD photometry and the increase in computing power made it possible to apply proper statistical tools to the problem of *inverting* the observed CMDs to infer the underlying physical properties. Patenaude (1978) is an example in the attempt at setting up empirical isochrones for open clusters for an easier comparison with theoretical ones. In fact, observers have been fond of establishing the so-called "semi-empirical methods" whereby the separation between characteristic points in the CMD are calibrated with theoretical models. Hence, we have, for instance, the vertical separation between the turn-off (TO) and the horizontal branch, or the horizontal separation between the TO and some point in the sub-giant branch. These distances are in fact ill-defined, as photometric errors, binaries, and the intrinsic sampling contribute to a dispersion that is difficult to quantity. Detailed attempts at calibrating these "empirical" distances (e.g., Meissner and Weiss, 2006) must be superseded by a proper modern statistical framework.

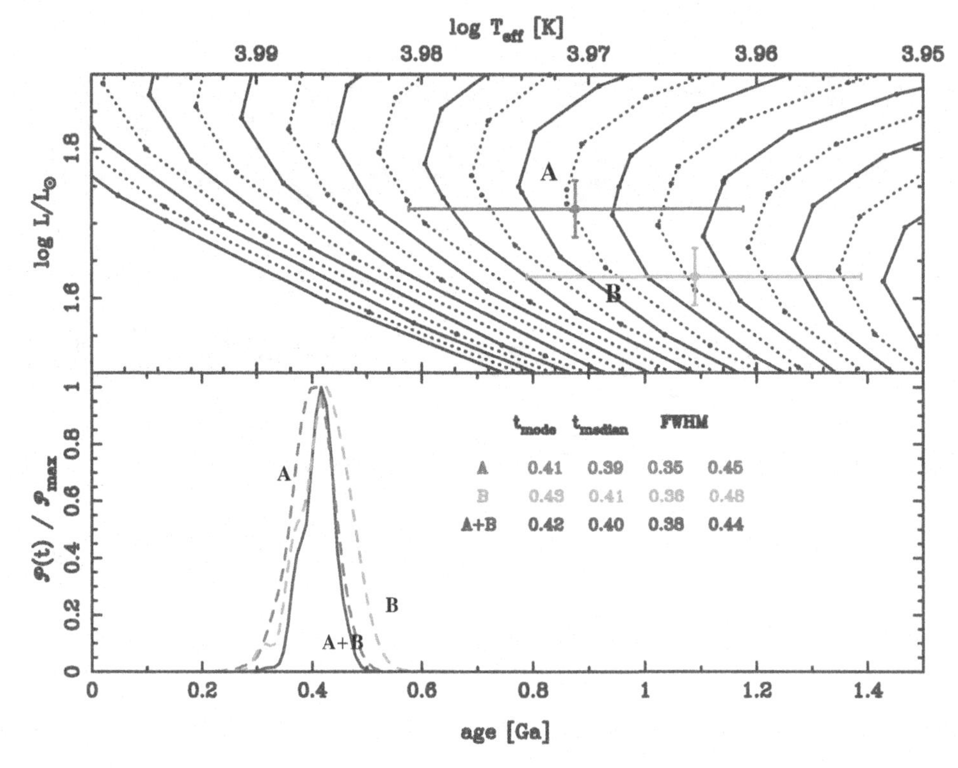

FIG. 6.8. Posterior PDF for the age of β Aur, using effective temperatures and absolute luminosities, assumed to be uncorrelated. As the bolometric corrections at these temperatures are small, it is not surprising to get results similar to those using absolute magnitudes (Figure 6.6), although the distributions appear to be wider.

Producing synthetic CMDs used to be costly, but now there are many tools available based on a different set of evolutionary tracks to create isochrones, luminosity functions, and integrated magnitudes (Table 6.1).

Naylor and Jeffries (2006) proposed a maximum likelihood method whereby a simulated underlying distribution function with some a priori information is made, and then they maximize a geometric likelihood to assess a goodness of fit. The advantage of this approach is that it allows them to include populations of (unresolved) binaries, where both the fraction of binaries and the mass ratio distribution can be accounted for. Their code is available at the website www.astro.ex.ac.uk/people/timn/tau-squared.

The cross entropy technique has also been used in solving the optimization problem (Monteiro *et al.*, 2010), although in this case the authors used a geometrical likelihood, weighted through Monte Carlo realizations, which, in effect, reduce to the Bayesian formulation, although in a rather convoluted way.

We can use `BayesGM` to infer, for example, the age of a coeval stellar population just as we did for a binary system. We form the combined likelihood of the N_* stars with n

TABLE 6.1. Some popular codes for generating isochrones and CMDs

`CMD v2.3`	stev.oapd.inaf.it/cgi-bin/cmd
Victoria-Regina	www3.cadc-ccda.hia-iha.nrc-cnrc.gc.ca/community/VictoriaReginaModels
Darmouth	stellar.dartmouth.edu/models
`IAC-STAR`	iac-star.iac.es/cmd/index.htm

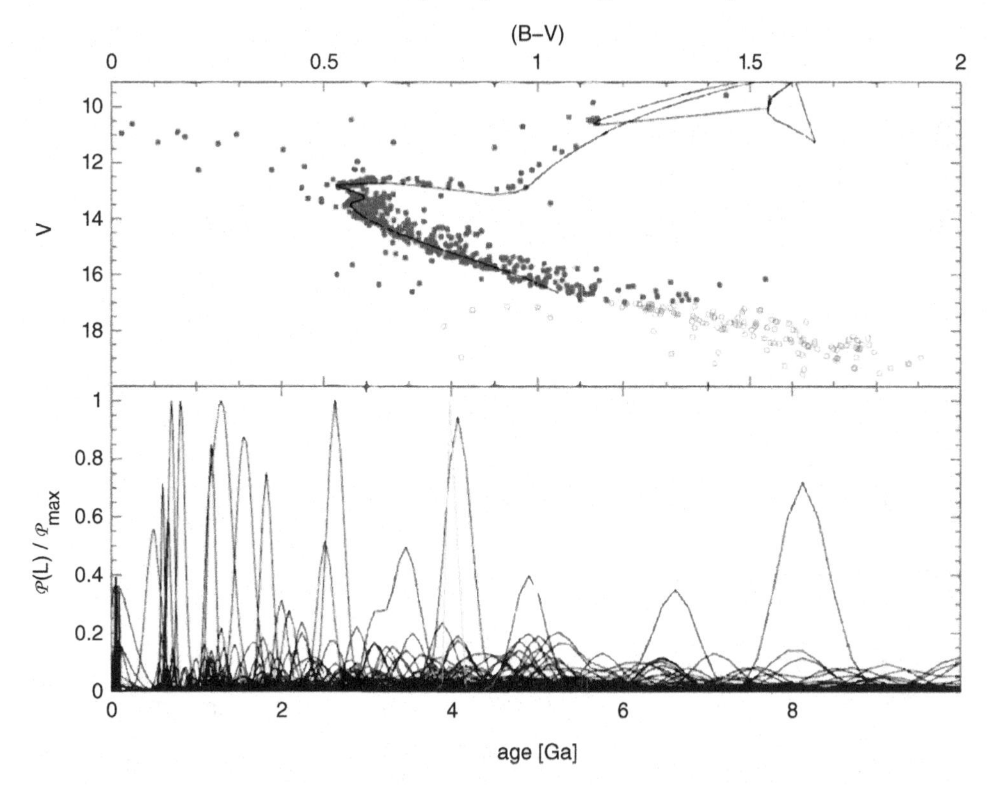

FIG. 6.9. `BayesGM` can be used to determine the ages of clusters in the same way as in detached binaries, under the assumption of coevality of its stellar populations, in this case M67 (NGC 2682). Each star produces its age PDF (black lines in the lower panel), whose product yields the cluster's age PDF (blue curve). A prior on metallicity of [Fe/H] $= 0.0 \pm 0.1$ is assumed, given the spectroscopic measures of both dwarf and giant stars (Santos *et al.*, 2009). Only stars with probabilities of membership larger than 50% based on proper motions and radial velocities have been used (Yadav *et al.*, 2008). Only stars brighter than $V = 17$ are considered, as some sets of tracks appear to present bluer isochrones than observed (Yadav *et al.*, 2008) at fainter magnitudes. The mode of the marginalized PDF on distance modulus is $\mu = 9.65$.

observables, say, $n = 2$ with (V, B-V), as

$$\mathcal{L}_{combined}(V, B - V|\vec{\vartheta}) = \prod_{j=1}^{N_*} \mathcal{L}_{phys}(V_j, (B - V)_j|\vec{\vartheta}), \tag{6.24}$$

where $\mathcal{L}_{phys}(\vec{\mathbf{D}}|\vec{\vartheta})$ is given in Equation 6.19 and we have the equivalent posterior PDF (Equation 6.20). Note that, at this stage, we are not imposing a priori that the stars must be coeval.

The results of the exercise for the M67 open cluster are shown in Figure 6.9 where the age PDF has been obtained by marginalizing over the other parameters (in this case, distance, reddening, metallicity). Each star yields its PDF in the parameters, and the lower panel shows the full range of PDFs reached by the ensemble of stars in M67. Clearly some stars cannot possibly be members of the cluster, for they have widely discrepant age distributions. This technique allows one to make a further selection (besides proper motions) for true members of the cluster. Note also that some distributions may be affected by some of the underlying assumptions made: some stars are certainly unresolved binaries and clearly the isochrones cannot possibly fit the blue stragglers present which will, nevertheless, produce unsensical PDFs unless they are filtered out.

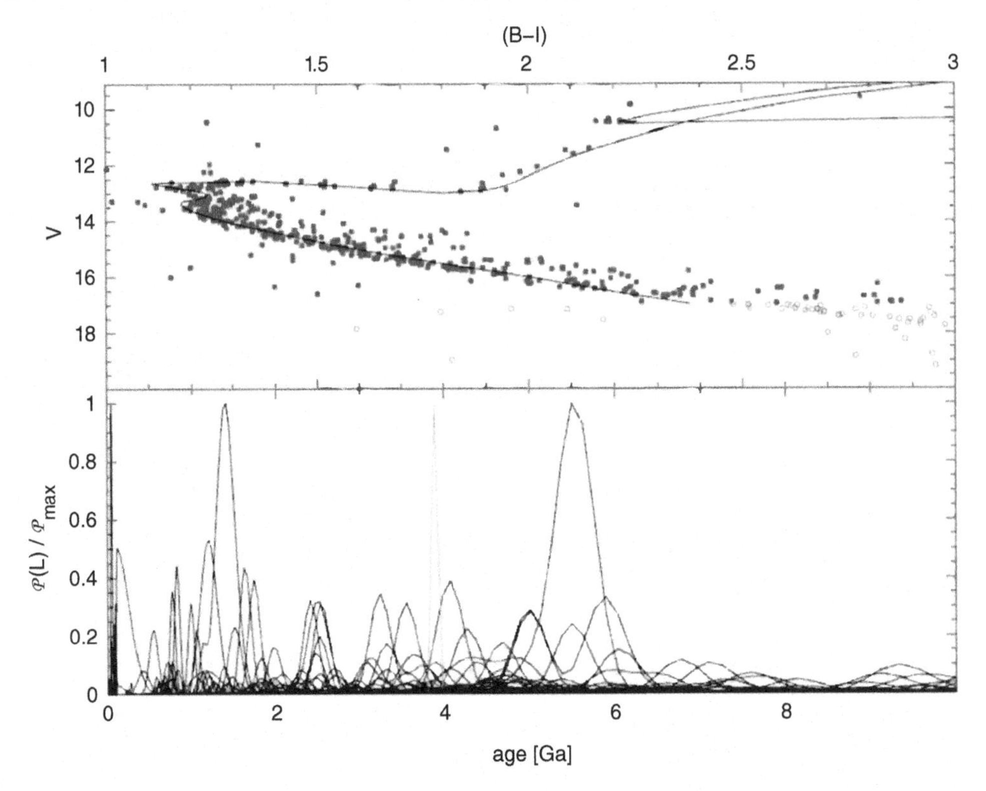

FIG. 6.10. The same data set for M67 as in Figure 6.9 but this time using the B, V, and I photometry yields the same modal age, but a mode of the posterior PDF in distance of $\mu = 9.53$. The difference cannot be accounted for by extinction and shows the level of systematics at $\delta\mu \sim 0.1$.

The product of all the marginalized PDFs yields the PDF of the age of the ensemble as $\pi(t) = \delta(t - t_*)$ and in this case its mode lies at 3.89 Gyr, with a sharply defined PDF. The corresponding isochrone is indicated in the upper panel of Figure 6.9. The open circles in the figure are stars fainter than $V = 17$ and which appear to be too blue for the set of isochrones (this problem is also found in a different context by An *et al.*, 2007). Could this be caused by chromospheric activity or is it a problem in the evolutionary tracks?

To check this issue, we can take the set of (B, V, I) independent measures and re-do the analysis. Figure 6.10 shows that in this case the mode of the posterior age distribution peaks at 3.98 Gyr, and the PDF is fully consistent with the one inferred from the (B, V) set. However, the modes of the distance modulus PDFs are significantly different, 9.65 and 9.53, and cannot be accounted for by the uncertainties in the reddening (which was also marginalized out). The outcome of the analysis is that there is a level of systematics that can only be explored using the fully N-dimensional PDFs, to assess correlations between the parameters and possible causes of inconsistencies.

We can now see an example of the multi-dimensional PDFs case by applying `BayesGM` to globular clusters, with flat priors.[5] Figure 6.11 shows the CMD of the old globular cluster NGC 6681, observed with HST, assuming the standard transformations to the B, V system. Here the modes of the posterior appear well defined as well, and the right panel of the figure shows the PDF marginalized over all parameters but age and metallicity.

[5] Note, however, that the set of stellar tracks used allowed only three different values for $[\alpha/\text{Fe}]$, namely, +0.0, +0.2, and +0.3.

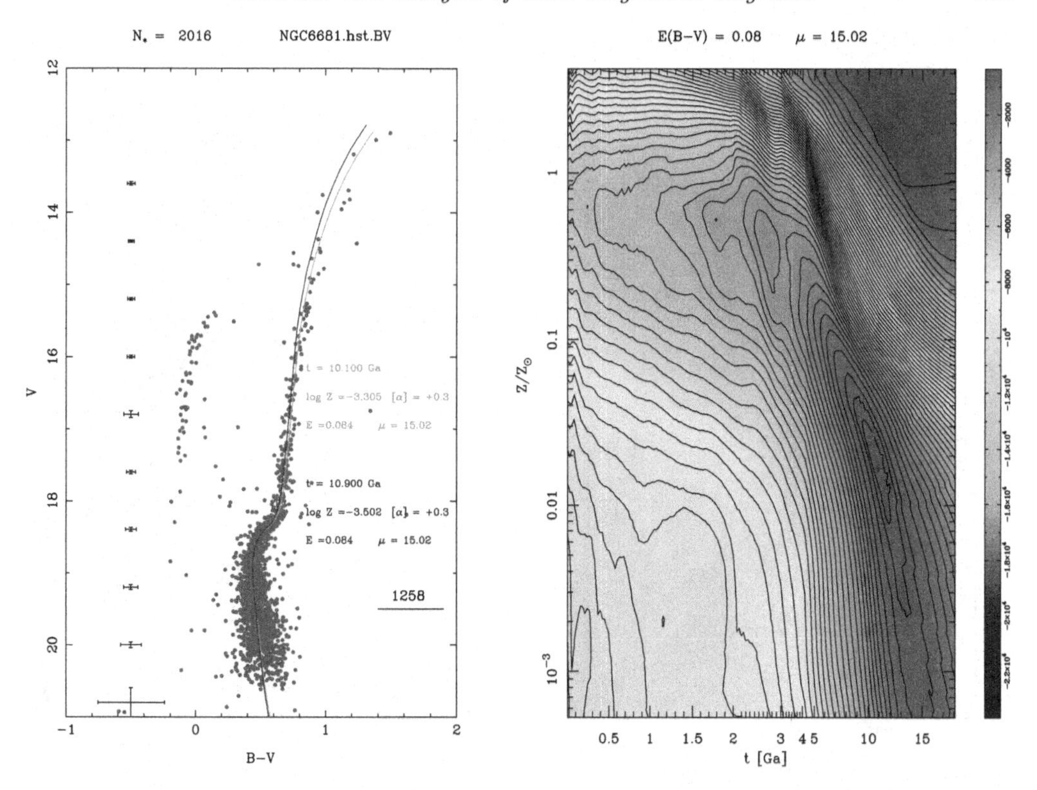

FIG. 6.11. Lifting the geometrical age-metallicity degeneracy. The left panel shows HST observations of NGC 6681, suitably transformed to the UBVRI system. The typical error bars at each magnitude are indicated on the left. Only the 1,258 stars brighter than $V = 19.5$ are used. `BayesGM` yields a posterior PDF on metallicity of $\log Z = -3.502$ with $[\alpha/\mathrm{Fe}] = +0.3$, a color excess $E(B-V) = 0.084$, distance modulus $\mu = 15.02$, and age $t = 10.9$ Gyr (black isochrone, plus sign on the right panel). The solution depends critically on the possible membership of three stars at the top of the RGB. Removing these three stars yields the black isochrone, whereas assuming that they are members produces an age that is older by 0.8 Gyr, the remaining posteriors being the same. In contrast with Figure 6.2 where the locus of maximum likelihood followed a long stretc.h, here the inclusion of the evolutionary speed along each isochrone creates closed contours (right panel), and hence lifts the degeneracy.

The probability contours have the same shape as the ones we saw in the age-metallicity degeneracy (Figure 6.2), except that this time we can lift the degeneracy entirely: only a tiny area has the maximum probability.

The technique is very powerful, yet subject to some interesting systematics. Figure 6.11 also shows the resulting isochrone for the modal age (and metallicity, distance, etc.) when the top three stars are removed. These are very bright stars, and one may wonder whether they belong to the cluster at all. In this case, the modal age shifts by 0.8 Gyr to younger ages (still within the top-most inner probability contour), and the modal posterior metallicity moves -0.2 dex. This is not, however, a sign of degeneracy because the data set is different, it just shows the sensitivity of some parameters to outliers (the distance and reddening are, quite rightly, unaffected by their presence).

Outliers and possible non-members do not always perturb the modes. Figure 6.12 shows the ground-based CMD of NGC 6397 and a sharply peaked marginalized two-dimensional PDF with modes at 16.9 Gyr and $Z = -4.2$ dex. In this case, removing the four brightest stars shifts the modes to 17.5 Gyr and $Z = -4.41$ dex. Taken at face value, this globular cluster appears older than the age of the universe as inferred from a set of completely

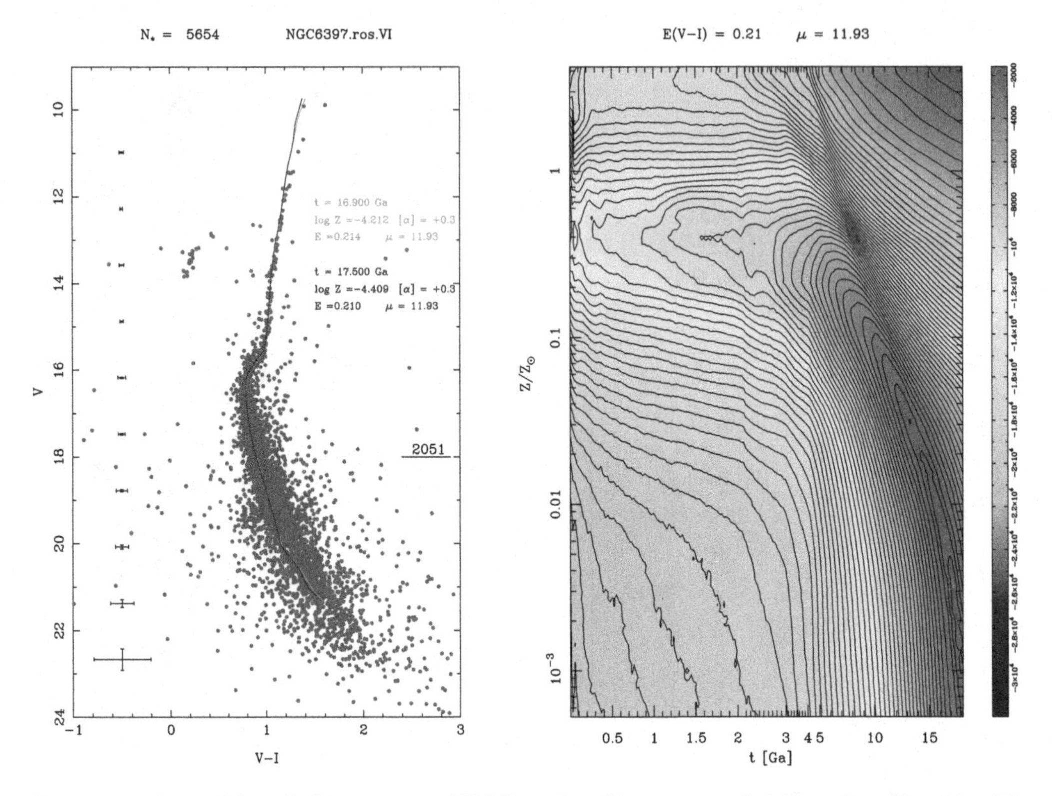

FIG. 6.12. Ground-based observations of NGC 6397 yield 2051 stars brighter than $V = 18$, with average photometric uncertainties as indicated on the left. The age-metallicity degeneracy is lifted (right panel), although in this case the issue of whether the four top-most stars in the RGB belong to the cluster does not change the mode of the posterior PDF of the age: in both instances the ages appear to be larger than 16.5 Gyr, which may point to a limitation in the stellar evolutionary tracks at these low metallicities.

independent measures (CMB fluctuations and the expansion rate), but in fact points to a systematic in this set of tracks at these very low metallicities.[6]

An important caveat is that an increasingly large number of mainly spectroscopic (but also some photometric) observations is revealing that many globular clusters have multiple, not simple stellar populations. The spread and anti correlation of Na and O, for example, may be accounted for in a self-pollution scenario, where the ejecta from the old population led to a composition of the younger one enriched in He, N, Na, Al, but depleted in C, O, Ne, and Mg. However, the Bayesian technique can, in fact, assign individual probabilities of membership to one or another of the population, precisely because their different abundances may lead to differential evolution and hence positions in the CMDs.

6.4 Composite resolved stellar populations

A variable star formation rate and chemical evolution history give rise to a composite stellar population, which is a mixture of stars of different ages and chemical abundances. Disentangling which populations were formed in this scenario is the main goal of the general inverse problem: Given an observed CMD, what are the star formation and chemical enrichment histories?

[6] Because the data set has been modified as some stars has been excluded, note that the evidence changes, and hence the posteriors must be compared properly normalized.

The problem is far from trivial, in part due to the apparent age-metallicity degeneracy, and was limited for a long time to a qualitative comparison between the observations and synthetic CMDs with prescribed SFR and $Z(t)$ histories. Aparicio *et al.* (1990) and Tosi *et al.* (1991) were the early pioneers to attempt the statistical inversion of the CMDs based on the comparison of number counts in suitably-defined areas of the CMD and were a step beyond a qualitative analysis. There soon were many other attempts (see, e.g., Gallart *et al.*, 2005, for a review) in this direct approach, including Ng *et al.* (2002) who used an optimization technique with a genetic algorithm. The thorny issue, as we will see, is to compare properly simulated CMDs with observations (Section 6.6).

Dolphin (1997) (later refined and applied in Dolphin (2002)'s `MATCH` code at americano.dolphinsim.com/match), Olsen (1999), and Harris and Zaritsky (2001) (`StarFish` code www.noao.edu/staff/jharris/SFH) proposed to decompose a generic CMD into a linear combination of "elemental" CMDs produced by coeval populations of well-defined ages and metallicities. This is also the method used by Makarov and Makarova (2004) with their `StarProbe` code. The observed CMD is thus posited to be produced by the linear combination of "partial" CMDs as

$$N(i,j) = \sum_k R_k \, n(i,j)_k, \tag{6.25}$$

where $N(i,j)$ is the number of stars in the bin (i,j) of the observed CMD, made up of the weighted sum of k partial CMDs with counts $n(i,j)_k$ in that bin. If the partial CMDs were computed for a nominal $SFR = 1\mathrm{M}_\odot\,\mathrm{yr}^{-1}$, then the weights R_k provide the star-formation rate contributed by the k-th partial CMD. One can add a foreground CMD to model any underlying contamination, completeness, etc.

Different statistics have been used to infer the values of R_k for the set of partial CMDs, although none of them is entirely correct (cf. Section 6.6). Imposing the non-negativity constraint that $R_k \geq 0$ $(\forall k)$ allows one to determine them through iterative or steepest descent methods (in some cases, claims of a unique solution have been made; see Dolphin, 2002). For a grid of, say, 10×10 CMDs spanning 10 ages and 10 metallicities, this is a problem of finding the absolute minimum in a parameter space with $K = 100$ dimensions, a non-trivial task given the likely presence of many secondary minima. For this reason, the parameter space can be efficiently explored using genetic algorithms to find the absolute maximum corresponding to the best fit (Ng *et al.*, 2002; Aparicio and Hidalgo, 2009), as many secondary minima do exist in this highly dimensional optimization problem. Their code, `IAC-POP`, is available at www.iac.es/galeria/aaj/iac-pop_eng.htm.

Tolstoy and Saha (1996) pioneered a Bayesian formulation of the problem by pondering on the method to compare data sets drawn from simulations with the actual observed CMD, while Cignoni and Shore (2006) used the Richardson-Lucy technique to deconvolve the observed CMD in order to produce a "reconstructed" CMD which can then be compared with simulations.

In fact, the proper way to formulate the problem is to realize that this is, in statistical parlance, an inverse problem that is *ill-defined* and may have multiple solutions (Craig and Brown, 1986). Yet claims have been made that unique solutions can nevertheless be found, not only to the functionals defining the star formation and chemical enrichment histories, but also the shape of the IMF, the mass ratio distribution of the unresolved binaries, etc. (Vergely *et al.*, 2002; Wilson and Hurley, 2003). To some extent, this may perhaps be true, but as in all ill-posed this comes at the prize of a compromise between accuracy/smoothness and resolution. To see this, write the probability that a given star with data set $\vec{\mathbf{D}}_i$ comes from a star-formation episode at age t when the star-formation rate was $SFR(t)$ and the metallicity $Z(t)$:

$$\mathcal{P}\left[SFR(t), Z(t)|\vec{\mathbf{D}}_i\right] = \pi(t)\,\pi(Z) \int_{t_0}^{t_1}\int_{Z_0}^{Z_1} SFR(t)\,\mathcal{L}_{phys}\big(\vec{\mathbf{D}}_i|\vec{\vartheta}, Z(t)\big)\,dt\,dZ, \tag{6.26}$$

where the limits of the integrals come from our prior knowledge $\pi(t)$ that the star must have an age between these limits (could be a least informative prior between 0 to 15 Gyr), and $\pi(Z)$ describes our prior probability on metallicity (could be flat between −5 and 1, or limited to some range if we have previous measures of Z for these stars). For an ensemble of N_* stars the combined probability that their full CMD arises from the episodes of star formation described by $SFR(t)$ and chemical enrichment given by $Z(t)$ becomes

$$\mathcal{P}_{CMD}\left[SFR(t), Z(t)|\vec{\mathbf{D}}\right] = \prod_{i=1}^{N_*} \mathcal{P}\left[SFR(t), Z(t)|\vec{\mathbf{D}}_i\right]. \quad (6.27)$$

Clearly, given some precision in the photometric data, the fine-grained details of the functions $SFR(t)$ and $Z(t)$ are impossible to constrain, hence the infinity of possible solutions, and the ill-posed nature of the problem. To regularize the problem, one seeks not to maximize this probability but rather the log of the probability to which one adds a regularization term, which depends on the type of solution one seeks (Craig and Brown, 1986). The choice varies between terms that penalize variations (i.e., imposes constant solutions to the functions), or terms that penalize gradients or large deviations (i.e. imposes a more or less strict smoothness to the solutions).

The general problem, as far as we are aware, has not yet been solved, and only partial solutions have been explored. For instance, in the case where our prior in metallicity is peaked at some well-defined value Z_* (as is the case in some dwarf galaxies, whose dispersion in metallicity appears to be quite small), one can assume that $\pi(Z) = \delta(Z - Z_*)$, and the problem reduces to solving for only one function, $SFR(t)$. One could impose a parametric form to this function (say, in terms of a piece-wise constant function, or a series of Gaussian bursts) and solve for the parameters of the assumed function. Instead, Hernandez *et al.* (1999) realized that seeking the maximum of the probability function is equivalent to asking that the variation of the function is zero, ensuring that it is an extremum. In turn, using the Euler-Lagrange equation, $\delta\mathcal{P}_{CMD}[SFR(t), Z_*] = 0$ implies a set of N_* coupled differential equations that can easily be solved. In this way a fully non-parametric solution of $SFR(t)$ was built, with no prior information on shape, form, or amplitude. The technique was applied to some nearby dwarf galaxies (Hernandez *et al.*, 2000a), whose HST-based CMDs were well-measured,[7] as well as to the solar neighborhood probed by Hipparcos (Hernandez *et al.*, 2000b) with a wealth of detail in spite of the few stars used. This was also explored independently by Cignoni *et al.* (2006) with an entirely different method, and there are many more results in both clusters and nearby galaxies using a variety of methods, most of which are described in recent review articles (Tolstoy *et al.*, 2009; Cignoni and Tosi, 2010).

Interestingly, even if the general problem of determining *both* functions independently can be solved, we know from chemical evolution theory that there must be some coupling: an enhanced episode of star formation will lead to an enrichment on some timescale that depends on the element and on the details of the nucleosynthetic yields, among other things. Yuk and Lee (2007) propose to compute in a self-consistent way the chemical enrichment history while inferring at the same time the star-formation rate history. However, this involves imposing some prior information, and it is far from clear if, for instance, a closed box model or the assumed recycling are correct. At some point, one can imagine the infall of unenriched gas, or else the accretion of a satellite bringing enriched material. There is ample scope for progress in this field.

[7] Dolphin (2002) rightly noted that some zero point offsets that were adopted were wrong, shifting the solutions by a small amount. Much work remains to be done on systematics.

6.5 Unresolved populations and pixel CMDs

The extreme case of composite populations is reached when their stars cannot be resolved. This is the case, for instance, when we are analyzing the integrated spectra of galaxies or clusters: we have access only to luminosity-weighted estimates of the quantities of interest (age, Z, etc.). Given the shape of the luminosity function, where a handful of bright stars can dominate the flux of the population (which in fact is dominated in number and mass by the less massive stars), one can easily see the major biases inherent in these techniques.

An intermediate case, which is essential to understand these stellar populations, is the one in the pixels of images of nearby galaxies. There we no longer have to deal with billions of stars (integrated spectra) but rather some $10^2 \cdots 10^4$ stars or so, depending of course on the distance and the size of the pixels. A formalism was proposed Renzini (1998) and observations were pioneered by Bothun (1986) with ground-based observations. Abraham *et al.* (1999) discussed, in a forward modeling approach, the interpretation of the four-band images of galaxies in the Hubble Deep field, and the limitations produced by the extinction-color-metallicity degeneracy. Not limited by the seeing, HST-based studies have dominated the field and are becoming an essential tool for understanding galaxy evolution (e.g., Conti *et al.*, 2003; Eskridge *et al.*, 2003; Kassin *et al.*, 2003; Lanyon-Foster *et al.*, 2007; Lee *et al.*, 2011).

A proper formulation of the inverse problem is still lacking, but the forward modeling techniques -which have so far been used- assume a full sampling of the underlyng stellar populations. This may be appropriate for integrated properties, but not in pixels, where the number of stars, while important, is not large enough to ensure the statistical convergence in the properties. This is also the case in the régime where the star formation rate is low, and, in general, in places where the number of stars is small enough to create stochastic variations in the properties such as luminosities and colors. The importance of these fluctuations is essential also to assess whether the IMF is universal: could the stochastic variations be consistent with samples drawn from the same IMF?

This stochasticity is equivalent to a lack of convergence in the properties and can be quantified in a simple way, either assuming quasi-Poisson counts (e.g., Cerviño *et al.*, 2002; Cerviño and Valls-Gabaud, 2003), and through the concept of the lowest luminosity limit, a limit ensuring that statistical fluctuations become unimportant (Cerviño and Luridiana, 2006; Cerviño and Valls-Gabaud, 2009). Another approach, more costly in CPU time, is to create Monte Carlo samples. For example, Popescu and Hanson (2009) proposed a code, `MASSCLEAN` (available at www.physics.uc.edu/~bogdan/massclean.html) to carry out multi-color simulations that do not assume a full sampling of the IMF, and confirm by and large the analytical predictions in the quasi-Poisson régime discussed earlier. This approach is likely to change entirely the interpretation of the integrated properties of stellar clusters (Popescu and Hanson, 2010a,b; Popescu *et al.*, 2012). In a similar way, new tools have been put forward to produce these stochastic variations in stellar populations. da Silva *et al.* (2012) present a code, `SLUG` (sites.google.com/site/runslug) that is also likely to revisit many results obtained thus far in galaxy evolution.

6.6 Best-fit solutions and uncertainties

An issue that is seldom addressed, if at all, is whether the best-fit solution found (by any method) is also a good fit. There is no guarantee whatsoever that the best straight-line fit to a parabola is a good fit, even if it is the best possible within the assumption of a straight line. In some cases the problem is not even tackled. Sometimes a χ^2 criterion is used (which is not appropriate, since number counts in any CMD cell or bin follow Poisson statistics), or a comparison with a *single* model realization is performed (for instance, analyzing the residuals in a Hess diagram, which is just a binned version of the CMD, as first used by Hess (1924) in a different context). None of these approaches is satisfactory: one has to deal *both* with Poisson counts and with the intrinsic variability in the *model* predictions (as different realizations of the model will invariably yield different results and hence residuals).

Some statistics that have been used in the context of comparing model CMDs (with $\{m_i\}$ stars) with observed CMDs (with $\{s_i\}$ stars) in a selection of B bins or boxes include the following:

(i) Pearson's χ^2_P. This takes into account the variability of the model counts $\{m_i\}$ only:

$$\chi^2_P = \sum_i^B \frac{(s_i - m_i)^2}{m_i}. \tag{6.28}$$

(ii) Modified Neyman's χ^2_N which adopts the form

$$\chi^2_N = \sum_i^B \frac{(s_i - m_i)^2}{\max(s_i, 1)}, \tag{6.29}$$

but only encapsulates the variability of the observed counts $\{s_i\}$.

(iii) Dispersion. Kerber *et al.* (2001) minimize the dispersion defined as

$$\mathcal{S}^2 = \sum_i^B (m_i - s_i)^2. \tag{6.30}$$

While clearly it has the correct behavior, it does not take into account the Poisson distribution of both model and observed counts. Could the minimum $\mathcal{S}^2$ value reached be different should a different realization of the model $\{m_i\}$ be used for the comparison?

(iv) Percentile position. Kerber *et al.* (2001) also use the percentile position of $\mid s_i - m_i \mid$ within the distribution of $\mid m_i = m_{ij} \mid$ with $j = 1, \cdots N_{sim}$ and m_{ij} being Poisson realizations with parameter m_i in each bin i. Then the statistic

$$\text{pss} = \sum_i^B (1 - p_i) \tag{6.31}$$

is minimized. This tackles the part of the intrinsic variability of the model outcomes, but not the one from the observations.

(v) χ^2 statistic for Poisson variables. Both Ng (1998) and Mighell (1999) point out that the number counts in the bins the CMD has been divided into follow Poisson statistics, not Gaussian ones (unless we are in the large numbers limit, which will not be the case in sparsely populated bins). The statistic proposed is

$$\chi^2_\gamma = 2 \sum_i \frac{[s_i + \min(n_i, 1) - m_i]^2}{s_i + 1}, \tag{6.32}$$

and many codes have implemented this to assess the goodness (or otherwise) of their fits. However, comparing Poisson counts is tricky, as all high energy physicists know. While the sum of Poisson variates is also Poisson distributed (see, e.g., Cerviño and Valls-Gabaud, 2003, for some consequences in the context of stellar populations), the difference of Poisson counts, such as $m_i - s_i$ which one could naively use, do not follow Poisson statistics but are distributed following a Skellam distribution (the fact that the variate may become negative is a clear hint).

(vi) Poisson likelihood ratio. Dolphin (2002) correctly argued that the proper analogy with the χ^2 ratio (which only applies to Gaussian statistics) for Poisson variates is the ratio

$$-2\ln \mathrm{PLR} = 2\sum_i \left(m_i - s_i + s_i \ln\left(\frac{s_i}{m_i}\right)\right), \tag{6.33}$$

as proposed by Baker and Cousins (1983). However, using a likelihood ratio has some constraints that makes this quantity unsuitable for seeking a proper comparison (see the following).

The likelihood of a model is obviously a *relative* probability. If we have two models A and B giving likelihoods $\mathcal{L}_A$ and $\mathcal{L}_B$, model A is $\mathcal{L}_A/\mathcal{L}_B$ times as likely as model B. This is also known as the bookmakers' odds (Syer and Saha, 1994). To compare two likelihoods or use a likelihood ratio, however, two conditions are essential (Protassov *et al.*, 2002), but are too often overlooked:

(i) The models must be nested.
(ii) The values of the parameters must not reach zero.

The first condition implies that, for example, one cannot use a likelihood ratio for comparing a fit using a polynomial and another one using an exponential function. It can be used to compare the fits obtained by a polynomial of degree k with another fit using a polynomial of degree $j \neq k$. The second condition, of strict positivity, implies that one cannot use a likelihood ratio when decomposing a CMD into a linear combination of partial CMDs, as quite a few odd partial CMDs are unlikely to contribute at all (say, young populations to a globular cluster). Necessarily some $R_k = 0$ will happen in Equation 6.25 and the likelihood ratio cannot be applied.

At any rate, χ^2-like statistics such as the ones discussed depend on the size of the bins the CMD is divided into, and care must be taken when assessing their significance in comparison with CMDs with different cell/bin sizes. In addition, one has to take into account the *intrinsic* variability of the model predictions, so a comparison between a *single* realization and an observed CMD makes very little sense. Obviously, one can only take into account a Poisson dispersion in the observed counts, but models have not only this intrinsic variability but also one associated with them, even for a fixed set of values for the parameters.

A nearly size-independent statistic was suggested by Bell *et al.* (2008) where the rms deviation of the data with respect to the model is minimized, and takes explicitly into account the Poisson variability of the model. For a set of B bins, one forms the *distribution* of

$$\sigma/\mathrm{total} = \sqrt{<\sigma^2>}\left[\frac{1}{B}\sum_i^B s_i\right]^{-1}, \tag{6.34}$$

where

$$<\sigma^2> = \frac{1}{B}\left[\sum_i^B (s_i - m_i)^2 - \sum_i^B (m_i' - m_i)^2\right]. \tag{6.35}$$

This statistic takes into account the variability of the model, as $\{m_i'\}$ are Poisson realizations of the model with expectation values $\{m_i\}$, but *not* the uncertainties in the observed counts $\{s_i\}$. It is, in fact, *designed* to detect the fluctuations in the observed number counts with respect to the models.

Should we know the distribution function of the model, we could apply the standard statistical tools to infer the proper credibility intervals. This is obviously not the case in CMDs, but we can generate samples from the model (a sample of infinite size would be as good as the model). Different samples (different realizations) are likely to populate regions of the CMD that may not coincide with the observed ones; hence, we need to smooth out the model and data for a proper comparison. The simplest way of doing this is binning, with the underlying assumption that the model distribution function is constant within the bin (hence, small bins are preferred, but they should be large enough that bins contain both model and observed stars as much as possible).

If we divide the CMD into B cells, of arbitrary sizes and shapes, each containing $\{s_i\}$ observed stars (with a total of $S = \sum_i^B s_i$ stars), and $\{m_i\}$ model stars (with a total of $M = \sum_i^B m_i$ stars), each cell has a probability w_i of having the appropriate number of stars (model or observed) constant within the bin. The probability distribution function for the bin occupancies, multivariate distribution of counts, will be given by a multinomial distribution

$$P(s_i, m_i \mid w_i) = M!\, S! \prod_{i=1}^{B} \frac{w_i^{m_i+s_i}}{m_i! s_i!}, \tag{6.36}$$

where the weigths $\{w_i\}$ of the distribution function are unknown. We only have the constraint, given by the normalization, that $\sum_i^B w_i = 1$. We can treat the weights as nuisance parameters and marginalize over them using the identity

$$\left(\prod_{i=1}^{B} \int w_i^{n_i}\, dw_i\right) \delta\left(\sum w_j - 1\right) = \frac{1}{(N+B-1)!} \prod_{i=1}^{B} n_i!, \tag{6.37}$$

and we get

$$P(s_i, m_i) = \frac{M!\, S!\, (B-1)!}{(M+S+B-1)!} \mathrm{ZAp}, \tag{6.38}$$

which is equal to $P(s_i \mid m_i)\, P(m_i)$. And likewise with $P(m_i)$, and so Bayes's theorem gives us

$$P(s_i \mid m_i) = \frac{S!\, (M+B-1)!}{(M+S+B-1)!} \mathrm{ZAp};. \tag{6.39}$$

For a fixed number of bins B, model stars M, and observed stars S, the first term is a constant and hence

$$\mathrm{Prob} \propto W = \mathrm{ZAp}. \tag{6.40}$$

This statistic was first proposed by Saha (1998) and has been widely used, for example, in comparisons of N-body simulations with discrete data sets (Sevenster *et al.*, 1999; Beaulieu *et al.*, 2000) or indeed in inversions of CMDs (Hernandez *et al.*, 2000a,b; Kerber *et al.*, 2002), correlations between the SFR history and the glaciation epochs (de La Fuente Marcos and de La Fuente Marcos, 2004), and in setting constraints on the properties of clusters (e.g., Rengel *et al.*, 2002; Kerber and Santiago, 2005). Contrary to some baseless claims (Dolphin, 2002), it *does* allow a proper comparison between data and models. Note that it takes model counts and observed counts on the same footing. If M is arbitrarily large, the bins can be made small enough to contain one observed star at most, so that the probability goes as $\prod_k(1+m_k)$ where k is the running index for boxes with $s_k = 1$, and with $m_k \gg 1$ we get the same result as in the continuous distribution case.

More properly in the context of multi-dimensional Poisson counts, if M and S are not fixed but are the expectation values of the totals when s_i and m_i are drawn from a Poisson process, we get a modified W statistic (Saha, 2003) as

$$P(s_i|m_i) = \frac{e^{-M}\, M^M\, e^{-S}\, S^S\, (B-1)!}{(M+S+B-1)!} \mathrm{ZAp}. \tag{6.41}$$

It is clear that, normalizations aside (which can be fixed if indeed B, M, and S are kept fixed), Saha's W statistic (Equation 6.41) can be used to perform either

- *Parameter fitting*: Just compute the distribution of W for *a given data set* and different model parameters. That is, fix $\{s_i\}$, vary model parameters (hence $\{m_i\}$), then read off parameter estimates and confidence intervals for the parameters.
- *Goodness of fit*: Here we want the distribution of W for *fixed model parameters* and various simulated data sets. Fix model parameters (so $\{m_i\}$ are fixed), then vary simulated $\{s_i = m_i'\}$. Then compare the *distribution* of W with the *distribution* obtained from the actual data. The extent to which both distributions (actually, samples of W) can be drawn from the same underlying (and unknown) distribution function gives a proper measure of the goodness of the fit.

It is therefore the statistic of choice to be used in the context of CMD models.

REFERENCES

Abraham, R. G., Ellis, R. S., Fabian, A. C., Tanvir, N. R., and Glazebrook, K. 1999. The star formation history of the Hubble sequence: spatially resolved color distributions of intermediate-redshift galaxies in the Hubble Deep Field. *MNRAS*, **303**(Mar.), 641–658.

An, D., Terndrup, D. M., Pinsonneault, M. H., Paulson, D. B., Hanson, R. B., and Stauffer, J. R. 2007. The distances to open clusters from main-sequence fitting. III. Improved accuracy with empirically calibrated isochrones. *ApJ*, **655**(Jan.), 233–260.

Aparicio, A. and Hidalgo, S. L. 2009. IAC-pop: finding the star formation history of resolved galaxies. *AJ*, **138**(Aug.), 558–567.

Aparicio, A., Bertelli, G., Chiosi, C., and Garcia-Pelayo, J. M. 1990. CCD UBVR photometry of the old rich open cluster King 2 – comparison with theoretical models. *A&A*, **240**(Dec.), 262–288.

Bailer-Jones, C. A. L. 2011. Bayesian inference of stellar parameters and interstellar extinction using parallaxes and multiband photometry. *MNRAS*, **411**(Feb.), 435–452.

Baker S. and Cousins, RD. 1984. Clarification of the use of chi-square and likelihood functions in fits to histograms. *Nucl. Instrum. Methods*, **221**(Feb.), 437–442.

Bayes, T. 1763. An essay toward solving a problem in the doctrine of chances. *Phil. Trans. Roy. Soc.*, **53**, 370–418.

Beaulieu, S. F., Freeman, K. C., Kalnajs, A. J., Saha, P., and Zhao, H. 2000. Dynamics of the Galactic Bulge Using Planetary Nebulae. *AJ*, **120**(Aug.), 855–871.

Becker, S. A. and Mathews, G. J. 1983. A comparison between observed and theoretical H-R diagrams for the young LMC star cluster NGC 1866. *ApJ*, **270**(July), 155–168.

Bell, E. F., and 17 colleagues. 2008. The accretion origin of the Milky Way's stellar halo. *ApJ*, **680**(June), 295–311.

Bothun, G. D. 1986. Two-color CCD mapping of the luminous Type I irregular galaxy NGC 4449. *AJ*, **91**(Mar.), 507–516.

Breddels, M. A., and 23 colleagues. 2010. Distance determination for RAVE stars using stellar models. *A&A*, **511**(Feb.), A90.

Brott, I., and 8 colleagues 2011. Rotating massive main-sequence stars. I. Grids of evolutionary models and isochrones. *A&A*, **530**(June), A115.

Bruntt, H. and Southworth, J. 2008. A new level of photometric precision: WIRE observations of eclipsing binary stars. *Journal of Physics Conference Series*, **118**(Oct.), 012012.

Burnett, B. and Binney, J. 2010. Stellar distances from spectroscopic observations: a new technique. *MNRAS*, **407**(Sept.), 339–354.

Casagrande, L., Ramírez, I., Meléndez, J., Bessell, M., and Asplund, M. 2010. An absolutely calibrated T_{eff} scale from the infrared flux method. Dwarfs and subgiants. *A&A*, **512**(Mar.), A54.

Casagrande, L., Schönrich, R., Asplund, M., Cassisi, S., Ramírez, I., Meléndez, J., Bensby, T., and Feltzing, S. 2011. New constraints on the chemical evolution of the solar neighbourhood and Galactic disc(s). Improved astrophysical parameters for the Geneva-Copenhagen Survey. *A&A*, **530**(June), A138.

Cerviño, M. and Luridiana, V. 2006. Confidence limits of evolutionary synthesis models. IV. Moving forward to a probabilistic formulation. *A&A*, **451**(May), 475–498.

Cerviño, M. and Valls-Gabaud, D. 2003. On biases in the predictions of stellar population synthesis models. *MNRAS*, **338**(Jan.), 481–496.

Cerviño, M. and Valls-Gabaud, D. 2009. On the initial cluster mass distribution inferred from synthesis models. *Ap&SS*, **324**(Dec.), 91–94.

Cerviño, M., Valls-Gabaud, D., Luridiana, V., and Mas-Hesse, J. M. 2002. Confidence levels of evolutionary synthesis models. II. On sampling and Poissonian fluctuations. *A&A*, **381**(Jan.), 51–64.

Chanamé, J. and Ramírez, I. 2012. Toward precise ages for single stars in the field. Gyrochronology constraints at several Gyr using wide binaries. I. Ages for initial sample. *ApJ*, **746**(Feb.), 102.

Charlier C.V.L. 1889. *Publ. Astronomischen Gesell*, **19**, 1.

Cignoni, M. and Tosi, M. 2010. Star formation histories of dwarf galaxies from the color-magnitude diagrams of their resolved stellar populations. *Advances in Astronomy*, **2010**.

Cignoni, M. and Shore, S. N. 2006. Restoring color-magnitude diagrams with the Richardson-Lucy algorithm. *A&A*, **454**(Aug.), 511–516.

Cignoni, M., Degl'Innocenti, S., Prada Moroni, P. G., and Shore, S. N. 2006. Recovering the star formation rate in the solar neighborhood. *A&A*, **459**(Dec.), 783–796.

Collins, G. W., II and Smith, R. C. 1985. The photometric effect of rotation in the A stars. *MNRAS*, **213**(Apr.), 519–552.

Collins, G. W., II and Sonneborn, G. H. 1977. Some effects of rotation on the spectra of upper-main-sequence stars. *ApJS*, **34**(May), 41–94.

Conti, A., and 8 colleagues. 2003. The star formation history of galaxies measured from individual pixels. I. The Hubble Deep Field North. *AJ*, **126**(Nov.), 2330–2345.

Cousin R. 1995. *Am. J. Phys.*, **63**, 398.

Craig, I. J. D. and Brown, J. C. 1986. *Inverse Problems in Astronomy: A Guide to Inversion Strategies for Remotely Sensed Data.* Bristol, England: Adam Hilger.

da Silva, L., and 8 colleagues. 2006. Basic physical parameters of a selected sample of evolved stars. *A&A*, **458**(Nov.), 609–623.

da Silva, R. L., Fumagalli, M., and Krumholz, M. 2012. SLUG–stochastically lighting up galaxies. I. Methods and validating tests. *ApJ*, **745**(Feb.), 145.

Dale A. I. 1982. *Arch. Hist. Exact Sci.*, **27**, 23.

de La Fuente Marcos, R. and de La Fuente Marcos, C. 2004. On the correlation between the recent star formation rate in the Solar Neighbourhood and the glaciation period record on Earth. *New A*, **10**(Nov.), 53–66.

Devorkin, D. H. 2000. *Henry Norris Russell: Dean of American Astronomers.* Princeton, N.J.: Princeton University Press.

Dolphin, A. E. 1997. A new method to determine star formation histories of nearby galaxies. *New A*, **2**(Nov.), 397–409.

Dolphin, A. E. 2002. Numerical methods of star formation history measurement and applications to seven dwarf spheroidals. *MNRAS*, **332**(May), 91–108.

Dose, V. 2003. Bayesian inference in physics: case studies. *Rep. Pro. Phys.*, **66**(Sept.), 1421–1461.

Duerbeck, H. W. ed. 2006. *Organizations and Strategies in Astronomy, Vol. 7.* Berlin: Springer Verlag.

Eskridge, P. B., and 10 colleagues. 2003. Ultraviolet-optical pixel maps of face-on spiral galaxies: clues for dynamics and star formation histories. *ApJ*, **586**(Apr.), 923–938.

Flannery, B. P. and Johnson, B. C. 1982. A statistical method for determining ages of globular clusters by fitting isochrones. *ApJ*, **263**(Dec.), 166–186.

Gallart, C., Zoccali, M., and Aparicio, A. 2005. The adequacy of stellar evolution models for the interpretation of the color-magnitude diagrams of resolved stellar populations. *ARA&A*, **43**(Sept.), 387–434.

Gennaro, M., Prada Moroni, P. G., and Tognelli, E. 2012. Testing pre-main-sequence models: the power of a Bayesian approach. *MNRAS*, **420**(Feb.), 986–1018.

Gleissberg W. 1940. *Pub. Istanbul Obs.*, **13**, 2.

Gregory, P. C. 2005. *Bayesian Logical Data Analysis for the Physical Sciences: A Comparative Approach with "Mathematica" Support.* Cambridge, UK: Cambridge University Press.

Haffner, H. and Heckmann, O. 1937. Das farben-helligkeits-diagramm der praesepe auf grund neuer beobachtungen. *Veroeffentlichungen der Universitaets-Sternwarte zu Goettingen*, **4**, 77–95.

Harmanec, P. and Prša, A. 2011. Call to adopt a nominal set of astrophysical parameters and constants to improve the accuracy of fundamental physical properties of stars. *PASP*, **123**(Aug.), 976–980.

Harris, J. and Zaritsky, D. 2001. A method for determining the star formation history of a mixed stellar population. *ApJS*, **136**(Sept.), 25–40.

Hearnshaw, J. B. 1986. *The Analysis of Starlight: One Hundred and Fifty Years of Astronomical Spectroscopy. Cambridge: Cambridge University Press.*

Hermann D. B. 1994. *Ejnar Hertzsprung: Pionier der Sternforschung* Berlin: Springer-Verlag.

Hernandez, X. and Valls-Gabaud, D. 2008. A robust statistical estimation of the basic parameters of single stellar populations – I. Method. *MNRAS*, **383**(Feb.), 1603–1618.

Hernandez, X., Gilmore, G., and Valls-Gabaud, D. 2000a. Non-parametric star formation histories for four dwarf spheroidal galaxies of the Local Group. *MNRAS*, **317**(Oct.), 831–842.

Hernandez, X., Valls-Gabaud, D., and Gilmore, G. 1999. Deriving star formation histories: inverting Hertzsprung-Russell diagrams through a variational calculus maximum likelihood method. *MNRAS*, **304**(Apr.), 705–719.

Hernandez, X., Valls-Gabaud, D., and Gilmore, G. 2000b. The recent star formation history of the Hipparcos solar neighbourhood. *MNRAS*, **316**(Aug.), 605–612.

Hertzsprung, E. 1911. Ueber die verwendung photographischer effektiver wellenlaengen zur bestimmung von farbenaequivalenten. *POPot*, **63**.

Hess, R. 1924. *Probleme der Astronomie. Festschrift fur Hugo v. Seeliger.* Berlin: Springer Verlag.

Hobson, M. P., Jaffe, A. H., Liddle, A. R., Mukeherjee, P., and Parkinson, D. 2010. *Bayesian Methods in Cosmology.* Cambridge: Cambridge University Press.

Hurley, J. and Tout, C. A. 1998. The binary second sequence in cluster color-magnitude diagrams. *MNRAS*, **300**(Nov.), 977–980.

Javiel, S. C., Santiago, B. X., and Kerber, L. O. 2005. Constraints on the star formation history of the Large Magellanic Cloud. *A&A*, **431**(Feb.), 73–85.

Jeffery, E. J., von Hippel, T., DeGennaro, S., van Dyk, D. A., Stein, N., and Jefferys, W. H. 2011. The white dwarf age of NGC 2477. *ApJ*, **730**(Mar.), 35.

Jeffery, E. J., von Hippel, T., Jefferys, W. H., Winget, D. E., Stein, N., and De Gennaro, S. 2007. New techniques to determine ages of open clusters ussing white dwarfs. *ApJ*, **658**(Mar.), 391–395.

Jørgensen, B. R. and Lindegren, L. 2005. Determination of stellar ages from isochrones: Bayesian estimation versus isochrone fitting. *A&A*, **436**(June), 127–143.

Kassin, S. A., Frogel, J. A., Pogge, R. W., Tiede, G. P., and Sellgren, K. 2003. Stellar populations in NGC 4038/39 (the Antennae): exploring a galaxy merger pixel by pixel. *AJ*, **126**(Sept.), 1276–1285.

Kerber, L. O. and Santiago, B. X. 2005. Physical parameters of rich LMC clusters from modeling of deep HST color-magnitude diagrams. *A&A*, **435**(May), 77–93.

Kerber, L. O., Girardi, L., Rubele, S., and Cioni, M.-R. 2009. Recovery of the star formation history of the LMC from the VISTA survey of the Magellanic system. *A&A*, **499**(June), 697–710.

Kerber, L. O., Javiel, S. C., and Santiago, B. X. 2001. Constraints on thick disc and halo parameters from HST photometry of field stars in the Galaxy. *A&A*, **365**(Jan.), 424–430.

Kerber, L. O., Santiago, B. X., Castro, R., and Valls-Gabaud, D. 2002. Analysis of color-magnitude diagrams of rich LMC clusters: NGC 1831. *A&A*, **390**(July), 121–132.

Lachaume, R., Dominik, C., Lanz, T., and Habing, H. J. 1999. Age determinations of main-sequence stars: combining different methods. *A&A*, **348**(Aug.), 897–909.

Lanyon-Foster, M. M., Conselice, C. J., and Merrifield, M. R. 2007. Structure through color: a pixel approach toward understanding galaxies. *MNRAS*, **380**(Sept.), 571–584.

Laplace P.S.. 1812. *Théorie analytique des probabilités.* Paris: Courcier.

Lastennet, E. and Valls-Gabaud, D. 1996. A systematic study of the effects of unresolved binaries and rotation in open clusters. *The Origins, Evolution, and Destinies of Binary Stars in Clusters*, **90**, 464.

Lastennet, E. and Valls-Gabaud, D. 2002. Detached double-lined eclipsing binaries as critical tests of stellar evolution. Age and metallicity determinations from the HR diagram. *A&A*, **396**(Dec.), 551–580.

Lastennet, E., Fernandes, J., Valls-Gabaud, D., and Oblak, E. 2003. Disentangling discrepancies between stellar evolution theory and sub-solar mass stars. The influence of the mixing length parameter for the UV Psc binary. *A&A*, **409**(Oct.), 611–618.

Lastennet, E., Valls-Gabaud, D., Lejeune, T., and Oblak, E. 1999. Consequences of HIPPARCOS parallaxes for stellar evolutionary models. Three Hyades binaries: V 818 Tauri, 51 Tauri, and theta (2) Tauri. *A&A*, **349**(Sept.), 485–494.

Lebreton, Y. 2000. Stellar structure and evolution: deductions from Hipparcos. *ARA&A*, **38**, 35–77.

Lee, J. H., Kim, S. C., Park, H. S., Ree, C. H., Kyeong, J., and Chung, J. 2011. Hubble Space Telescope pixel analysis of the interacting face-on spiral galaxy NGC 5194 (M51A). *ApJ*, **740**(Oct.), 42.

Ludwig, H.-G., Freytag, B., and Steffen, M. 1999. A calibration of the mixing-length for solar-type stars based on hydrodynamical simulations. I. Methodical aspects and results for solar metallicity. *A&A*, **346**(June), 111–124.

Luri, X., Torra, J., and Figueras, F. 1992. The proximity parameter. *A&A*, **259**(June), 382–385.

Maeder, A. 1974. Stellar evolution near the main sequence: on some systematic differences between cluster sequences and model calculations. *A&A*, **32**(May), 177–190.

Maeder, A. and Meynet, G. 2012. Rotating massive stars: from first stars to gamma ray bursts. *Reviews of Modern Physics*, **84**(Jan.), 25–63.

Maeder, A. and Peytremann, E. 1970. Stellar rotation. *A&A*, **7**(July), 120.

Makarov, D. I. and Makarova, L. N. 2004. Modeling the star population of resolved galaxies. *Astrophysics*, **47**(Apr.), 229–241.

Malkov, O. Y., Sichevskij, S. G., and Kovaleva, D. A. 2010. Parametrization of single and binary stars. *MNRAS*, **401**(Jan.), 695–704.

Mayne, N. J. and Naylor, T. 2008. Fitting the young main-sequence: distances, ages and age spreads. *MNRAS*, **386**(May), 261–277.

Meissner, F. and Weiss, A. 2006. Global fitting of globular cluster age indicators. *A&A*, **456**(Sept.), 1085–1096.

Meyer-Hofmeister, E. 1969. A theoretical Hertzsprung-Russell-diagram for the star cluster NGC 1866. *A&A*, **2**(June), 143–150.

Mighell, K. J. 1999. Parameter estimation in astronomy with Poisson-distributed data. I. The χ^2_γ statistic. *ApJ*, **518**(June), 380–393.

Monteiro, H., Dias, W. S., and Caetano, T. C. 2010. Fitting isochrones to open cluster photometric data. A new global optimization tool. *A&A*, **516**(June), A2.

Naylor, T. 2009. Are pre-main-sequence stars older than we thought? *MNRAS*, **399**(Oct.), 432–442.

Naylor, T. and Jeffries, R. D. 2006. A maximum-likelihood method for fitting color-magnitude diagrams. *MNRAS*, **373**(Dec.), 1251–1263.

Ng, Y. K. 1998. Stellar population synthesis diagnostics. *A&AS*, **132**(Oct.), 133–143.

Ng, Y. K., Brogt, E., Chiosi, C., and Bertelli, G. 2002. Automatic observation rendering (AMORE). I. On a synthetic stellar population's color-magnitude diagram. *A&A*, **392**(Sept.), 1129–1147.

Nielsen, A. V. 1969. *Centaurus*, **9**, 219.

Nordström, B., and 8 colleagues. 2004. The Geneva-Copenhagen survey of the Solar neighbourhood. Ages, metallicities, and kinematic properties of 14,000 F and G dwarfs. *A&A*, **418**(May), 989–1019.

Olsen, K. A. G. 1999. Star formation histories from Hubble Space Telescope color-magnitude diagrams of six fields of the Large Magellanic Cloud. *AJ*, **117**(May), 2244–2267.

Patenaude, M. 1978. Age determinations of open clusters. *A&A*, **66**(May), 225–239.

Perrin, M.-N., Cayrel de Strobel, G., Cayrel, R., and Hejlesen, P. M. 1977. Fine structure of the H-R diagram for 138 stars in the solar neighbourhood. *A&A*, **54**(Feb.), 779–795.

Protassov, R., van Dyk, D. A., Connors, A., Kashyap, V. L., and Siemiginowska, A. 2002. Statistics, handle with care: detecting multiple model components with the likelihood ratio test. *ApJ*, **571**(May), 545–559.

Pont, F. and Eyer, L. 2004. Isochrone ages for field dwarfs: method and application to the age-metallicity relation. *MNRAS*, **351**(June), 487–504.

Popescu, B. and Hanson, M. M. 2009. MASSCLEAN–Massive Cluster Evolution and Analysis Package: description and tests. *AJ*, **138**(Dec.), 1724–1740.

Popescu, B. and Hanson, M. M. 2010a. MASSCLEANcolors–Mass-dependent integrated colors for stellar clusters derived from 30 million Monte Carlo simulations. *ApJ*, **713**(Apr.), L21–L27.

Popescu, B. and Hanson, M. M. 2010b. MASSCLEANage–stellar cluster ages from integrated colors. *ApJ*, **724**(Nov.), 296–305.

Popescu, B., Hanson, M. M., and Elmegreen, B. G. 2012. Age and mass for 920 Large Magellanic Cloud clusters derived from 100 million Monte Carlo simulations. *ApJ*, **751**(June), 122.

Ramírez, I. and Meléndez, J. 2005. The effective temperature scale of FGK stars. II. T_{eff}:Color:[Fe/H] calibrations. *ApJ*, **626**(June), 465–485.

Reddy, B. E., Tomkin, J., Lambert, D. L., and Allende Prieto, C. 2003. The chemical compositions of Galactic disc F and G dwarfs. *MNRAS*, **340**(Mar.), 304–340.

Rengel, M., Mateu, J., and Bruzual, G. 2002. The determination of the age of globular clusters: a statistical approach. *Extragalactic Star Clusters*, **207**, 716.

Renzini, A. 1998. The stellar populations of pixels and frames. *AJ*, **115**(June), 2459–2465.

Renzini, A. and Buzzoni, A. 1983. Theoretical foundations of evolutionary population synthesis. A progress report. *Mem. Soc. Astron. Italiana*, **54**, 739–745.

Robertson, J. W. 1974. Core-helium stars in young clusters in the Large Magellanic Cloud. *ApJ*, **191**(July), 67–78.

Rosenberg H. 1910. *Astron. Nach.*, **186**, 71.

Rosenberg, H. 1929. Lichtelektrische photometrie. *Handbuch der Astrophysik*, **2**, 380.

Rosenberg, H. O. 1936. Darkening at the limb and color index of an eclipsing variable (u Cephei). *ApJ*, **83**(Mar.), 67.

Russell, H. N. 1912. *Proc. Phil. Soc. Amer.*, **51**, 569.

Russell, H. N. 1914a. Relations between the spectra and other characteristics of the stars. *Popular Astronomy*, **22**(May), 275–294.

Russell, H. N. 1914b. Relations between the spectra and other characteristics of the stars. *Popular Astronomy*, **22**(June), 331–351.

Russell, H. N. 1931. Notes on the constitution of the stars. *MNRAS*, **91**(June), 951–966.

Russell, H. N., Dugan, R. S., and Stewart, J. Q. 1927. Book Review: *Splendour of the Heavens, a Popular Authoritative Astronomy. Popular Astronomy*, 355.

Saha, P. 1998. A method for comparing discrete kinematic data and N-Body simulations. *AJ*, **115**(Mar.), 1206–1211.

Saha, P. 2003. Book Review: *Principles of Data Analysis* / Capella Archive, 2003. *The Observatory*, **123**, 398.

Salaris, M. and Cassisi, S. 2005. Evolution of Stars and Stellar Populations. *New York: Wiley.*

Santos, N. C., Lovis, C., Pace, G., Melendez, J., and Naef, D. 2009. Metallicities for 13 nearby open clusters from high-resolution spectroscopy of dwarf and giant stars. Stellar metallicity, stellar mass, and giant planets. *A&A*, **493**(Jan.), 309–316.

Schaltenbrand, R. A. 1974. Three-color photometry of SA 94 in the RGU system. *A&AS*, **18**(Oct.), 27.

Sevenster, M., Saha, P., Valls-Gabaud, D., and Fux, R. 1999. New constraints on a triaxial model of the Galaxy. *MNRAS*, **307**(Aug.), 584–594.

Shapley, H. 1960. *Source Book in Astronomy, 1900–1950. Cambridge: Harvard University Press.*

Siess, L., Forestini, M., and Dougados, C. 1997. Synthetic Hertzsprung-Russell diagrams of open clusters. *A&A*, **324**(Aug.), 556–565.

Smith, R. W. 1977. Russell and stellar evolution – his "Relations between the spectra and other characteristics of the stars". *Dudley Observatory Reports*, **13**, 9–13.

Soderblom, D. R. 2010. The ages of stars. *ARA&A*, **48**(Sept.), 581–629.

Southworth, J. 2011. Homogeneous studies of transiting extrasolar planets – IV. Thirty systems with space-based light curves. *MNRAS*, **417**(Nov.), 2166–2196.

Stello, D., and 24 colleagues. 2009. Radius determination of Solar-type stars using asteroseismology: what to expect from the Kepler Mission. *ApJ*, **700**(Aug.), 1589–1602.

Strand, K. A. 1968. Ejnar Hertzsprung, 1873–1967. *PASP*, **80**(Feb.), 51.

Syer, D. and Saha, P. 1994. Bookmakers' odds for the sky distribution of gamma-ray bursts. *ApJ*, **427**(June), 714–717.

Takeda, G., Ford, E. B., Sills, A., Rasio, F. A., Fischer, D. A., and Valenti, J. A. 2007. Structure and evolution of nearby stars with planets. II. Physical properties of 1,000 cool stars from the SPOCS Catalog. *ApJS*, **168**(Feb.), 297–318.

Tolstoy, E. and Saha, A. 1996. The interpretation of color-magnitude diagrams through numerical simulation and Bayesian inference. *ApJ*, **462**(May), 672.

Tolstoy, E., Hill, V., and Tosi, M. 2009. Star-formation histories, abundances, and kinematics of dwarf galaxies in the Local Group. *ARA&A*, **47**(Sept.), 371–425.

Torres, G. 2010. On the use of empirical bolometric corrections for stars. *AJ*, **140**(Nov.), 1158–1162.

Torres, G., Andersen, J., and Giménez, A. 2010. Accurate masses and radii of normal stars: modern results and applications. *A&A Rev.*, **18**(Feb.), 67–126.

Tosi, M., Greggio, L., Marconi, G., and Focardi, P. 1991. Star formation in dwarf irregular galaxies – Sextans B. *AJ*, **102**(Sept.), 951–974.

Trotta, R. 2008. Bayes in the sky: Bayesian inference and model selection in cosmology. *Contemporary Physics*, **49**(Mar.), 71–104.

Valenti, J. A. and Fischer, D. A. 2005. Spectroscopic properties of cool stars (SPOCS). I. 1,040 F, G, and K dwarfs from Keck, Lick, and AAT planet search programs. *ApJS*, **159**(July), 141–166.

VandenBerg, D. A., Bergbusch, P. A., and Dowler, P. D. 2006. The Victoria-Regina stellar models: evolutionary tracks and isochrones for a wide range in mass and metallicity that allow for empirically constrained amounts of convective core overshooting. *ApJS*, **162**(Feb.), 375–387.

VandenBerg, D. A., Casagrande, L., and Stetson, P. B. 2010. An examination of recent transformations to the $BV(RI)_C$ photometric system from the perspective of stellar models for old stars. *AJ*, **140**(Oct.), 1020–1037.

van Dyk, D. A., Degennaro, S., Stein, N., Jefferys, W. H., and von Hippel, T. 2009. Statistical analysis of stellar evolution. *Annals of Applied Statistics*, **3**, 117–143.

Vergely, J.-L., Köppen, J., Egret, D., and Bienaymé, O. 2002. An inverse method to interpret color-magnitude diagrams. *A&A*, **390**(Aug.), 917–929.

Vogt, H. 1926. Die beziehung zwischen den massen und den absoluten leuchtkraften der sterne. *Astronomische Nachrichten*, **226**(Jan.), 301.

von Hippel, T. 2005. From young and hot to old and cold: comparing white dwarf cooling theory to main-sequence stellar evolution in open clusters. *ApJ*, **622**(Mar.), 565–571.

von Hippel, T., Jefferys, W. H., Scott, J., Stein, N., Winget, D. E., De Gennaro, S., Dam, A., and Jeffery, E. 2006. Inverting color-magnitude diagrams to access precise star cluster parameters: a Bayesian approach. *ApJ*, **645**(July), 1436–1447.

Waterfield, R. L. 1956. Report of his observatory. *MNRAS*, **116**, 217.

Waterfield, R. L. 1956. *J. Brit. Astr. Assoc.*, **67**, 1.

Wilson, R. E. and Hurley, J. R. 2003. Impersonal parameters from Hertzsprung-Russell diagrams. *MNRAS*, **344**(Oct.), 1175–1186.

Yadav, R. K. S., and 9 colleagues. 2008. Ground-based CCD astrometry with wide-field imagers. II. A star catalog for M 67: WFI@2.2 m MPG/ESO astrometry, FLAMES@VLT radial velocities. *A&A*, **484**(June), 609–620.

Yıldız, M. 2007. Models of α Centauri A and B with and without seismic constraints: time dependence of the mixing-length parameter. *MNRAS*, **374**(Feb.), 1264–1270.

Yıldız, M., Yakut, K., Bakış, H., and Noels, A. 2006. Modeling the components of binaries in the Hyades: the dependence of the mixing-length parameter on stellar mass. *MNRAS*, **368**(June), 1941–1948.

Young, P. A., Mamajek, E. E., Arnett, D., and Liebert, J. 2001. Observational tests and predictive stellar evolution. *ApJ*, **556**(July), 230–244.

Yuk, I.-S. and Lee, M. G. 2007. Modeling star formation history and chemical evolution of resolved galaxies. *ApJ*, **668**(Oct.), 876–890.

Zwitter, T., and 25 colleagues. 2010. Distance determination for RAVE stars using stellar models. II. Most likely values assuming a standard stellar evolution scenario. *A&A*, **522**(Nov.), A54.

7. Tutorial: Modeling tidal streams using N-body simulations

J. PEÑARRUBIA

7.1 Introduction: exercise goals

The main goal of this practical course is to build up a theoretical representation (N-body model) of the observed properties of the stellar stream associated to the globular cluster Palomar 5. Our priors are (i) a static (simplified) representation of the Milky Way potential, (ii) the position on the sky of the cluster remnant core, (iii) its heliocentric radial velocity, and (iv) its heliocentric distance.

We use the position of the stellar stream as detected in the Sloan Digital Sky Survey (SDSS) (see Grillmair and Dionatos, 2006) as observational constraints on the free-parameters of our models, which in this simplistic exercise correspond to the 2D-tangential components of the current velocity vector (i.e., proper motions) of Pal 5. Note that there are available measurements of Pal 5 proper motions. However, measuring those quantities for stellar systems as faint $M_V = -4.77 \pm 0.20$ and distant ($D \simeq 21$ kpc) as Pal 5 is subject to large observational uncertainties that translate into poorly constrained Galactocentric orbital parameters. To illustrate this issue, we adopt the Galactocentric proper motions of Pal 5 (μ_α, μ_δ) as free parameters that we derive from fitting the orientation of the stellar stream on the sky, and compare their values with measurements available in the literature. The second main goal of the exercise is thus to inspect the reliability of the existing proper motion measurements for Pal 5.

The third and last goal of the exercise is to construct a theoretical model of Pal 5 (core plus associated stream) via N-body calculations of the best-fitting orbit. Here the mass distribution of the stream progenitor is fixed ab initio from existing theoretical works (e.g., Dehnen *et al.*, 2004). One of the main unknowns corresponds to the stream age or, equivalently, to the orbital time required to reproduce the length and shape of the observed tidal stream.

7.1.1 *Methodology*

Our construction of N-body models for stellar tidal streams follows a laborious, but straightforward method:

- We use a simple semi-analytic code to solve the equations of motion of a point-mass particle moving in a Milky Way-like gravitational potential. Since the adjoined mass of the remnant core plus the known parts of the stream is small ($\lesssim 10^5$ $M_\odot$) compared with that of the Milky Way, the dynamical friction force term can be safely neglected (e.g., Peñarrubia *et al.*, 2006). We also hold the parameters that defined the Milky Way potential fixed through the evolution of our clsuter models for simplicity, given that tidal streams are barely sensitive to the past evolution of the host potential (Peñarrubia *et al.*, 2006). This setup clearly simplifies the derivation of Pal 5 past orbit.
- The initial Galactocentric position vector of Pal 5 is set up as $\mathbf{r} = (X, Y, Z)_{\rm Pal5} = (X_\odot - D\cos l \cos b, D \sin l \cos b, D \sin b)$, where $D \approx 21$ kpc is the measured Heliocentric distance, $(l, b) = (0^\circ.837, 45^\circ.854)$ the Galactocentric coordinates, and $X_\odot = 8$ kpc is the distance from the Sun to the Galaxy center.
- The initial Galactocentric velocity vector of Pal 5 is $\mathbf{v} = (U, V, W)_{\rm Pal5} = (U_\odot + U_{\rm hel}, V_\odot + V_{\rm hel}, V_\odot + V_{\rm hel})$, where $(U, V, W)_\odot = (10.5, 225.0, 7.0)\,{\rm km\,s^{-1}}$ is the sun velocity vector in the Galactic Standard of Rest (Binney and Merrifield, 1998), and $(U, V, W)_{\rm hel}$ are the heliocentric measurements. Of the three velocity components, we

only fix the radial one, which we derive by projecting the heliocentric velocity vector along the line-of-sight in our study, i.e., $v_r \equiv \hat{\mathbf{r}}_{\rm hel} \cdot \mathbf{v}_{\rm hel} = -56 \pm 4\,{\rm km\,s^{-1}}$ (Smith, 1985), while keeping the tangential components as free parameters.
- We use test particles to integrate its orbit back in time for a period $t_{\rm age}$ in order to provide initial conditions for our N-body realizations of Pal 5. The parameter $t_{\rm age}$ is free in our study. We fix its value by demanding the stellar stream of Pal 5 to show a length and shape similar to that detected in the SDSS survey. Since future, deeper surveys may reveal pieces of the stream that to date remain undetected, our estimate of $t_{\rm age}$ provides a strict lower limit.
- We then construct N-body representations of a stellar cluster in dynamical equilibrium and integrate their orbit forward in time for a time period $t_{\rm age}$. The last suite of free parameters in our models correspond to those that define the initial mass distribution of the Pal 5 globular cluster. For simplicity, we adopt the best-fitting models of Dehnen *et al.* (2004), who use King (1966) models with a shape parameter $W_0 = 3.5$, which is a dimensionless measure for the depth of the gravitational potential, an initial mass $M_0 = 1.2 \times 10^4\ {\rm M}_\odot$, and a King tidal radius $R_t = 10\,{\rm kpc}$, to construct representative models for the Pal 5 remnant as well as for its associated tidal stream.

7.2 Numerical setup

The main numerical tools used in this course are (i) a fairly simple semi-analytic code that integrates the orbits of test particles in a Milky Way-like static potential and (ii) an algorithm that describes self-consistently the evolution of a Pal 5 N-body representation. Here we provide a brief overview of the Milky Way model as well as the numerical codes that were used during the tutorial.

7.2.1 *The Galactic potential*

The Galaxy is modeled as a static potential with three components: a Miyamoto and Nagai (1975) disk, a Hernquist (1990) bulge, and a *spherical* Navarro, Frenk, and White (1996, hereafter NFW) dark matter halo. In this section we outline our model parameters.

We assume that the Milky Way is embedded in a dark matter halo that follows an NFW density profile

$$\rho(r) = \frac{\rho_0}{(r/R_s)[1 + (r/R_s)]^2} \quad (r \leq R_{\rm vir}); \tag{7.1}$$

In cosmological simulations, the characteristic density ρ_0 and scale radius R_s are sensitive to the epoch of formation and correlate with the halo virial radius via the concentration parameter $C_{\rm vir} \equiv R_{\rm vir}/R_s$.

The virial radius is defined so that the mean over-density relative to the critical density is Δ,

$$\frac{M_{\rm vir}}{(4/3)\pi R_{\rm vir}^3} = \Delta\ \rho_{\rm crit}, \tag{7.2}$$

where $\rho_{\rm crit} = 3H_0^2/8\pi G$, and $H_0 = 100\,h\,{\rm km\,s^{-1}Mpc^{-1}}$ is the present day value of Hubble's constant.

The choice of Δ varies in the literature, with some authors using a fixed value, such as NFW, who adopted $\Delta = 200$, and others who choose a value motivated by the spherical collapse model, where (for a flat universe) $\Delta \sim 178\,\Omega_{\rm m}^{0.45}$ (Eke *et al.*, 1996). The latter gives $\Delta = 95.4$ at $z = 0$ in the concordance ΛCDM cosmogony, which adopts the following cosmological parameters: $\Omega_{\rm m} = 0.3$, $\Omega_\Lambda = 0.7$, $h = 0.7$, consistent with constraints from

CMB measurements and galaxy clustering (see, e.g., Spergel *et al.*, 2007, and references therein).

Also, in density profiles with outer slopes ≥ 3, the cumulative mass profiles diverges as $r \to \infty$. To avoid that problem, impose a truncation at $R_{\rm vir}$ that can be written as (Kazantzidis *et al.*, 2004)

$$\rho(r) = \frac{\rho_0}{C_{\rm vir}^{\gamma}(1 + C_{\rm vir}^{\alpha})^{(\beta-\gamma)/\alpha}} \left(\frac{r}{R_{\rm vir}}\right)^{\epsilon} \exp\left(-\frac{r - R_{\rm vir}}{R_{\rm dec}}\right) \quad (r > R_{\rm vir}); \tag{7.3}$$

where $R_{\rm dec}$ is a small quantity, whose value we choose as $0.1 R_{\rm vir}$.

To obtain a continuous logarithmic slope ϵ is defined as

$$\epsilon = \frac{-\gamma - \beta C_{\rm vir}^{\alpha}}{1 + C_{\rm vir}^{\alpha}} + \frac{R_{\rm vir}}{R_{\rm dec}}. \tag{7.4}$$

The potential associated to the above density profile is

$$\Phi_h(r) = -4\pi G \left[\frac{1}{r}\int_0^r \rho(r') r'^2 dr' + \int_r^{\infty} \rho(r') dr'\right]. \tag{7.5}$$

We choose a virial mass and a concentration that provides a reasonable description of the circular velocity curve of the Milky Way, namely, $M_{\rm vir} = 10^{12}$ M$_\odot$, $R_{\rm vir} = 258$ kpc and $R_s = 21.5$ kpc, assuming $\Delta = 101$ (Klypin *et al.*, 2002).

The Milky Way disk is simulated as force component $\mathbf{f}_d = -\nabla\Phi_d$ in the form of a Miyamoto and Nagai (1975) disk model

$$\Phi_d(R, z) = -\frac{GM_d}{\sqrt{R^2 + (a + \sqrt{z^2 + b^2})^2}}; \tag{7.6}$$

where M_d is the disk mass, $a = 6.5$ kpc and $b = 0.25$ kpc are the radial and vertical scale lengths.

The Milky Way bulge follows a Hernquist (1990) profile

$$\Phi_b(r) = -\frac{GM_b}{c + r}; \tag{7.7}$$

with a mass $M_b = 1.3 \times 10^{10}$ M$_\odot$ and a scale radius $c = 1.2$ kpc.

7.2.2 *The semi-analytical code*

To follow the cluster orbit, we have constructed a simple semi-analytic code that solves the equations of motion in a Milky Way-like potential. We treat clusters as point masses that move in a host halo potential whose parameters are shown. At each time step, our code solves the following equation of motion

$$\ddot{\mathbf{r}} = -\nabla(\Phi_h + \Phi_d + \Phi_b) \tag{7.8}$$

using a standard leap-frog technique. Subsequently we integrate its orbit back in time for a time interval $t_{\rm age}$ in order to provide initial conditions for our N-body realizations of the Pal 5 cluster.

7.2.3 *The N-body code*

We follow the evolution of N-body models of Pal 5 in the Galaxy potential using SUPERBOX, a highly efficient particle-mesh gravity code (see Fellhauer *et al.*, 2000, for details). SUPERBOX uses a combination of different spatial grids in order to enhance the numerical resolution of the calculation in the regions of interest. In our case, SUPERBOX uses three

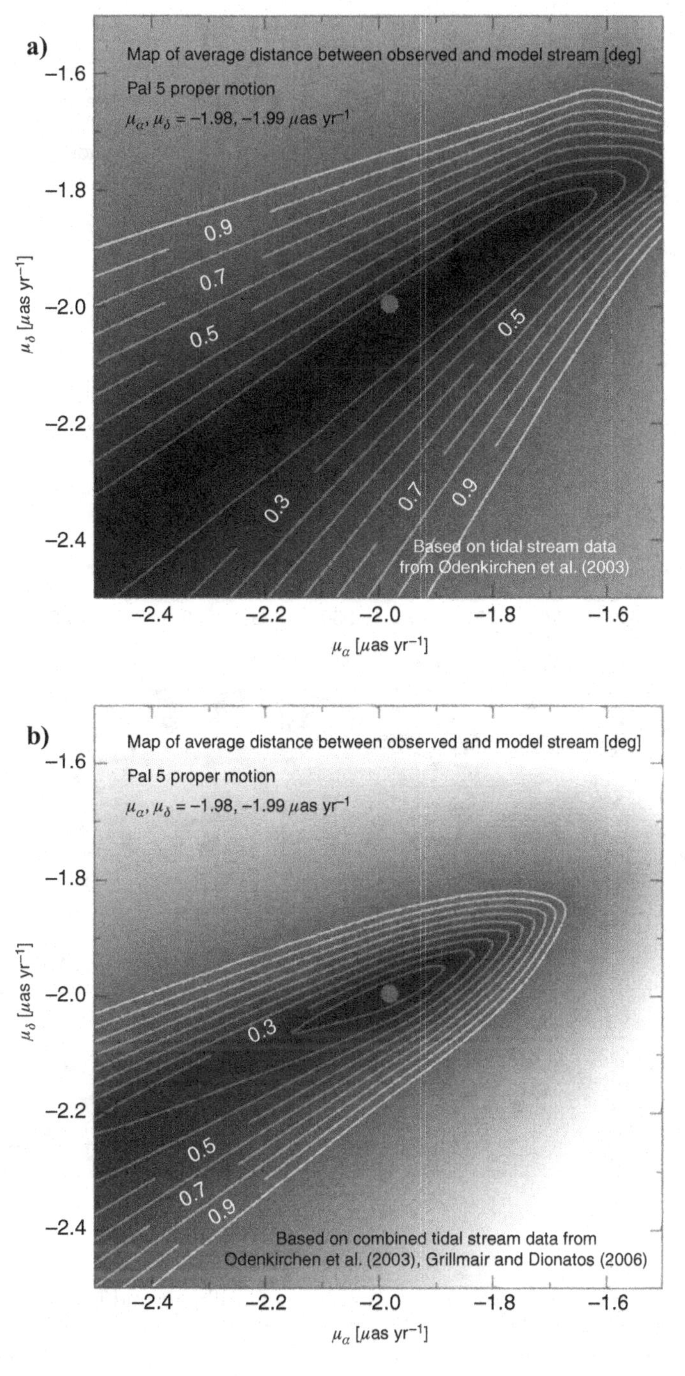

FIG. 7.1. Constraints on the proper motions of Pal 5 that fit to the trajectory of the tidal tails shown in Odenkirchen *et al.* (2003) only (a), and to Odenkirchen *et al.* (2003) and Grillmair and Dionatos (2006) (b). Note that both data sets lead to a best-fitting value of $(\mu_\alpha, \mu_\delta) \approx (-1.98, -1.99)$ mas/yr.

nested grid zones centered on the highest-density particle cell of the dwarf. This center is updated at every time step, so that all grids follow the cluster along its orbit.

Each grid has 128^3 cubic cells: (i) the inner grid has a spacing of $dx = 2R_h/126 \simeq 1.6 \times 10^{-2}\, R_h$, where R_h is the cluster half-light radius, and is meant to resolve the

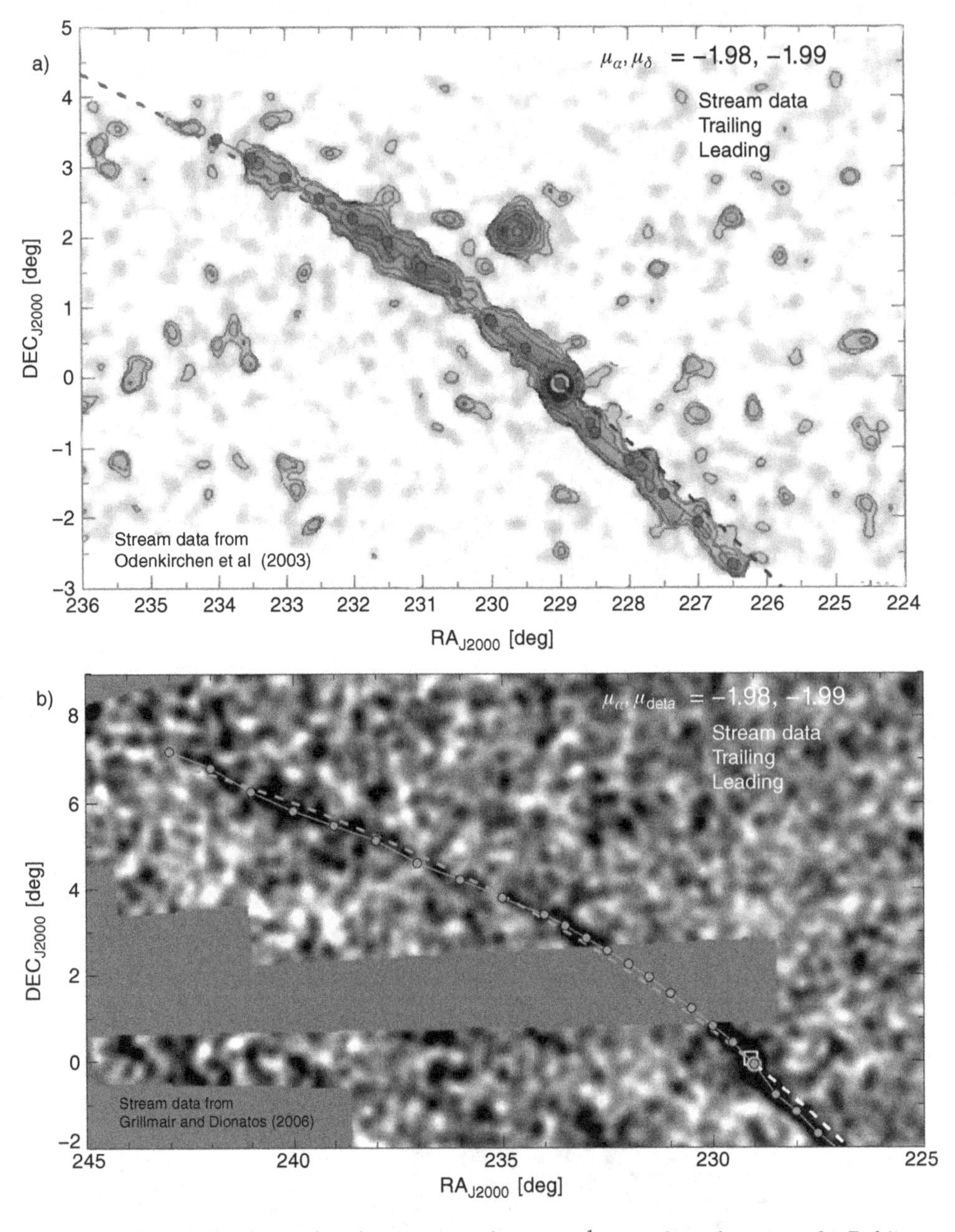

FIG. 7.2. Best-fit orbit $(\mu_\alpha, \mu_\delta) \approx (-1.98, -1.99)$ mas yr^{-1} over-plotted against the Pal 5 stream detections of Odenkirchen *et al.* (2003) (a), and those of Grillmair and Dionatos (2006) (b). The leading and trailing tails are denoted with blue and red dashed lines, respectively, whereas the stream and the remnant core of Pal 5 are respectively highlighted with solid and open symbols for ease of reference.

innermost region of the cluster particles. (ii) The middle grid extends well beyond the cluster tidal radius, with spacing $20R_h/126$. (iii) The outermost grid extends out to $1,000 \times R_h$ and is meant to follow particles that are stripped from the cluster and that orbit about the Galaxy.

SUPERBOX uses a leap-frog scheme with a constant time-step to integrate the equations of motion for each particle. We select the time-step according to the criterion of Power *et al.* (2003); applied to our cluster models (see Section 7.3.2), this yields $\Delta t = 4.6$ Myr.

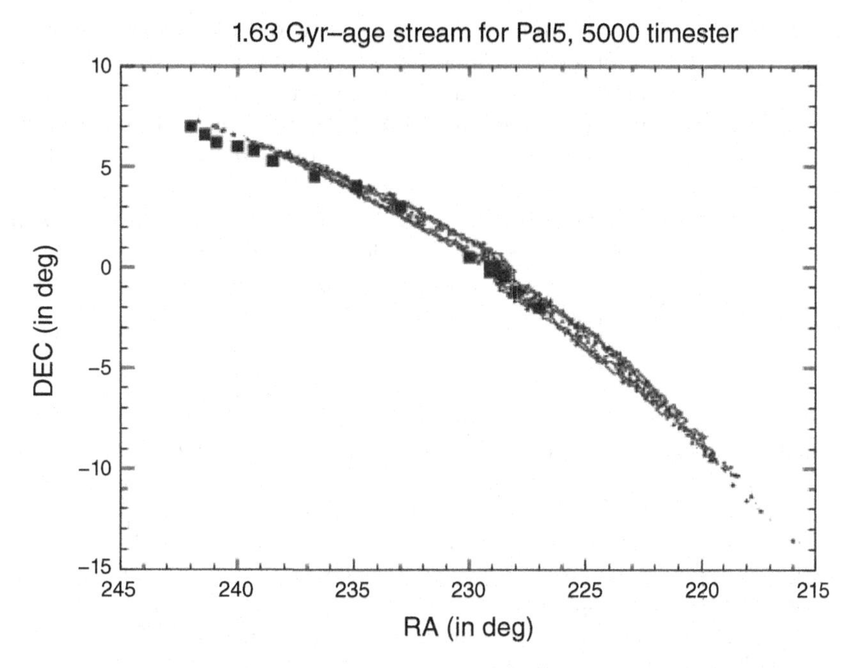

FIG. 7.3. N-body model for the tidal tail of Pal 5. Following Dehnen *et al.* (2004) we assume that the progenitor cluster follows a King (1966) density profile with an initial mass $M = 1.2 \times 10^4$ $M_{\odot}$, a central potential $W_0 = 3$, and a tidal radius $R_t = 10$ kpc. The length of the Pal 5 stream observed in the SDSS catalog Grillmair and Dionatos (2006) can be reproduced with a model that began to lose stars to tides approximately $t_{\rm age} = 1.63$ Gyr ago.

7.3 Results

7.3.1 *Constraints on the proper motions of Pal 5*

The fact that the tidal tails reveal the trajectory of the progenitor can be used to tightly constrain the possible range of proper motions of Pal 5. Figure 7.1 shows the results of a χ^2 fitting of the stream-projected location using the detections of Odenkirchen *et al.* (2003) (*upper panel*). Adding the detections of Grillmair and Dionatos (2006) (*lower panel*) leads to a tighter constraint on the proper motions of Pal 5, whose best-fitting value are close to $(\mu_\alpha, \mu_\delta) \approx (-1.98, -1.99)$ mas/yr.

It is interesting to compare these numbers against the estimates published in the literature. For example, Schweitzer *et al.* (1993) find $(\mu_\alpha, \mu_\delta) = (-2.47 \pm 0.17, -0.87 \pm 0.22)$ mas/yr, whereas Scholz *et al.* (1998) and K. Cudworth (unpublished result, listed in Dinescu *et al.*, 1999) find $(-1.0 \pm 0.3, -2.7 \pm 0.4)$ and $(-2.55 \pm 0.17, -1.93 \pm 0.17)$ mas/yr, respectively.

Figure 7.2 shows the best-fitting orbit of the leading and trailing tails (blue and red dashed lines, respectively) over-plotted against the Pal 5 stream detections of Odenkirchen *et al.* (2003) (*upper panel*), and those of Grillmair and Dionatos (2006) (*lower panel*). Clearly this Figure shows that none of the published proper motion estimates would provide a good match of the observed trajectory of the tails.

7.3.2 *Constraints on the stream age*

How long ago were the stars that currently distribute along the tidal tails stripped from Pal 5? To answer this question we run N-body calculations that simulate the disruption of this cluster. Following Dehnen *et al.* (2004) we adopt a progenitor model that follows a King (1966) density profile with an initial mass $M = 1.2 \times 10^4$ $M_{\odot}$, a central potential $W_0 = 3$, and a tidal radius $R_t = 10$ kpc. Subsequently, we explore models with a range of integration times ($t_{\rm age}$) until a model that describes the observed stream length is found.

Figure 7.3 shows that a reasonable match to the detected pieces of the stream in the SDSS catalog can be found for a cluster model that began losing stars to tides $t_{\rm age} = 1.63$ Gyr ago. We note, however, that this value is degenerated with the initial cluster mass: stars stripped from more massive progenitors would populate the associated tidal stream on a shorter time scale and thus reduce the time required to reproduce its length. As shown by Peñarrubia *et al.* (2006), a possible way to constrain the initial cluster mass would be to measure the velocity dispersion along the line-of-sight at different parts of the stream. However, given its expected low velocity dispersion (of a few km/s), low surface-brightness, and relative large heliocentric distance, this measurement may pose a strong challenge with the current instrumentation.

REFERENCES

Binney, J. and Merrifield, M. 1998. Galactic Astronomy. *Princeton, NJ : Princeton University Press.*

Dehnen, W., Odenkirchen, M., Grebel, E. K., and Rix, H.-W. 2004. Modeling the disruption of the globular cluster Palomar 5 by Galactic tides. *AJ*, **127**(May), 2753–2770.

Dinescu, D. I., Girard, T. M., and van Altena, W. F. 1999. Space velocities of globular clusters. III. Cluster orbits and halo substructure. *AJ*, **117**(Apr.), 1792–1815.

Eke, V. R., Cole, S., and Frenk, C. S. 1996. Cluster evolution as a diagnostic for Omega. *MNRAS*, **282**(Sept.), 263–280.

Fellhauer, M., Kroupa, P., Baumgardt, H., Bien, R., Boily, C. M., Spurzem, R., and Wassmer, N. 2000. SUPERBOX – an efficient code for collisionless galactic dynamics. *New A*, **5**(Sept.), 305–326.

Grillmair, C. J. and Dionatos, O. 2006. dA 22° tidal tail for Palomar 5. *ApJ*, **641**(Apr.), L37–L39.

Hernquist, L. 1990. An analytical model for spherical galaxies and bulges. *ApJ*, **356**(June), 359–364.

Kazantzidis, S., Mayer, L., Mastropietro, C., Diemand, J., Stadel, J., and Moore, B. 2004. Density profiles of cold dark matter substructure: implications for the missing satellites problem. *ApJ*, **608**(June), 663–679.

King, I. R. 1966. The structure of star clusters. III. Some simple dynamical models. *AJ*, **71**(Feb.), 64.

Klypin, A., Zhao, H., and Somerville, R. S. 2002. ΛCDM-based models for the Milky Way and M31. I. Dynamical models. *ApJ*, **573**(July), 597–613.

Miyamoto, M. and Nagai, R. 1975. Three-dimensional models for the distribution of mass in galaxies. *PASJ*, **27**, 533–543.

Navarro, J. F., Frenk, C. S., and White, S. D. M. 1996. The structure of cold dark matter halos. *ApJ*, **462**(May), 563.

Odenkirchen, M., and 9 colleagues 2003. The extended tails of Palomar 5: A 10° arc of globular cluster tidal debris. *AJ*, **126**(Nov.), 2385–2407.

Peñarrubia, J., Benson, A. J., Martínez-Delgado, D., and Rix, H. W. 2006. Modeling tidal streams in evolving dark matter halos. *ApJ*, **645**(July), 240–255.

Power, C., Navarro, J. F., Jenkins, A., Frenk, C. S., White, S. D. M., Springel, V., Stadel, J., and Quinn, T. 2003. The inner structure of ΛCDM haloes – I. A numerical convergence study. *MNRAS*, **338**(Jan.), 14–34.

Scholz, R.-D., Irwin, M., Odenkirchen, M., and Meusinger, H. 1998. New space motion of Galactic globular cluster Palomar 5. *A&A*, **333**(May), 531–539.

Schweitzer, A. E., Cudworth, K. M., and Majewski, S. R. 1993. Membership, photometry and kinematics of the globular star cluster PAL 5. *The Globular Cluster-Galaxy Connection*, **48**(Jan.), 113.

Smith, G. H. 1985. Spectroscopy of red giants in the globular cluster Palomar 5. *ApJ*, **298**(Nov.), 249–258.

Spergel, D. N., and 21 colleagues 2007. Three-year Wilkinson Microwave Anisotropy Probe (WMAP) observations: implications for cosmology. *ApJS*, **170**(June), 377–408.